Biologic Readings for Today's Students

Biologic Readings for Today's Students

Edited by

Gideon E. Nelson
James D. Ray, Jr.
University of South Florida

John Wiley & Sons, Inc.
New York London Sydney Toronto

Library of Congess Catalogue Card Number: 75 – 152500

ISBN 0 – 471 – 63155 – 8

Printed in the United States of America

10 9 8 7 6 5

Preface

This book of readings is intended for use in introductory biology courses. The articles were selected on the basis of readability, relevance to today's problems, presentation of new advances in biology, and unusually clear explanations of complex biological concepts.

The readings were chosen from fourteen periodicals, ranging from *Today's Health* and *The New York Times Magazine* to *Science* and *Science Journal.* The arrangement of the selections follows the subject-matter organization in *Fundamental Concepts of Biology* by Nelson, Robinson, and Boolootian.

Certain subject areas in this reader contain more articles than others simply because of the situation that we encountered in the literature. For example, there are almost no essays on topics such as cellular respiration suitable for use by general biology audiences. By contrast, articles on environmental abuse are quite abundant.

The issues and problems examined in these essays should extend beyond the classroom into the everyday affairs of the readers. We hope that this book will function significantly in the development of an informed and active community.

Gideon E. Nelson
James D. Ray, Jr.

Contents

I. Introduction 1
1. The Scientific Method
Henry Margenau and David Bergamini 2

II. Energy 10
2. The New Food Revolution
Gene Gregory 11
3. The Green Revolution: Cornucopia or Pandora's Box?
Clifton R. Wharton, Jr. 20
4. Aquaculture
John E. Bardach 29

III. Cells 43
5. At the Controls of the Living Cell
Lawrence Lessing 44
6. Lysosomes in Disease
A. C. Allison 55
7. The Chalones
W. S. Bullough 62

IV. Nervous System 69
8. The Search for the Memory Molecule
David Perlman 70
9. Hallucinogens of Plant Origins
Richard Evans Schultes 79
10. Cannabis
Andrew T. Weil 94
11. The Speed That Kills or Worse
Jonathan Black 103

V. Homeostasis 112
12. The Many Clocks of Man
John D. Palmer 113

VI. Behavior 123
13. On War and Peace in Animals and Man
N. Tinbergen 124
14. The Possible Biological Origins of Sexual Discrimination
Lionel Tiger 140

VII. Reproduction 153
15. The Reproductive Life Cycle
Sir Alan Parkes 154
16. The Mechanism of Bacterial Sexuality
William Hayes 163

VIII. Heredity 173
17. Heredity and the Embryo
D. H. Carr 174
18. Will the Baby Be Normal?
Robert W. Stock 182
19. Genetic Counseling Units in the United States
Henry T. Lynch 192
20. Genetic Engineering: Controlling Man's Building Blocks
Joshua Lederberg 195
21. Hopi Indians, Inbreeding, and Albinism
Charles M. Woolf and Frank C. Dukepoo 203

IX. Development 216
22. The Size of Life
John Tyler Bonner 217

X. Evolution 222
23. The Chemical Origin of Life
Cyril Ponnamperuma 223
24. Origin of Life on Other Planets?
Eugenia Keller 231
25. Human Origins
D. R. Pilbeam 239
26. Human Biological Adaptability
Gabriel W. Lasker 253
27. The Exploitation of Animals by Man
M. L. Ryder 266

XI. Environment 279
28. Nutrient Cycling
F. H. Bormann and G. E. Likens 280
29. Effects of Pollution on the Structure and Physiology of Ecosystems
G. M. Woodwell 290
30. Aquatic Weeds
L. G. Holm, L. W. Weldon and R. D. Blackburn 298
31. Population Resources and Technology
Harrison S. Brown 318
32. Beyond Family Planning
Bernard Berelson 323
33. Infectious Diseases and Human Ecology
A. Macdonald 346

XII. Epilogue 364
34. What We Must Do
John Platt 365
35. On the Life Sciences: Science to Wisdom
Jean Mayer 378

Biologic Readings for Today's Students

I. Introduction

All too often, basic science texts begin with a pedantic consideration of methodology that says little that is relevant for today's students. Our first essay, *''The Scientific Method''* is not the usual pedestrian introduction.

Henry Margenau, physicist and philosopher, and David Bergamini, science writer, have done an able job of treating scientific methodology with imagination, and the article is well worth more than a casual perusal.

1.

The Scientific Method

Henry Margenau, David Bergamini

It is, rather, an attitude and a philosophy, providing guidance by which dependable over-all concepts can be extracted from impressions that swarm in on man's senses from the outside world. AUTHORS

We have only to take a walk on a spring morning, or watch the waves break on a shore, or gaze at the stars over the back porch, to be reminded that the world around us is a tapestry intricate beyond compare. But it is one thing simply to admire the facade, and quite another to delve behind it. How does the scientist begin to see through it? How can he glimpse, beyond it, the law and order of the universe? What gives him his unique confidence? The answers lie in the scientist's particular style of inquiry, which he has developed over the ages in his quest for knowledge, and which has come to be known as the scientific method.

Few aspects of science are more difficult for the layman to grasp, yet none better repays the effort – for the scientific method goes to the very heart of all scientific activity. The term "scientific method" is itself somewhat of a misnomer. It is not a method in the sense of a formal procedure. It furnishes no detailed map for exploring the unknown, no surefire prescription for discovery. It is, rather, an attitude and a philosophy, providing guidance by which dependable over-all concepts can be extracted from impressions that swarm in on man's senses from the outside world.

Its practitioner may be the kind who seems to be forever turning over rocks to expose what lies beneath, smashing things open to count and catalogue their interior contents – always obsessed with facts. Or he may be someone we see as a dreamer, pursuing visions, spinning gossamers of equations high in an abstract sky – working not with facts but with ideas that are seemingly born of sheer creative fancy.

It is the great triumph of the scientific method that it enables these two extremes of talent, the data gatherers and theory makers, to complement each other. It bridges the chasm between facts and ideas. It gives the scientist the discipline to distinguish between ideas that are relevant and useful, and those that are empty and misleading. It enables him to exploit those fugitive moments of intuition and insight that are as indispensable to science as to any other mental endeavor.

With its virtues, the method has certain natural limitations. It cannot replace the inspiration of Archimedes discovering a basic law of hydrostatics while sitting in his bath. It cannot conjure up the good luck of Alexander Fleming chancing on penicillin. In short, it cannot create science automatically any more than the theory of harmony can write a symphony or a Navy manual can win a sea battle.

The scientific method is the offspring of a branch of philosophy called epistemol-

---

ogy (from the Greek, *episteme,*"knowledge," and logos, "theory"). From Plato on down, men have pondered the question of what knowledge consists of, whence it comes, and how we acquire it. In the seventeenth century the English epistemologist Francis Bacon, focusing on scientific knowledge in particular, attempted to provide a tidy blueprint for its acquisition. He offered the scientist a fourfold rule of work: observe, measure, explain, and then verify. By the nineteenth century a more sophisticated version of the method was being propounded: pose a question about nature; collect pertinent evidence; form an explanatory hypothesis; deduce its implications; test them experimentally; and then accept, reject, or modify the hypothesis accordingly.

But observe what? Measure what? How frame the hypotheses or explanations? Because there are obviously no pat answers to these questions, present-day epistemologists doubt that creative scientists can be hedged about by hard-and-fast instructions. One man may pick up a butterfly and notice the peculiarities of its wing markings. Another may wonder about the ratio of its flying surface to its body weight. The first may go on to enunciate a new theory of insect camouflage and the second a new theory of insect aerodynamics, but neither gains his insight by measuring all available evidence indiscriminately or by testing every arbitrary hypothesis that flits into his mind. One may arrive at the answer he seeks after a good night's sleep. The other may badger his inspiration for many years before he sees the light. Or then again, on a lazy day, both may simply appreciate the beauty of the butterfly and let it flutter off.

History may wait decades for an encounter of the right man and the right butterfly on the right day. No single investigator is likely to proceed from an initial observation to a finished theory about any significant subject in the course of his lifetime. When Newton drew up his fundamental laws of motion and gravitation in the 1680's, he capped an investigation that stretched far into the past: back through Galileo's experiments with falling cannon shot and Kepler's formulation of laws for the movements of the planets; back through the shrewd speculations about force and relativity made by a group of medieval thinkers known as the impetus school; back through the deductions of ancient Greeks about geometry and space; back through the stargazing of the Babylonians; all the way back to Stone Age observations on the motion of projectiles. When Darwin arrived at his concept of evolution, he answered questions about changing forms of life that had been posed by scientific thinkers as long ago as Aristotle. Astronomers today still consult the readings of star positions made by their ancient colleagues in Samarkand and Sumer.

Any program of creative research in science moves ahead through tentative, chancy, inspired steps taken by many men over many years. A theory dimly glimpsed in the nineteenth century may emerge into full view only after it has been framed in mathematical patterns of logic worked out in the twentieth century. An experiment recognized as vital may endure lengthy postponement while the apparatus is being devised to perform it. The physiologist Wilton Earle, who pioneered in growing live human tissues in bottles outside the body, was once held up for an entire year for lack of a flask stopper that would not poison his delicate sterile cultures. It took him months to ascertain that the stoppers he needed should be made of silicon, and months more to persuade a manufacturer to turn them out.

These obstacles that impede research, as well as the strokes of luck and intuition that help it on its way, are basic to all human enterprise. What is unique to scientific endeavor is the linkage it forges from ideas to facts, and from facts to ideas.

Nothing is more astonishing about science than its ability to make imaginative conjectures and then to convert them into tangible realities that no one had previously suspected. Out of Maxwell's equations of electromagnetism came radio and television. Out of Einstein's formulas on matter and energy came the atomic bomb. When the scientific imagination works in the opposite direction – crystallizing theories out of facts – the transformation is equally dazzling. Here, in one decade, lies a musty lot of old bones and odd birds, and there, in the next decade, stands a theory of evolution. Here is a pile of dirty pitchblende and an endless labor of refining to do, and there is radium and an awareness of the atomic nucleus.

To understand the remarkable interaction that science sets up between observable facts and abstract ideas, it is necessary to consider these questions: What constitutes a fact of science? What constitutes an acceptable theory? What happens intellectually in the process of bridging the gap between the two?

The facts with which science works are called data – those things that are given to us through our observation of the world around us and our absorption of its unceasing flow of sights, sounds, smells, tastes, and tactile sensations. It is this kind of outer experience that is employed in the making of science, rather than the kind of inner experience generated out of man's hungers, fears, loves, and other urges. In so far as he can, the scientist, when considering evidence, excludes emotion and bias from his thinking. He tries to discount his sense of righteousness or of sin, his simple pains or joys in living, his repugnance to ugliness or attraction to beauty. As takeoff points for his inquiry, the scientist accepts only those ingredients of experience that bear discernibly on objective knowledge – knowledge as distinct from hope, desire, contemplative enjoyment, or mystical rapture.

This restricted realm of experience in which the scientist operates is described by epistemologists as cognitive (from *cognoscere,* "to find out," "to know"). It admits only those perceptions that we receive directly through our senses, plus the purely logical processes by which we select, judge, and reason. It contains such items as the chest of drawers that we have seen and touched and therefore *know* to be there, even when we have turned our backs on it; the gallon of gasoline that we have tested and therefore *know* will run a car for a certain mileage; the sun that we *know* will rise again tomorrow because we have seen it rise so many times in the past.

The raw stuff of science lies strewn about on the surface of experience like so many nuggets on a sandbar. Scientists sometimes call the field from which it is taken the "protocol" plane. The word stems from a practice of Greek authors in classical times. When a writer planned a work, he would first paste on the manuscript a sheet of papyrus, listing the topics that he expected to explain in the text. This preliminary sheet was known as the *protokollon,* from *protos* "first," and *kolla,* "glue." Hence it seems appropriate to refer to the raw sensory materials of knowledge as the protocols, or forewords, of science.

When he selects his protocol facts, the scientist tries to define them as precisely as he can through measurement. The kinds of measure he employs vary a great deal. The astronomer classifies a star by it brightness and spectral color. The chemist discriminates between substances by their boiling points, acidity, and so forth. The physicist defines a push or a pull in terms of force that can be registered on a spring balance.

By such means a scientist can assign a numerical value to a protocol fact that heretofore had only sensory or aesthetic value. He changes it from a qualitative fact, which is subjective and private, into a quantitative fact, which is objective and public – in other words, communicable.

Sensations of hotness and dampness, for example, can be converted into universally understood quantities like 90°F. or 70 percent relative humidity.

Through the use of standards and gauges – through measurement, in short – the scientist distills from his private sensations an essence that all men can inspect, verify, and use. The Stone Age primitive who put his finger in boiling water may have voiced his equivalent of "Hot!" Today, having invented thermometers and scales of temperature, we have a mutual understanding of how hot "hot" really is.

Once measured, protocol facts provide grist for the mill of the theorist. Examining various numbers arrived at by measurement, he can seek out relationships between them, which he expresses in formulas. Readings of temperature and pressure made on the air in a tire, for example, can be converted into the basic law of gases.

Readings on gauges do not automatically produce laws of basic understanding, They may lead nowhere – unless the scientist sees a pattern in them. And such patterns are difficult to discern unless measured values are expressed in suitable units. Many of our present units, our weights and measures, are legacies from the early years of scientific inquiry, and do not suggest any truly fundamental ideas. The metric system is based not on natural units, which are meaningful throughout the cosmos, but on eighteenth-century measurements of the circumference of the earth. The Fahrenheit temperature scale, based on a zero point defined by a haphazard mixture of salt and ice, does not suggest the true nature of temperature, which we now know to be molecular agitation.

Through increasingly refined measurements, the scientists seeks and finds ever more fundamental "numbers" in nature. It is by this process that he has discovered such crucial quantities as absolute zero, the speed of light in a vacuum, the mass of the hydrogen atom. Picking out fundamental numbers from a welter of measurements, coming to have a familiar feel for them, and eventually evolving from them some concept about nature, constitute the most difficult and creative part of a scientist's work. All too easily, he may pursue blind alleys or go not quite far enough in his analysis. Before Newton, for instance, Galileo, Kepler, and William Gilbert, the great Elizabethan student of magnetism, all came close to grasping the essentials of gravitation, but veered off through various misconceptions concerning what physicists now call force.

In entering the domain of concepts, the scientist mentally shifts gears, as it were. Whereas in dealing with protocol facts and measurements he has used inductive reasoning, he now applies deductive reasoning. In reasoning inductively, he has worked his way logically from particulars to generalities. "All the swans I have ever seen are white; therefore all swans are white." Or: "I have never seen a fire-breathing reptile; therefore dragons do not exist." But these generalizations that he makes are never entirely certain. After all, swans have been found in Australia that are black, and someone might conceivably turn up a reptile that had learned the sideshow trick of swallowing fire.

After enough tests, however, if a generalization seems unexceptionable, the scientific theorist proceeds to treat it as true – at least tentatively. He takes it as a premise and reasons out possible consequences that might arise from it. If his premise is true, the conclusions he derives from it will be true too. In applying this type of reasoning – deductive logic – he can draw on the symbols and patterns devised by mathematicians, the purest scientists of all. When Einstein envisaged a new theory of space and time, he was able to take over a mathematical structure previously conceived by a school of nineteenth-century non-Euclidean geometers. When modern biochemists theorize about DNA molecules, they shape their information with tools already hammered out by probability theorists and topologists.

The deductive stage of scientific investigation is sometimes called the "C-domain," the "C" representing concepts. In this domain the scientist constructs theories more or less at will and explores their implications. Each twist he gives to his hypotheses puts him into a new orbit of possible consequences. Each fresh go-round broadens his perspective on the protocol facts from which he started. After enough orbits, he may begin to discern new factual possibilities not apparent to him in his original observations or measurements.

As he formulates his final theory, the scientist subjects it to intensive criticism. Seeking to make it as useful as possible, he asks himself: Is this proposed a universal throughout the extent of space and the passage of time? Does it lead anywhere? Does it predict one state of affairs as arising out of another? Could it be transposed from one frame of reference to another and still remain valid? And finally, because of his innate passion for orderliness, his aesthetic appreciation of things that are meet and fitting, he asks: Is this theory as elegant as possible? Could I formulate it more succinctly?

Now comes the moment of verification and truth: testing the theory back against protocol experience to establish its validity. If it is not a trivial theory, it suggests the existence of unknown facts that can be verified by further experiment. If the theoretical predictions do not jibe with observable facts, then the theorist has to forget his disappointment and start all over again.

If a theory survives all tests and is accepted into the canon of scientific law, it becomes a fact in its own right and a foundation for higher spires of thought. Abstract though it may be, a proved theory can suggest new hi-fi sets or hybrid cattle just as surely as do experiments with electricity or stockbreeding. It serves as a starting point for new theories just as surely as any experience on the plane of protocols. Galileo's formula for the increasing speed at which a body falls freely near the surface of the earth became a single example of Newton's law of gravitation. Newton's law, in turn, became a single special case in Einstein's theory that gravitation is a manifestation of the geometry of space and time. At this moment some child in a hamlet in Maine may be preparing himself for the work of constructing a "unified-field theory" of both atom and cosmos, in which Einstein's sweeping concepts of relativity will appear as mere details.

In full bloom, rooted in protocols, stemmed with measurements and headed with theories, a science becomes a complete organism. Every part of it communicates with every other part wherever located – on the P-ground of protocol fact, in the M-air of measurement, or higher still in the C-space of concepts. Vital juices flow from experimental instruments to mathematical symbols and back again. Past ideas remain as germinal in present science as do the cells of childhood in the body of an adult.

The astonishingly effective tie-in that the scientific method creates between literal-mindedness and imaginative symbolism has produced problems by its own luxuriant success. Today the roots of the tree of science delve so deep into the subatomic underworld, and the branches reach so high into the real of the galaxies, that scientists are not always sure what it is that they are dealing with.

Is an evanescent atomic particle, artificially created and artificially detected, a protocol fact or a concept? Is the expansion of the universe, deduced from spectra of faint heavenly bodies, a reality of nature or a description imposed on nature by the mind? Are the human senses and intelligence, developed for a middlesized world of tables and chairs, capable of exploring an atom many billions of times smaller than man, or a starry galaxy hundreds of millions of trillions of times larger? Is it possible that, in the words of the British geneticist J.B.S. Haldane, the universe "may not only be queerer than we imagine, but queerer than we can imagine"?

Obviously after their recent successes in both atomic and cosmic matters, scientists are not about to give up as hopeless their epic quest for knowledge. Nonetheless, there has been a marked increase in scientific modesty during recent decades. Many scientists of the nineteenth century believed that the number of scientific facts to be learned was finite. They often felt that they would someday achieve absolute truth and ultimate understanding. Today, however, their successors speak only of reaching "partial understanding," of continually approaching truth but never grasping it completely.

The difficulties that now beset the scientific imagination have revived an issue which is over two thousand years old. What is reality? Which is more real and dependable: an idea generated by the mind, or an impression of something outside, generated through the senses? Rationalist philosophers like Plato and Descartes have inclined to the view that ideas exist independently. Empiricist philosophers like Bacon and Hume have felt that ideas do not exist except as they are suggested by observable facts. Taking a position somewhere between the two schools, the great eighteenth-century German metaphysician Immanuel Kant once mused: "Concepts without factual content are empty; sense data without concepts are blind . . .The understanding cannot see. The senses cannot think. By their union only can knowledge be produced."

Under the aegis of the scientific method the rationalists and empiricists of science – the theorists and the experimentalists – have got on so successfully together that, for the most part, they have left philosophy to the philosophers. But they have come to acknowledge that the age-old debate about the nature of reality is fundamental to science today. Ideas do, indeed, lead a life of their own that is independent of observation and experiment. The concept of a circle, for instance, might still have its own validity even if the universe were devoid of any circular shapes. The ratio of the circumference to the diameter of a circle, designated as π by the ancient Greeks, and known as pi to all school children since, has been calculated to 100,000 decimal places, yet no one, to this day, has ever seen a circle perfect enough to exemplify this ratio beyond about six decimal places. The mathematical concept of a circle, in short, is more dependable and certain than any circle that can be drawn with a compass or observed in nature.

That the ideas of science are created by the mind, as well as discovered in the outside world, was brought home dramatically to twentieth-century scientists when they realized that they had been attributing to subatomic particles – such as the electron – a number of qualities that have meaning in our everyday world but little or no meaning in the microcosm of the atom.

Scientists think of the electron as the carrier, within the atom, of negative electrical charge. The rest of us, when we think of it at all, usually envisage streams of pellets in power lines. The electron is too small to touch, taste, smell, hear, or see even with a microscope but we nonetheless imagine it to be essentially concrete and palpable. After all, electrons can be shot from the nozzle end of a particle accelerator as if they were so many bullets; the fragments of their targets can be traced in a photographic emulsion almost as vividly as if they were splinters from a door frame shattered by machine-gun fire.

But do electrons really have anything in common with pellets or bullets? Imagine what would happen if we could shrink a little red ball and make it as small as we wished. When its diameter got down to one hundredth of an inch, the ball would cease to be visible to the naked eye. It would still have color, roundness, and size, however, as could be determined through a magnifying glass. When the magnifying glass no longer sufficed, a microscope would still serve as a legitimate extension of human vision to follow the ball's shrinkage. But no matter how superbly effective the instrument of observation, the ball would begin to lose its color when its diameter shrank to less than 25 millionths of an inch; since this is about the wavelength of red light, anything smaller cannot reflect light as we know it. The ball would become not black, but simply unimaginable in terms of color.

As it continued to shrink, the ball would remain detectable by shorter waves, which can be received by instruments other than the eye. Ultraviolet rays would continue to bounce off it until it was less than one millionth of an inch in diameter. Eventually the ball would lose any appreciable shape, size, or position. At 40 trillionths of an inch in diameter – the size of an electron – it could no longer withstand the impact of the most lightweight ray or particle that could be hurled in to rebound from it. Any probe, devised experimentally or imagined theoretically, would succeed, not in "seeing" the ball, but in knocking it galley-west.

This is precisely the position in which the electron itself is found – that is, in no position at all. The very act of finding an electron is synonymous with dislocating it. It is impossible to know, at one and the same time, where an electron was when you found it and where it is when you become aware of the finding. It is, in sum, impossible to pinpoint the place and speed of movement of any one electron at any one instant.

The scientist attributes to the electron many valid descriptive qualities – spin, charge, mass – which are deduced rather than observed directly. If he says, however, that the electron has no position or definite size, or that it behaves sometimes like a particle, sometimes like a wave, he is using words that refer to sensory qualities meaningful in our own world but altogether irrelevant in the world of the microcosm. To look for color in an electron is like trying to smell a sunbeam.

The dilemma posed to the human senses by the individual electron – or by any other subatomic particle – was met head on in 1927 by the quantum physicist Werner Heisenberg, when he enunciated the so-called "uncertainty principle." According to this principle, we can never justifiably attribute to the electron, at any one time, all the everyday descriptive qualities of common-sense objects. The very act of finding it traveling at a given speed will throw its whereabouts into question. The very act of determining its whereabouts will give it a speed or direction that is unknown. Locating electrons in space and time can be achieved only by applying the laws of probability. As with human beings, the electron population is statistically predictable, but the behavior of capricious individuals within the group is not.

The concept that chance governs the behavior of electrons has played hob with traditional notions that a particular cause produces a particular effect. In the subatomic world, one event is not a *necessary* consequence of another event, but merely a *possible* consequence – with a calculable probability.

Many scientists feel that Heisenberg's principle of uncertainty is intrinsic to the whole universe; that the intimate details of the cosmos have a fuzziness which science will never be able to clarify or overcome. Other scientists, perhaps a minority, feel that the uncertainty may be the result of human limitations rather than of any unpredictability on the part of nature itself. In effect, Heisenberg's dictum may simply be a reminder, to the human being who seeks to capture the spirit of the cosmos, that attempts to visualize its minutiae are futile. The Biblical injunction against the making of graven images may still hold.

More and more, scientists regard their concepts as "models" that are analogies of nature rather than photographic likenesses of it. Furthermore, different models may be used to represent the same concept. The atom has sometimes been pictured as a solid sphere armed with hooks; at other times it has been thought of as a pudding of positive electric charge filled with negatively charged raisins – the electrons.

In the model of the atom that has proved to be most widely useful – the one conceived by Niels Bohr in 1913 – it is compared to a miniature solar system. If we were to magnify such a solar system 10 trillion times, we would get a wild picture of the sort of wild picture that scientists must contend with. The "sun" in this system – the nucleus of the atom – is as big as a marble, but unimaginably denser. The nearest "planet" (electron) orbits a third of a mile away. It is lighter, fluffier, and larger than the "sun" – about the relative size of a fifty-foot gas balloon. If an earthling were placed beside the same model and magnified the same number of times, he would be 10 billion miles tall, and his head would be far beyond the outermost planets.

The incredible complexities of the electron, though, they may stagger the imagination of the layman, have proved immensely instructive to the scientist. He continues to look for manifestations of nature. But when he finds them he knows they are simply manifestations – facets of reality which happen to suit a particularly useful, and often beautiful, idea that his mind has created. His method, in short, is a means of creating new ideas and then of selecting the ones that best fit the appearances of nature.

In a manner of speaking, the scientist stands in a house of mirrors – mirrors that he himself has built, and that he himself keeps improving so they give increasingly detailed and accurate reflections. Reality continues to elude him, but he can see it darkly in one or another of the flawed glass surfaces before him. Some of the mirrors give one distortion, some another; all offer only reflections. By piecing together these reflections, the scientist seeks ever better approximations of truth, but he no longer expects to see truth naked.

II. Energy

Basic to an understanding of life is the knowledge of energy, its pathway through living systems, conversion into usable forms, and utilization. These aspects are usually well presented in the initial chapters of biology texts. However, there are many problems of concern beyond the traditional presentations for today's student.

As long as man was a member of natural ecosystems and depended upon them for his consumable products, few problems developed. But as agronomic societies grew into technological ones, many problems appeared. Crop surpluses in the developed countries and mass malnutrition among developing nations are but two indications. A further acknowledgment was the election of Norman Borlaug for the 1970 Nobel Peace Prize. This ''dirty-handed'' agronomist was cited for his success in applying technology to the solution of world food problems.

The role of the green revolution in dealing with man's food problems is considered in two of three essays in this unit. Gene Gregory's ''The New Food Revolution'' is an optimistic overview of man's recent and projected developments in providing a greater worldwide supply of food. Gregory, a chemical engineer, spent many years in southeast Asia as a journalist and editor. Clifton Wharton, economist and president of Michigan State University, in his paper, ''The Green Revolution: Cornucopia or Pandora's Box,'' gives a viewpoint tempered with skepticism. His words are not unlike those of India's Prime Minister Indira Gandhi who, in advising her chief ministers, said, ''The warning of the times is, unless the green revolution is accompanied by a revolution based on social justice, the green revolution may not remain green.''

''Aquaculture'' by John E. Bardach of the School of Natural Resources, University of Michigan, states the roll of aquatic animal culture in the production of high-grade protein food.

2.

The New Food Revolution

Gene Gregory

Nature has been "reasonably impartial" in distributing her riches, both natural and human. We have the knowledge of technology, and hopefully, the understanding of human nature necessary to assure that the world will in the future by well fed **AUTHOR**

Until relatively recently, all too few practical measures had been taken to come to grips with the problem of producing more food of the kind that was needed in the places it was needed most.

Much attention has been given to throttling, by medicinal or other artificial means, the population explosion – ironically enough, due largely in the first place to the introduction of modern medicine in less-developed countries. But one serious estimate, by the U.S. scientist and educator Harrison Brown in "The Challenge of Man's Future," holds that the earth could – with technology now within our reach – sustain a population of 50 thousand million, which even at the uncontrolled rate of present population growth would not be attained for four or five generations.

Since the evidence to date is that birth control methods simply are not working, and are not likely to be fully effective for another generation or two, more immediate solutions to the world's food problem must be sought elsewhere. Highest priority must be given to the expansion of food production, within each of the lesser-developed countries.

Food aid, while beneficial as a stopgap measure to relieve famine caused by drought or flood, tends to attach the symptoms rather than the causes. While necessary in times of distress, even when food aid is maintained for a few years only, it breeds an unwholesome attitude of dependence among the recipients and makes it very difficult for local agriculture ever to get properly organized in face of competition from 'free food."

Even where shipments of agricultural goods have been used to generate local currencies which in turn become part of the overall funds available for economic aid programmes, those funds have too often been diverted into the development of new non-agriculture-related industries rather than improving and increasing food production.

The possession of a steel mill, a national airline and a polyethelene plant have been considered as symbols of progress – but were in fact monuments to the misconceived notion of planners that somehow these investments constituted a short cut to the solution of underdeveloped problems.

Even though as much as 80 per cent of the population of the less-developed countries is engaged in agricultural pursuits, according to a recent review of the Development Assistance Committee of the Organization for Economic Co-operation and Development(OECD), only 12 per cent of the total aid flow (including food aid) went to agriculture. As a result of misplaced emphasis on industrialization,

Reproduced from the March 1969 issue of the *Unesco Corrier* with permission.

agriculture was deprived of needed local investment inputs at all levels – irrigation, manufacture of fertilizers and pesticides, research, education, processing industries.

In a sense, the fearful spectre of a world food crisis was a blessing in disguise. There came a widespread recognition of the truth that while massive emergency food aid could alleviate much of the immediate suffering, a change of emphasis in foreign aid programmes was long overdue.

But perhaps even more important, the urgency of immediate action on the part of the governments of lesser-developed countries provoked a general up-grading of agricultural priorities for investments as well as a lifting of some of the impedimenta that have prevented private foreign capital from investing in industries for the manufacture of badly needed agricultural inputs and the processing of farm products for the home and export markets.

But, as yet, in the lesser-developed countries, all of them agricultural countries, there has been almost no agricultural research of higher agricultrual education. Three exceptions are examples for all governments to imitate:

Mexico, which had for years imported half of the wheat consumed in the country, after an extensive wheat research programme begun in 1944 and conducted by the Rockefeller Foundation, was self-sufficient in this grain by 1956, and in 1964 exported a half million tons.

At Los Banos in the Phillipines, research sponsored jointly by the Rockefeller and Ford Foundations developed new varieties of rice, the IR 5 and IR 8, which increase yields up to 8.5 tons per acre – or about 15 times the average yield of traditional varieties. As a result, in the short span of about three years the Phillipines has been transformed from a rice deficit to a rice surplus country.

American aid agronomists adapted improved varieties of Guatemalan corns to Indonesia and Thailand which did, in fact, produce something of an agricultural revolution in the latter. In eight years the crop in Thailand increased from non-existent production to an export of $104 million, the fourth largest corn export in the world.

Extended to other critical food deficit areas, the results of these research efforts, largely privately financed, have produced a veritable "green revolution" in the world's most food deficient regions which just two years ago were threatened by widespread and seemingly continuing famine.

In India, the area planted with new varieties of wheat, rice, millet and sorghum increased from about 23,000 acres in 1965-66 to nearly 4 million acres one year later. As a result, India harvested 95.6 million tons in 1967-68, and expects to achieve self-sufficiency in 1970-71 (see text opposite).

Pakistan, which imported large quantities of Mexican wheat seed, will harvest a record 5.5 million ton grain crop this year; and by 1970 should be producing twice as much food grains as in 1965, enough to meet the country's requirements.

Particularly noteworthy results have been obtained by development teams of the Republic of China in African countries where they have introduced their Taiwan Native 1 rice species. Extensive experiments in the Cameroons, Gabon, Gambia, Malawi, Niger, Senegal, Chad and Togo by these teams have more than doubled production obtained with traditional rice strains.

As a result of these and other developments, farming has suddenly become both respectable and lucrative in many countries.

In Rome, early last October, Addeke H. Boerma, Director-General of FAO, expressed cautious optimism that "the long lag in food production which plagued the past decade may be coming to an end." And in Paris, he told delegates to a Unesco international scientific conference on the use and conservation of resources of the biosphere, "I believe that the agriculture of the developing countries is now reaching the point of "take-off". (See the "Unesco Courier," January 1969.)

But the coming agricultural revolution is not confined to progress in the developing countries. New but untried ideas will displace a whole raft of outdated notions about food production – first in the advanced industrial countries, but eventually profoundly affecting the developing countries.

For the first time in history the prospect is that there will be enough food produced to feed, and feed well, all mankind – despite the fact that the population will grow in the next 31 years as much as it has in the last 1,000,000 years. Predictions for the remaining three decades of this century, based upon a two-year study of U.S. agriculture just completed, include:

Wheat harvests will jump to 300 bushels per acre, more than ten times present yields.

Corn output will reach 500 bushels per acre, against today's average 75 bushels.

A single cow will produce as many as 1,000 calves in its lifetime, as compared to the present 10, through the use of hormones which keep the entire cycle of ovulation and pregnancy in cows and other domestic animals under exact control. Embryos will be transferred from special breeder animals to ordinary "incubator" animals with substantial increases in prolificacy of a herd and a radical improvement in quality animal production.

Milk production will soar to 30,000 pounds per cow from today's 8,000. Synthetic milk will also be manufactured from carrot tops and pea pods.

Today's farm labourers will be increasingly replaced by "farm engineers," scientists and technicians trained in fields ranging from electronics to air conditioning.

Farming will become automated and computerized, with tilling of the soil, planting and harvesting crops and regulating the growing process controlled from a centre equipped with computers, radar and remote control devices.

Fields will be covered by huge plastic domes, or by other means enabling the constant control of environment affecting crop growth.

Remote-controlled harvesters will not only pick but also grade, package, freeze and deliver food to wholesale depots.

Space satellites will collect up-to-the-minute data on soil conditions, insect damages, crop growth patterns throughout the world, as well as make long range weather and production forecasts.

But this is just a beginning, In the United States alone there are presently some 30,000 research projects under way, and spending for agricultural research is more than $850 million annually.

While today 7,000,000 Americans live on farms producing many times the volume of food 31,000,000 farmers produced a century ago, by the year 2,000 a scant 2,000,000 of the 300,000,000 people in the U.S. will be "farmers." The average U.S. farmer now produces enough food to feed 40 people, four times the number he could feed 30 years ago; in the next 30 years, productivity will increase another four-fold based upon technology already known and within reach.

But in agriculture, as in other fields that are increasingly science – and technology – based, half of what is known today will be obsolete by 1980. Or, to put it another way, at the present rate of technological advance, half of the agricultural technology of 1980 has not yet been discovered.

One of the most immediate effects of the coming agricultural revolution, both in the advanced and in the developing countries, will be an increase in capital intensive farming. Post World War II increases in mechanization in the advanced countries have already far surpassed the "tractor stage":

Airplanes are increasingly used to spray test-tube "miracles" over crops and pasture lands.

Computers determine when and how to breed calves by artificial insemination.

Devices such as the "autosow," which acts as a kind of mechanical mother for baby pigs, are turning pig farming into a mass production industry much like the poultry business.

The "Bacon Bin," a recently devised total confinement production facility – complete with automatic feeding and manure disposal systems – can raise and fatten 1,000 hogs at once, from birth to maturity, in record time.

Major international food companies continue to expand their large-scale farming activities, primarily in developing countries. But since foreign-owned "plantations" smack of "colonialism," and for a host of other impedimenta common to the newly independent countries, there have been limited opportunities for private business to contribute directly to the development of badly needed food industries through this kind of direct investment.

With the change of attitudes in the developing countries toward the development of the agricultural sector, the day may not be far off when direct foreign investments in agricultural enterprises may again be welcomed, but for the moment the opportunities for investment in the developing countries lie in other equally interesting sectors vital to increasing food production.

Someone must supply the new varieties of wheat, rice, corn, millet and sorghum in massive and continuing quantities at relatively low prices. And the new "miracle" seed varieties are not enough by themselves to insure the fruition of a green revolution in the developing countries. Suitable quantities of fertilizers, insecticides, water and farm machinery are necessary. Credit, marketing, and other services of modern agricultural systems must be made available.

The way the new seed varieties were introduced into the Philippines, India and Pakistan had as much to do with their success as does the quality of the seeds themselves. Rice "kits" consisting of the seed, fertilizer, insecticides needed to grow a half-acre of one of the seed varieties were sold or distributed together. This novel distribution concept presents a special challenge to abilities of international corporate enterprises to develop the required joint marketing facilities.

The governments of developing nations, local private business and the farmers cannot sustain this effort without outside help. They lack the means to do the adaptive research, the facilities needed to train farmers in modern agricultural techniques, the public finances for large-scale irrigation and rural electrification projects, or for adequate storage and modern transport facilities. They often lack the required technology, capital and managerial personnel necessary to build new plants to produce the required seed, fertilizer, insecticide and other essential inputs.

To meet these needs, governments of OECD countries have appreciably increased their aid to agricultural programmes of developing countries in recent years. The World Bank announced in October 1968 that it would quadruple loans to speed the development of world agriculture. The Inter-American Development Bank now commits about 27 per cent of its funds for agricultural development purposes, and the Asian Development Bank – formed just two years ago – is presently engaged in a country-by-country survey to assess priority projects to be considered for assistance by its projected Agricultural Development Fund.

But producing food is business. And many of the resources needed by developing countries to get the nascent agricultural revolution off the ground are the almost exclusive possessions of private enterprise in western industrial countries.

To mobilize the resources of private industry for food production in developing countries, the FAO has created the FAO/Industry Co-operative Programme, a joint effort by the FAO and international business to resolve problems involved in the establishment of industries in the developing countries to manufacture fertilizers, pesticides and farm implements or to process food and other agricultural products. At its March 1968 meeting, representatives of 37 major international firms met at FAO headquarters in Rome to work out a widely acceptable code of conduct for the protection of private foreign investors in developing countries.

The most tangible sign of the effectiveness of this programme has been in India's crucial fertilizer industry. Until the drought of 1965-66, administrative and political problems had prevented an effective involvement in international private business in major new ventures in India. After the creation of the programme, the FAO actively assisted the search for ways to eliminate obstacles to new investments in critical food or food-related industries.

As the urgency of increased agricultural production grew, the climate for investment in India in these industries began to change for the better. With the 50 per cent jump in fertilizer consumption in 1967 having contributed substantially to the rapid recovery from near-famine conditions, top priority has been given to the development of 15 major fertilizer factories.

But fertilizer production is just one field of immediate opportunity for international business. Almost all developing countries have undertaken more or less extensive agricultural training programmes, which hold the key to the introduction of new farming techniques including the mechanization of farm production. The South American Farm Mechanization Training Centre at Buga, Columbia, illustrates how such programmes can be effectively developed through co-operation among industry, the FAO and governments.

With FAO's guidance through the Freedom from Hunger Campaign, the Colombian National Training Service and a British farm machinery manufacturer set up the centre to train technicians from Colombia and other Latin American countries as instructors in farm machinery operation and repair and as government extension officers.

With wide experience in such training programmes in the United Kingdom, France, the Federal Republic of Germany and each of the Scandinavian countries, the British firm was in a position to provide the necessary instructional staff, as well as a broad range of tractors, equipment for workshops and teaching aid. The company's donation to this project, which will eventually amount to approximately $300,000, represents a far-sighted investment in South American agricultural development.

Another field of potential industry-government co-operation involves one of the most widely discussed developments in the world food industry, a process developed by a British petroleum company at its French refinery for producing protein from crude oil. The final product, a yellowish, odourless powder, is comparable to fishmeal in composition. But, like fishmeal, it is not yet suitable for direct human consumption. It is, however, suitable for feeding to animals and poultry, which in turn convert it to the traditional and still most acceptable food forms – meat, milk and eggs. (See article page 21).

While some experts say that it will be at least ten years before we will see human food derived directly from protein produced from petroleum or from natural gas, several oil and chemical companies consider the potential demand for the product great enough to begin their own development programmes. India has already started producing proteins from petroleum hydrocarbons with pilot plants under the direction of the Indian Institute of Petroleum at Dehradun and the Regional Research Laboratory at Jorhat.

A number of high protein concentrates have also been developed from various vegetable and fish sources. Dr. Norman W. Pirie, head of the biochemistry department at Rothamsted Experimental Station in England, has developed one of the most recent – a process for extracting protein from green leaves, such as pea cannery waste.

A Swedish company is now producing a fish protein concentrate which mixed with water and cocoa powder, has been used recently in Nigeria and Biafra.

At the American University of Beirut, a new supplementary – called Aubina – has been developed from wheat, Egyptian chick-pea, bone-ash and skim milk, and tests have shown that children find it good to eat.

Both soya and cotton-seed have long been sourcea of protein concentrate. But while the day may not be far off when meat, bread and even bacon made from these materials will be part of man's normal diet, so far protein- rich foods have received an extremely cool reception.

While there is a certain compelling logic in the argument that protein concentrates are "good for you," there still remains the problem of how the stuff tastes. And experience shows that unless people are really at the starvation level, they are particular about the taste of the food they eat.

This was the sum and substance of a number of scientific papers presented to the combined FAO/WHO/UNICEF Protein Advisory Board last September. According to Dr. O. Ballarin, an executive of the Brazilian subsidiary of a Swiss food company, all efforts by private enterprise in South America to sell and promote high-protein produce have met with disappointing results, and in some cases failed completely. In Peru, production of a high-protein fish flour produced from anchoveta was discontinued after sales sank to a disastrous low. Similar difficulties have been met with in Africa.

Specialists give these reasons for consumer resistance to something that is obviously "good for them":

Traditions regarding food are stronger in developing countries than in developed ones. New foods, delivered in a new form or a new package, are looked upon with suspicion.

Most people ignore the meaning of the word "protein," which has no commercial appeal.

People shun high-protein products for social reasons, regarding them as "the food of the poor."

In Ethiopia, some of these obstacles were overcome by the nutritionists, who called their product Faffa (meaning "grow big and strong" in the local language). Faffa was a success. But in many other countries, where illiteracy ranges up to 85 per cent, and people cannot even read the name of the product, promotion campaigns have to be carried out through films, radio and television.

Nevertheless, specialists such as Dr. Husain Ali Bhimjee Parpia, Director of the Central Food Technological Research Institute in Mysore, India, are firmly convinced that protein malnutrition can be overcome in much of the developing world if the available resources of oil-seed protein are utilized more efficiently. As soon as acceptable products can be perfected and attractively presented through regular commercial channels, vegetable protein sources will certainly make a major contribution to solving present nutritional problems.

Weaning foods, biscuits, ice cream, beverages and candies are the preparations which are now under consideration by major international food processors. Three

American companies are now apparently convinced they are on the brink of a major breakthrough in this field. One is planning to manufacture and sell throughout Asia a highly popular Hong Kong soybean drink with a satisfactory amino acid content. A major soft drinks bottling firm has developed a chocolate-flavoured, soybean-based high-protein beverage, fortified with vitamins; this potent concoction has already been launched on the Brazilian market. In El Salvador, a third company has just completed consumer testing on a new high-protein soft drink made from corn meal, cottonseed and nut meal.

But if the development of synthetic nutrients provides the basis for an entire new sector of the food industry, prospects for expansion in the exploitation of the world's marine resources is undoubtedly more immediate. The sea is, for the foreseeable future, an almost limitless source of high-protein content food. Annual production of plants within the oceans, according to the FAO's Fish Stock Evaluation Branch Chief, John Allan Gulland, is approximately 130 thousand million tons – hardly any of which is used for human consumption. And of the 23 thousand million tons of fish and shellfish the ocean grows per year, the present world-wide catch is about 60 million tons.

While much of the plant life – the phytoplankton – is generally microscopic and neither easy to harvest nor attractive to use, there are vast possibilities for expanding the fishing industry. At present the world catch of fish, rising annually by from 6 to 8 per cent is growing faster than the world's population. World fisheries production increased from 19 million metric tons in 1948 to 60 million tons in 1967, or at a rate much faster than food production from the land.

But present fishing methods are still traditional and primitive. Production might well be doubled in the next ten years through the use of new electronic equipment which detects the presence of schools of fish, speeds the movement of nets, and even lures fish into the nets.

While most of the world's fishing grounds are being quite intensively worked, the Indian Ocean – one of the world's least known oceans - may well be one of the richest and most underdeveloped. Recent estimates submitted to the FAO's Indian Ocean Fishery Commission indicate that the yearly catch of fish in the Indian Ocean could be increased about ten times, from its present level of 2 million to 20 million metric tons annually – without damaging the stocks. In the more immediate future, current production could be raised to 10 million tons a year, putting the ocean's fisheries on a par with those of the North Atlantic, which yield approximately 11 million tons a year.

Still, if fishing remains confined to the types of fish now exploited, experts estimate that the world catch would probably approach its physical limit within the useful working life of fishing vessels now being built. Alternative sources of natural production include both fish and other animals more or less unfamiliar to most fish consumers.

For instance, squid of various types appear to be abundant in many parts of the world, but exploiting the very large quantities of this creature – now eaten readily only in Japan and Mediterranean countries – depends as much on changing the tastes of the consumer as on being able to catch the animals economically.

Several developments over the past decade in the conversion of fish into seafoods for indirect or direct human consumption have been little short of spectacular. Large quantities of fish are now converted into meal, which is in turn indirectly consumed in the convenient protein form of meat, eggs and chicken.

Fish protein concentrate (FPC), a high protein powder made from otherwise inedible fish, is also highly suitable for direct consumption by people. FPC is dry, can be stored indefinitely and transported conveniently without wastage. Both tasteless and odourless it is suitable as an additive to other foods such as breakfast cereals. Added to animal foods, it will lower the production cost of meats, poultry and eggs by shortening the growing period and improving their quality.

Even more important investigations are currently being made in the U.S.S.R. and Japan into the possibilities of exploiting the vast quantities of Antarctic krill, *Euhausia superba* .This small shrimp, about five centimetres long, is the main food of the balloon whale. Rough calculations, based partly on estimates of the quantities that must have been eaten by whales before whaling drastically reduced their number, suggest that the annual yield of krill could be at least 50 million tons – or about the equivalent of the entire world fish catch at present.

Soviet experiments show that catching the krill in quantities may be quite easy and that an acceptable form of meal for animal feeding can be produced. A more interesting prospect is that, if the technical problems of separating the meat from the shell can be resolved, krill may be used for direct human consumption. Properly presented, in the form of imitation lobster tails, for example, it would appeal to the rapidly growing number of consumers of the much similar prawns, shrimps and other crustaceans.

Much has been made these days of advancing fishing from the primitive stage of hunting to the stage of agriculture. All the many forms of farming – pastoral, ranching, intensified gardening and factory farming – have their marine equivalents. Oysters, mussels and shrimp have long been raised in shoreline ocean beds and on ropes suspended from special rafts, just as the Japanese systematically have for centuries raised and harvested large quantities of *nori* (a highly nutritious edible seaweed) in their bays and inlets. And recently in Britain there has been an increasing interest in intensive marine farming of battery plaice and sole.

The importance of sea farming is primarily that fish can be raised with relatively cheap food. Unlike chicken, which can have no more than about 5 per cent of fish meal without tasting of fish, the diet of plaice or sole can presumably contain much more fish meal or other products – including perhaps meal from krill.

Last autumn, a subsidiary of an Anglo-Dutch food-detergents manufacturer, put the first crop from its fish farm at Loch Ailort in the western Scottish Highlands of the U.K. market. In its initial stage of fish-raising, the firm is now concentrating on raising rainbow trout – but with a difference. Trout farming in fresh water is a well- established industry in many countries. But Loch Ailort trout are raised in sea water and the largest – a little over two years old – tip the scale at 5.5 lbs. each. Not only do the salt water trout get bigger faster, but due to the careful regulation of their diet, the trout have delicate pink flesh resembling salmon.

Understandably, these pioneer efforts in fish farming have been concentrated on the highest priced varieties, but with a fish farm of eight square miles able to produce more fish than are presently caught in the entire North Sea, the business could soon take on massive proportions including a wide variety of popularly priced seafoods.

This emergent *aquicultural* revolution has its parallel in British *agriculture,* and Britain's "convenience" food industry has played a major role in both. The experimental and developmental fish farming was a logical outgrowth of the heavy investment in the British frozen foods industry.

To meet the tight standard specifications, the introduction of new varietal strains and improved growing techniques have also produced substantial yield increases in all crops grown under contract for the convenience food industry. Pea yields, for example, have risen by more than 40 per cent while other crops such as spinach and broad beans have increased even more.

Forward planning of the industry, with its ability to forecast sales and plan production tonnages, has brought a measure of stability to vegetable growing never before known. Likewise, as a result of the development of convenience foods, average output of poultry in Britain more than doubled during the past ten years.

But a comparison of per capita British consumption of frozen foods in 1967 (11.4 lbs.) with consumption in the U.S. (68.0 lbs.) indicates that the convenience food revolution in the U.K. has just really got under way. And the same conclusions can be drawn for Sweden (22.7 lbs.), Switzerland (15.4 lbs.), and Denmark (11.4 lbs.). In all other European countries and the lesser-developed countries it has hardly begun. What is true for frozen foods, largely holds for the entire range of convenience foods and the catering industry.

Consider further that the U.S. is on the threshold of still another stage in the revolutionary development of the convenience food industry. What are considered convenience foods now will be dwarfed by what will be accepted as convenience foods only a decade from now.

In spite of the critical dimensions of the world food situation at present, food production is already the world's greatest success story. Most developed countries have solved their basic problems – with the possible exception of the problem of mounting surpluses.

This great achievement of the past two decades can be attributed to reliance, not on grandiose theories, but on treating food production as a business – from the farm, to the processing stage and distribution – all requiring respected skills and capital inputs which are essential to ultimate success.

FAO Director-General Addeke H. Boerma recently said that he saw no intrinsic reason why yields in most developing countries should not grow at the same rate as those in the advanced and largely temperate areas.

Nature has been "reasonably impartial" in distributing her riches, both natural and human. We have the knowledge of technology, and hopefully, the understanding of human nature necessary to assure that the world will in the future be well fed. Through the close co-operation of governments and private business, of all nations, the last three decades of the twentieth century can go down in history as the age of the revolution that achieved this monumental feat.

3.

The Green Revolution: Cornucopia or Pandora's Box?

Clifton R. Wharton, Jr.

The Green Revolution offers him (the quiet, passive peasant) the dramatic possibility of achieving his goal through peaceful means. It has burst with such suddenness that it has caught many unaware. Now is the time to place it in its long-range perspective and to engage in contingency planning so that we may respond flexibly and quickly as the Revolution proceeds. AUTHOR

The application of science and technology to traditional agriculture has begun to produce dramatic results, above all in Asia. The rapid expansion of certain food grains in the developing world is being particularly widely heralded, and justly so, as the "Green Revolution." The discussion of the phenomenon tends to cluster around two views. On the one hand, some observers now believe that the race between food and population is over, that the new agricultural technology constitutes a cornucopia for the developing world, and that victory is in sight in the "War on Hunger." Others see this development as opening a Pandora's box; its very success will produce a number of new problems which are far more subtle and difficult than those faced during the development of the new technology. It is important to give careful attention and critical analysis to both interpretations in order to be optimistic about the promise of the Green Revolution where justified, and at the same time to prepare for the problems that are now emerging. The Green Revolution offers an unparalleled opportunity to break the chains of rural poverty in important parts of the world. Success will depend upon how well the opportunity is handled and upon how alert we are to the inherent consequences.

It is now generally known that major technological breakthroughs in food production are believed to have lifted the spectre of famine in the immediate future and to have postponed the prospect of Malthusian population disaster. Startling developments have been accomplished in wheat, rice and corn – major food staples in much of the developing world. The possibilities for doubling or even tripling production are based upon new high-yield varieties coupled with adequate supplies of water, fertilizer, pesticides and modern equipment. Overnight, the image of agriculture in the developing countries has changed from that of an economic backwater to that of a major potential contributor to overall development. The new varieties are rapidly spreading both within countries and across national boundaries. A recent estimate of the International Agricultural Development Service of the U.S. Department of Agriculture reveals that in Asia alone the estimated acreage planted with these new high-yield varieties rose from 200 acres in 1964-65 to 20

Reprinted by permission from *Foreign Affairs*, April 1969. Copyright 1969 Council on Foreign Relations, Inc., New York.

million in 1967-68. Traditional food-importing national like the Philippines and Pakistan are becoming self-sufficient and have the prospect of becoming net food exporters.

It will be no easy task to achieve the potential increased production offered by the new technology, particularly when it involves millions upon millions of diverse farms and farmers scattered over the countryside. If the increased production is in fact obtained, this will automatically produce a whole new set of second-generation problems which must be faced if development is to be sustained and accelerated. Therefore, two considerations need to be borne in mind. First, there is reason to believe that the further spread of new varieties will not be as fast as early successes might suggest. Second, the new problems arising out of the spread of the new technology, whatever its speed, need to be foreseen and acted upon now. The probable developments in each case have the greatest significance for economic growth and for the conduct of international relations.

The reasons for believing that the new technology will not in fact spread nearly as widely or as rapidly as supposed and predicted include, first, the fact that the availability or irrigated land imposes at least a short-run limit to the spread of the new high-yield varieties. Most of these require irrigation and careful water control throughout the growing cycle. In most Asian countries about one-fourth to one-half of the rice lands are irrigated; the remainder are dependent upon monsoons and seasonal rains. The speed with which additional land can be converted to the new technology depends on the rapidity with which new irrigation facilities can be constructed; and here the high capital costs are likely to be a retarding factor.

Large-scale irrigation projects can seriously strain the investment capacity of developing nations. For example, the massive Mekong River development scheme, involving Laos, Cambodia, Viet Nam and Thailand, has been estimated to require a capital investment over the next 20 years of about $2 billion, roughly 35 percent of the annual national income of the four countries involved and exceeding the annual net new investment of all the countries of Southeast Asia combined. Further, significant additional costs are involved in converting existing irrigation systems to the requirements of modern agriculture. Many of the old gravity irrigation systems were not designed to provide the sophisticated water controls demanded by the new varieties. (For example, each plot must be controlled separately throughout the growing season.)

Second, there are doubts about the ability of existing markets to handle the increased product. Storage facilities and transport are inadequate and crop grading often deficient. Not only must the marketing system be expanded to handle a larger output; there also is an increased need for farm supplies and equipment. Fertilizers, pesticides and insecticides must be available in the right quantities, at the right times, and in the right places. Given the inadequacy of the agricultural infrastructure, the need to expand and modernize marketing systems is likely to reduce the pace of the Revolution.

Because many of the new varieties, especially rice, do not appeal to the tastes of most consumers, it is difficult to calculate the size of the market. Some argue that until newer varieties which are closer to popular tastes are developed, the market will be limited.

Third, the adoption of the new technology is likely to be much slower where the crop is a basic food staple, grown by a farmer for family consumption. Such farmers are understandably reluctant to experiment with the very survival of their families. Peasant producers are obviously far more numerous in the developing world than are commercial farmers and the task of converting them to a more modern technology is considerably more difficult. So far, spectacular results have been achieved primarily among the relatively large commercial farmers. Some semi-subsistence farmers have begun to grow the new varieties, but the rate at which they adopt them may be slower.

Fourth, farmers must learn new farming skills and expertise of a higher order than was needed in traditional methods of cultivation. The new agronomic requirements are quite different as regards planting dates and planting depths; fertilizer rates and timing; insecticide, pesticide and fungicide applications; watering and many others. Unless appropriate extension measures are taken to educate farmers with respect to these new farming complexities the higher yields will not be obtained.

Fifth, many of the new varieties are non-photosensitive and the shorter term will allow two or three crops per year instead of one. Multiple cropping is good, but there may be difficulties if the new harvest comes during the wet season without provision having been made for mechanical drying of the crop to replace the traditional sun drying. In addition, there may be resistance if the new harvest pattern conflicts with religious or traditional holidays which have grown up around the customary agricultural cycles.

Sixth, failure to make significant institutional reforms may well be a handicap. There is evidence in several Latin American countries that a failure to make needed changes in policies now detrimental to agriculture, or a reluctance to effectuate the institutional reforms required to give real economic incentives to small farmers and tenants, has been primarily responsible for the very slow spread of Mexico's success with new varieties of wheat and corn to its neighbors to the south.

From all this one may deduce that the "first" or "early" adopters of the new technology will be in regions which are already more advanced, literate, responsive and progressive and which have better soil, better water management, closer access to roads and markets – in sum, the wealthier, more modern farmers. For them, it is easier to adopt the new higher-yield varieties since the financial risk is less and they already have better managerial skills. When they do adopt them, the doubling and trebling of yields mean a corresponding increase in their incomes. One indication of this is the large number of new private farm-management consultant firms in the Philippines which are advising large landlords on the use of the new seed varieties and making handsome profits out of their share of the increased output.

As a result of different rates in the diffusion of the new technology, the richer farmers will become richer. In fact, it may be possible that the more progressive farmers will capture food markets previously served by the smaller semi-subsistence producer. In India, only 20 percent of the total area planted to wheat in 1967-68 consisted of the new dwarf wheats, but they contributed 34 percent of the total production. Such a development could well lead to a net reduction in the income of the smaller, poorer and less venturesome farmers. This raises massive problems of welfare and equity. If only a small fraction of the rural population moves into the modern century while the bulk remains behind, or perhaps even goes backward, the situation will be highly explosive. For example, Tanjore district in Madras, India, has been one of the prize areas where the new high-yield varieties have been successfully promoted. Yet one day last December, 43 persons were killed in a clash there between the landlords and their landless workers, who felt that they were not receiving their proper share of the increased prosperity brought by the Green Revolution.

Other experts argue that the new technology's stimulus to production and income cannot be stemmed. It is true that the rapidity with which the new seed varieties have spread in country after country belies the customary view of an inert, unresponsive peasantry. In 1965, India began a program of high-yield varieties which set a goal of 32.5 million acres by 1970-71; last year's crop season saw 18 million acres already planted, which contributed to the most successful year in recent Indian agricultural history (some 100 million tons of food grains, 11 million over the previous record year of 1964-65). Self-sufficiency in food grains is predicted in three or four years. Other countries are experiencing similar situations where the demand for the new seeds is outstripping the available supplies and black markets are even developing in seeds and fertilizer.

Nevertheless, if we assume that the new varieties will continue to live up to expectations and spread rapidly and widely, the increased production will in turn lead to a new set of difficulties. First, large tracts planted in one of the new varieties may be susceptible to disease and infestation which could cause massive losses. Heretofore, reliance upon seed selected by individual farmers meant that neighboring farms growing the same crop usually planted two or more different varieties or strains. This heterogeneity provided a built-in protection against widespread plant diseases, since not all varieties are equally susceptible. But where a single variety is introduced, covering large contiguous areas, the dangers of pathologic susceptibility are multiplied. For example, the new wheat introduced from Mexico into the Indo-Gangetic belt in India and Pakistan has involved a small range of genotypes – and the same has been true in Iran, Turkey and certain Middle Eastern countries. Any change in the spectrum of races of wheat rust in any of these countries could threaten the wheat crop on a massive scale, since it would involve the entire area.

Two steps are necessary to avoid these dangers: first, a diversified breeding program which can continually produce new varieties; second, an able and well-organized plant protection service which can quickly identify dangerous outbreaks and initiate prompt steps to combat them. Both activities must rely primarily upon national organizations rather than the regional or international ones. Both demand a skilled, well-trained staff. Some nations have recognized these dangers and are taking steps to meet them, but others still have not been made sufficiently aware. Aid givers – public and private – who are responsible for promoting the new varieties bear an equal responsibility to promote indigenous research and plant protection services. The outbreak of any major disease which wipes out the harvest of thousands of farmers is far more likely to be blamed on the producers and spreaders of the miracle seed than on Fate. Agricultural development could be set back several decades.

Second, it is vitally important to expand the entire complex of services and industries required to achieve the higher production. Any government or foreign-aid agency which distributes the "miracle" seed but fails to provide the insecticide and fertilizer in the appropriate quantities when and where needed is courting political disaster; unless these inputs are available and used, some local, traditional varieties will outyield the new ones. A seed industry, agricultural chemical plants, processing and storage firms, factories producing hand sprayers, dusters, water pumps and engines – these are just a few of the agriculturally related industries which must develop if the Revolution is to take hold.

The skills and the capital needed cannot be provided solely by the public sector. Private capital must also be utilized. In a few countries the spread of the new technologies has already forced an abrupt departure from the previous practice of having government agencies serve as the major or sole distributor of the required inputs. Private industry, especially American, has stepped in to provide a new, more dynamic pattern of distribution. In the Philippines, for example, ESSO has become a major distributor of fertilizer and agricultural chemicals. Frequently, such ventures have involved links with local firms. In India, the International Minerals and Chemicals Corporation, with the Standard Oil Company of California, built a fertilizer plant with a yearly capacity of 365,000 tons; the U.S. firms provide the management but control is held by an Indian firm. Storage silos, seed multiplication firms and even integrated farm-to-retail firms are just a few of the activities where private U.S. resources are being harnessed to serve the Green Revolution.

Equally important are the increased farm services which are required, particularly agricultural credit. For example, from studies conducted at the International Rice Research Institute, it is estimated that whereas the total cash costs of production for the average Filipino rice farmer using traditional methods and varieties is about $20 per hectare, the cost rises to $220 when the new, high-yielding IR-8 is grown. Although the yield may increase threefold, leading to a net return four times greater than with traditional varieties, the farmer must have access to substantially greater credit to finance his operations. Especially for the poorer farmers with low cash reserves, who may want to adopt the new varieties, the village moneylender and merchant will not be adequate unless they in turn have access to additional funds. Indeed, the Green Revolution must be accompanied both by an increase in the amount of credit available and by the expansion and modernization of credit institutions and mechanisms. Tapping the capital markets in the modern urban sector must be encouraged, and ways must be found at the village level to mobilize local capital, especially the increased savings which are possible from higher farm incomes. The Green Revolution will generate increased cash which, if properly marshalled, can contribute to capital formation and agricultural progress.

Third, much more attention must be devoted to marketing the increased output. Where there has been semi-subsistence agriculture, the impact of the new technology upon the *marketed* product is even greater than on total production. If the crop is a food staple and if the peasant farm family traditionally consumes some 70 to 80 percent of its total product each year, a doubling of output does not lead to a doubling in the amount retained for family consumption. Some modest increase in consumption is likely, but the bulk of the increased production will enter the market. Thus a doubling in yields in a semi-subsistence agriculture usually leads to much more than a doubling of the amount sold.

The impact of this explosive increase upon the traditional marketing network and storage capacity can be calamitous. The case of India is illustrative. During the past crop year, India experienced a marvelous increase in food-grain production, but the marketing network and storage facilities were not prepared to cope with it. The result can be seen in the mountains of food-grain stored in schools and in the open air under conditions which are apt to reduce if not negate the gains. The food-deficit psychology which underlies the failure of planners and policy-makers to anticipate these results is not limited to the developing nations. Aid givers were equally surprised. Strangly, the lessons of the Indian experience do not yet seem to have affected the thinking and planning of other nations which are promoting the new technology.

Fourth, the slowness with which the food-deficit psychology dies also has an important consequence in terms of government pricing policies. The fact that agriculture, even semi-subsistence agriculture, does respond to price, is only gradually becoming accepted. But the shock which quantum jumps in food production may have on domestic prices has not been sufficiently appreciated. The downward pressure on prices, especially where transport is deficient and storage is inadequate, may in fact be so severe as to have a disincentive effect upon producers. Unless adequate attention is given to developing a sound pricing policy to prevent excessive dampening of incentives, the spread of the new technology may in fact be cut short before any "takeoff" has occurred. Premature discouragement could produce a reversion leading to a slowing up in food production or even a rejection of the new technology.

It has been amply demonstrated throughout the world that peasant and subsistence farmers are responsive to favorable prices, provided the return in real and they receive the benefit. For example, from 1951-53 through 1961-63, the farmers of Thailand in response to favorable prices increased their exports of corn at an average annual compounded rate of 35.8 percent; casava, 25.0 percent; and kenaf, 43.8 percent. Filipino farmers responded to a governmental price-support program for tobacco by changing from native to Virginia tobacco and then booming production from 3 million kilos in 1954 to over 30 million kilos in 1962. This list of crops where peasant farmers have responded to favorable prices is large – rubber, oil palm, coffee, jute, wheat, barley, sorghum, millet, gram, cotton. Thus, if the full potential offered by the new technology is to be realized, every effort must be made in insure that there is in fact a significant return to the producer and that the rapid rise in output does not lead to a counter-productive slump in prices.

Fifth, the goals of increased food production are frequently couched in terms of some desirable, minimal standards of nutrition. Such nutritional goals are commendable, but they can be attained only by individuals who have the income with which to purchase the better diet. Effective demand for food depends upon both the income of the demanders and the price of the food. If the increased production leads to lower costs and prices, then consumers will be able to increase their food purchases and hopefully raise their levels of nutrition. Equally important is the need to increase incomes so that the greater production entering the market can be purchased. The food problem in a developing world is both a problem of production and supply and a problem of demand and income. Unless the higher levels of effective demand materialize, the prospect will be market gluts, price depression and, in certain cases, shifts by the farmers away from the higher-yielding varieties. Hence, every effort must be made both to reduce the unit costs of the increased food output and to augment the incomes of consumers who purchase food; otherwise, the second bowl of rice will not be bought – despite the technical feasibility of producing it.

Sixth, one of the major avowed aims of most nations which are eagerly promoting the Green Revolution is to achieve self-sufficiency in food production. In Southeast Asia, for example, the Philippines already claims to have become self-sufficient. Malaysia predicts that she will be self-sufficient by 1971; Indonesia by 1973. Some believe that these target dates are overly optimistic. But if the rice-deficit nations of the region such as Malaysia, Indonesia and the Philippines eventually become self-sufficient by successfully adopting the new technology, what will happen to the rice-surplus nations like Burma and Thailand whose economies are heavily dependent upon rice exports? To whom will they sell their rice? Self-sufficiency will not only be detrimental to the rice-exporting nations, but will reduce one of the few areas of economic interdependence in the region. Unless action is taken in advance to offset the predictable impact of the new technology, hopes of promoting regional economic integration will be substantially reduced. Whether or not one agrees with the goal of self-sufficiency for these nations, the policies have been adopted and will be pursued. Many developing nations spend some 30 percent of their foreign exchange on food imports and wish to eliminate this drain as well as the irritation of chronic deficits in domestic production. We should anticipate the predictable consequences of these policies – in this case major economic dislocations in trade – so that we can be equally ready with developmental efforts or foreign assistance to reduce the dimensions of the problems. Unless the exporting nations take immediate stock of their prospects and seek to diversify their agriculture, the impact of such trade distortions could have major consequences for their economies and pace of development.

Seventh, a critical question is whether these technological developments are a "once-and-for-all" phenomenon. How likely is it that new technological improvement will continue to be made? The application of science to agriculture over the last 300 years has resulted in a tenfold increase in yield per acre on the best farmed lands in the temperate zone. This expansion is what led to the production controls introduced by the surplus nations, such as the United States, to keep demand and supply in reasonable balance. Today's Green Revolution is the result of a similar application of science to agriculture in the developing world. But it should be noted that the institutionalized application of science is largely concentrated at present in food crops. Before World War II, primary attention in agricultural research in the developing world was devoted to the major crops – rubber in Malaysia, sugar in the Philippines, coffee in Kenya, palm oil in Nigeria, coffee in Brazil, bananas in Honduras. Staple food crops were either ignored or received scant attention. Thus the successes of the recent application of science to peasant agriculture could be interpreted as an exploitation of a "technical gap" in food crops left by years of neglect. If current developments merely represent a "catching-up," then as soon as population overtakes current developments, we are back to "square one."

Much will depend upon whether or not the necessary manpower is trained in each country to provide a continuing human resource which can produce a constant stream of new technology. The manpower trained in the Rockefeller Foundation's Mexican program has always been a greater contribution, in my view, than the new varieties. Successful adoption should not deflect attention from the importance and role of continuous agricultural research. The development of indigenous competence to engage in agricultural research is critical and becomes even more critical as the new varieties are adopted. The target should be not *a* new technology but ever-new technology, and this requires skilled manpower.

These are only a few of the possible consequences of the successful spread of the new technology. There are several broader consequences and issues which can be raised only as questions in this brief presentation:

To what extent will the diffusion of the new technology accentuate the displacement of rural people and heighten the pace of migration to the cities? If higher yields per acre, multiple cropping plus mechanization, force surplus manpower out of agriculture, what are the prospects for increased employment in industry and services to absorb this manpower?

For the average developing nation the Green Revolution means that instead of devoting two-thirds to three-fourths of its agricultural resources to food production, these resources may now be shifted to other higher paying crops. The question then becomes, what crops and for what markets?

If agriculture becomes more modern, dynamic and wealthy, will the non-agricultural sector allow agriculture to retain a significant share of this increased income or merely follow the previous patterns of taxing agriculture for non-agricultural development?

What will be the political significance of these changes if successful adoption of the new technology leads to an economically invigorated and strengthened rural population – almost invariably a large majority in developing nations? Will rural-based political parties and movements emerge to alter the recent dominance of urban centers?

What will be the global effect of a food explosion in the tropical and sub-tropical world? Will such developments lead to an improved reallocation of productive specialization among the developed and developing world, or will nationalistic trade barriers continue to flout natural comparative advantages?

One final danger lies in assuming that there is no longer an urgent need for measures to reduce rates of population growth. Quite the contrary. While the new developments are a splendid gift of time to allow a holding operation, effective population measures continue to be essential. Whether one assumes a growth rate of 2.5 or 3 percent, the inexorable fact is that, give or take a few years, the population of the developing world will double in about 25 years.

The significance of the food-population problem is more than humanitarian and developmental; it also has critical implications for the conduct of international relations. Relations between nations are often profoundly affected by long-run forces over which men can exercise only limited control in the short run. The food-population race is an excellent example of such a set of forces. Predictions regarding both population and food, as well as their interaction over varying lengths of time, must be taken into account in the conduct of developmental assistance, not only by aid-giving nations and international organizations, but by the governments of developing nations themselves. Policies and programs designed to win the race between food and population may have unintended, though often predictable, consequences which may have a very broad impact.

Charles Malik once said that "one of the principal causes of both international conflict and internal strife is unfounded expectations. These are based ultimately either on deception or on a belief in magic."[1] What we have in hand seems to many people to approach magic; let us hope that it does not become the source of deception.

To speak of the possible consequences and problems associated with the next phase of the Green Revolution should not be misinterpreted as a plea for the suppression of the Revolution because, like Pandora's box, it will lead to even greater problems than those it was designed to eliminate. On the contrary, I would strongly argue that the list of second-generation problems is a measure of what great opportunities exist for breaking the centuries-old chains of peasant poverty. They also demonstrate how closely interrelated are the various factors which impel or retard agricultural development. This complex interrelationship makes interdisciplinary research and coöperation vital if the current problems are to be solved and future ones anticipated. The most realistic prediction is that each country is likely to experience a different set of these problems and that there will be variations among countries between the two extremes of optimistic and pessimistic prognoses.

The quiet, passive peasant is already aware of the modern world – far more

[1]Charles H. Malik, "What Shall It Profit A Man?", Columbia Journal of World Business, Summer 1966.

than we realize – and he is impatient to gain his share. The Green Revolution offers him the dramatic possibility of achieving his goal through peaceful means. It has burst with such suddenness that it has caught many unawares. Now is the time to place it in its long-range perspective and to engage in contingency planning so that we may respond flexibly and quickly as the Revolution proceeds. Perhaps in this way we can ensure that what we are providing becomes a cornucopia, not a Pandora's box.

4.

Aquaculture

John E. Bardach

Aquaculture ranges in intensity from simple weeding of natural stands of algae to complete husbandry of domesticated fish like trout or carp. AUTHOR

Optimistic forecasters of the world food supply in the year 2000 predict that serious regional food shortages will exist; pessimists-warn of widespread famine as populations in many countries increase more rapidly than agricultural production. The one bright spot in this otherwise gloomy picture is the rate of growth of 8 percent per year of the world fish harvest during the last decade (*1*). The annual harvest now amounts to 53 million metric tons (*2*) and is expected to increase for some time to come, though estimates of the possible total sustained yield differ widely, ranging from 120 to 2000 million metric tons. All are based on extrapolations from scanty data on primary productivity throughout the world's seas and on incomplete knowledge of the food relations of many harvested organisms.

Aquatic harvests supply almost nothing but proteins, since the bulk of aquatic plants are plankton algae that are uneconomical to harvest (*4*) and hold no promise for furnishing carbohydrate staples. A breakdown of the total yield from aquatic ecosystems into marine, fresh, and brackish water moieties shows that the yields from fresh and brackish water make up about 14 percent of the tonnage (*2*). Much of man's fishing has been in estuaries and bays of the sea, and in ponds, lakes, reservoirs, and streams, where he has long practiced aquaculture (*5*).

Aquaculture resembles agriculture rather than fisheries in that it does not rely on a common property resource but presumes ownership or leased rights to such bases of production as ponds or portions of, or sites in, bays or other large bodies of water. Products of aquaculture must compete successfully with those of fisheries and of animal husbandry; in Western food economy, aquaculture products such as trout, oysters, and shrimp bring good returns because they fall in the luxury class, whereas in developing countries various kinds of raised fish command a high price (*6*), since animal protein, including that derived from marine catches, is generally scarce. Although subsidized small home or village ponds may be justified in certain underdeveloped areas to help alleviate malnutrition, aquaculture, wherever it it practiced, should be examined primarily as a commercial enterprise that must compete with other protein supplies to be successful.

The organisms now being raised in aquaculture comprise several bivalve mollusks (mainly oysters of the genera *Ostrea* and *Crassostrea*), a few crustaceans (predominantly shrimp, in particular *Peneus japonicus*), and a limited number of fish species (*6*). Among the fish species, the

Reprinted from *Science*, vol. 161, pp. 1098-1106, 13 September 1968. Copyright 1968 by the American Association for the Advancement of Science. By permission of the author and publisher.

carp, *Cyprinus carpio*, and selected other members of the same family, Cyprinidae (minnows), are the most important. Trout and salmon are also important in aquaculture, especially the rainbow trout, *Salmo gairdneri* as are the Southeast Asian milkfish (*Chanos chanos*), and mullets, especially *Mugil cephalus*, and yellowtail (*Seriola quinqueradiata*), in Japan. Also noteworthy are the channel catfish industry (*Ictalurus punctatus*) in the southern United States and the use of *Tilapia* as pond-fish, mainly in Africa (*6*). Most of these species are adjusted to life in fresh or brackish water, but the culturing of some marine fishes is being attempted, notably in Great Britain with plaice (*Pleuronectes platessa*) and sole *(Solea solea) (7)* and with pelagic (high seas) schooling species in Japan and the United States, among them the Pacific sardine and mackerel *(Sardinops caerula, Pneumatophorus diego) (8)* and the pompano *(Trachinotus carolinus) (9)*. Some attached algae are also produced under semi-cultivation, both in temperate and tropical waters and certain phytoplankton species are cultured as food for oyster and shrimp larvae. These represent a special crop and are of minor or indirect nutritive value; they are omitted from this article, in which the potential of aquaculture for supplying high-quality protein is assessed.

Aquaculture furnishes the world with over 2 million metric tons, mainly fish, Mainland China alone reports annual production of 1.5 million metric tons of carp and carplike fishes (*6*). Two million metric tons represent nearly 4 percent of the total world catch and far exceed the United States, food fish harvest, although they are produced from a fraction of 1 percent of the world's waters.

Aquaculture ranges in intensity from simple weeding of natural stands of algae to complete husbandry of domesticated fish like trout or carp. It is sometimes difficult to distinguish intensive management from culture. The term, as used here, comprises practices that subject organisms to at least one, and usually more than one, manipulation before harvest. In addition, as in agriculture, the harvest in aquaculture takes most, if not all the organisms tended. Most often only one species is raised, although a few to several compatible species may be cultured simultaneously.

To be productive for husbandry, aquatic animals should have the following characteristics. (i) They should reproduce in captivity or semi- confinement (for example, trout) to make selective breeding possible or yield easily to manipulations that result in the production of their offspring (for example, carp). Failing ease of breeding, their larvae or young should be easily available for gathering (for example, oysters). (ii) Their eggs or larvae or both should be fairly hardy and capable of being hatched or reared under controlled conditions. (iii) The larvae or young should have food habits that can be satisfied by operations to increase their natural foods, or they should be able to take extraneous feeds from their early stages. (iv) They should gain weight fast and nourish themselves entirely or in part from abundantly available food that can be supplied cheaply, or that can be readily produced or increased in the area where the cultured species live.

Few aquatic organisms have all these attributes, and substantial expansion of the aquacultural crop depends in part on how biological and engineering skills can make the missing characteristics less crucial; other constraints are economic. I discuss here several operations and problems common to the raising of aquatic organisms (*10*), and I attempt to appraise realistically the potential of aquaculture on a world scale.

Selective Breeding of Aquatic Stock

Even before Jacob tended Laban's flocks, livestock had been subjected by man to selection for one or another desirable attribute, and breeding of domestic birds and animals has produced spectacular results. The first treatise on fish culture was written in China by Fan Li in 475 B.C. (*11*), but there are still only two aquatic animals over which genetic control has been exercised. These are carp and several species of trout; trout has a shorter and less varied history of breeding than carp. No true breeding programs exist with invertebrates, though oyster culture is advancing so rapidly that experiments in oyster genetics are likely to begin soon.

The breeding of aquatic animals, compared with terrestrial animals, has peculiar problems. Spawning habits often make the isolation of pairs difficult; isolation of numerous offspring requires many replicate ponds (aquariums are too unlike nature); and there is rarely more than one mating a year. Moreover, the environment has an overriding influence on the growth of poikilothermous animals; consequently, many different-sized animals of the same age are found together. Many aquatic animals require special environmental or social conditions for mating and reproduction, which are not easily duplicated under human control. Manipulations of water temperature or flow have triggered spawning; however, the development during the last two decades of the practice of hypophyzation, or treatment with pituitary hormone (*12*), to make some fishes spawn helps alleviate contraints on breeding for some species. This practice has influenced fish culture all over the world, from catfish growers in the southern United States and sturgeon breeders in the Ukraine, to the fishpond coöperatives of mainland China, where it is of paramount importance in making common carp produce eggs three times a year and in facilitating the propagation of its cyprinid pond mates, whose eggs were difficult to collect in rivers before the process of hypophyzation was developed. Use of pituitary material may also produce advances with the breeding of two species of fish important in brackish water culture – the milkfish and the gray mullet. Aquatic animals have one advantage over terrestrial animals from the breeder's point of view – a pair have large numbers of offspring, which permit mass selection.

Carp are readily adaptable to selective breeding because their eggs are large for fish eggs; they are not too delicate; and they are easily secured. Carp have been bred for fast growth, a body shape with more flesh than is found on the wild type, reduction of scale cover for greater ease in preparing the fish for the table, resistance to disease, and resistance to crowding and to low temperature (*13*). With such breeding practices as progeny testing (selection of parents according to the performance of their offspring) and diallele analysis (a system of mating that determines separately the genotypes of each parent) (*14*), further improvements on already well-domesticated strains may be expected. It is necessary to prevent reversion to the wild type. That these can occur rapidly is illustrated by the fate of carp introduced to America. After being brought to the New World in 1877 (*15*), carp was allowed to escape into lakes and rivers where indiscriminate mixtures of its prolific stock resulted in bony and scaly fish which soon became a nuisance in waters used for game fishing. There was no incentive for carp culture in the United States, where protein was abundant from land livestock. However, since carp has become a prized angling trophy in Western Europe, and because of the rapid eutrophication of American lakes and rivers (a process which favors carp) and the predicted narrowing of the gap, even in America, between the supply of terrestrial protein and the demand, it is not farfetched to think that carp may be cultured in the United States.

Trout, at least in America, were until recently raised mostly for stream stocking; consequently, disease resistance was the main concern in hatcheries not equipped for experiments in fish genetics. Demonstration of what may reside in the trout's gene pool has come mainly from two sources: the Danish table trout industry (*16*) and the experimental trout and salmon breeding program of the University of Washington in Seattle (*17*) where specially fed rainbow trout stock, continuously graded by selection during 30 years, grows to as much as 3 kilograms in 18 months while a wild rainbow trout in a lake at that age rarely weighs 200 grams (*18*); these fish tolerate higher temperatures than their wild congeners.

Trout and salmon eggs are larger than those of the carp; they develop slowly and are hardy, combining several advantageous properties. Salmon permit the establishment of hatcheries on suitable streams because they return to spawn to the stream with the odor of which they are imprinted as fry. In such hatcheries inadvertent selection from the spawning run of the largest – fastest growing – brood fish has produced strains that returned to the hatchery 1 year earlier than the offspring of their wild congeners. Salmonid fishes can be selected for higher fecundity, large egg size, and better survival and faster growth of fry, and for exact timing of their return to the parent stream (*19*). These breeding potentials should be used to increase the abundance of salmon especially since improved techniques now feasible in United States salmon hatcheries could produce about ten times as many young fish as are now released (*20*).

Salmon-fishing regulations are still based on propagation potentials in natural streams and require that 50 percent of the sum be allowed to escape the fishery. Salmon runs will increasingly depend on hatcheries that program their fish to return for stripping and the raising of a well-protected progeny whose rate of survival at the time of release is many times greater than that attained in nature, where maximum fish mortality takes place during the first few months of life. Thus, the salmon harvest of certain river mouths may almost be doubled in view of the fact that hatchery-dependent runs need only a few fish to supply the next generation. Salmon are highly valuable fish [$65 million for the United States catch in 1965 (*21*)], and it may be worthwhile to press for regional revision of escapement regulations and to examine the economic requirements and consequences of hatchery improvements.

Another advantage of breeding fishes is the ease with which many of them hybridize (*22*). At the University of Washington at Seattle, male steelhead (that is, seagoing rainbow) trout were crossed with fast-growing freshwater rainbow females. The growth rate of the offspring was intermediate between that of the parents, their shape was more fusiform than that of the female, and they migrated to sea. They had a voracious appetite and adopted parent streams to which they returned as 2-year-olds, weighing 2 to 3 kilograms on the average and occasionally as much as 5 kilograms (*23*). They would probably not breed true in the second generation, and they should therefore be hatchery-produced, but they represent an interesting use of the sea's unused fish food.

Difficult as it may be to raise the progeny of one pair of parents of carp and trout, to do so with oysters is still more complicated. Mass spawning is usually done on oyster beds, and although the female of the genus *Crassostrea*, to which the American oyster belongs, retains the eggs inside her shells until after fertilization, paternity on the oyster bed is impossible to ascertain. Although there are thousands of eggs for each carp or trout, there can be up to 100 million for each female oyster (*24*). This fact, however, has aided in mass selection. Progressive growers of Long Island oysters raise the larvae in warmed water and use cultured algal food. They also give proper attention to stirring and other manipulations simulating planktonic conditions. The many eggs and improved survival of free-floating larvae permit a filter screen to be used to select only 20 percent of the largest, most rapidly growing early larvae. These larvae exhibit good growth throughout life (*6*). But to achieve true selective breeding, growers of Long Island oysters now plan to rear single oyster progeny; since oysters reverse their sex from male to female halfway through their adult lives, the possibility of freezing sperm from a functional male is being tested, and it may be possible to use it to fertilize the same oyster later when it becomes a female (*25*).

The Raising of Aquatic Larvae

Many aquatic animals go through larval stages which do not resemble their adult phase; some larvae, including those of shrimp or oysters, and of many fishes, are planktonic and minute and feed on the smallest organisms. More than with domestic birds or mammals, nursing them through their early lives poses difficult technical and nutritional problems to growers. In British experiments with raising plaice and sole larvae in captivity, as much as 66 percent survival through the stage of metamorphosis has been accomplished. Ultraviolet radiation of the water decreased the danger of bacterial infection; tanks without corners minimized encounters with solid obstacles; and salinity, temperature, and *p*H were controlled. The size of the first food offered was geared to the tiny mouthparts of the larvae, but was increased with their capacity to take larger live food. Nauplii of the barnacle (*Balanus balanoides*) were used at first; they were replaced by nauplii of the brine shrimp (*Artemia*) with subsequent admixtures of small oligochaetes (*Enchytrea*) when the small fishes had metamorphosed and were resting on the bottom. Finally, chopped mussels (*Mytilus*) were used. Since plaice larvae, just before settling, consume 200 brine shrimp nauplii per day, the production of several hundred thousand young plaice posed serious technical problems in continuous food culture (*7*).

Obtaining and correctly supplying food was a significant part of the experiments at the U.S. Bureau of Fisheries at La Jolla, California, with larvae of high seas schooling species such as Pacific sardines and mackerels. In these experiments very small food organisms had to be supplied at the precise time of complete yolk absorption, and in sufficient quantities to allow larvae of limited mobility to find food in all parts of the aquarium. Because sardine larvae search in only about 1 cubic centimeter of water per hour, at the onset of feeding, but require a minimum of four food organisms per hour to replace energy lost in swimming and body functions, the rearing of 2000 larvae in 1,800,000 cubic centimeters of water (500 gallons) meant replacing 7,200,000 food organisms removed by larval predation each hour or approximately 86,400,000 food organisms during a 12-hour day.

The large quantities of food organisms in varying sizes needed for these experiments were collected mostly at night. A 1000-watt underwater lamp connected to a submersible pump was suspended several feet below the surface of the sea. Copepods were attracted from a wide distance and concentrated near the pump where they were sucked up with water and transported to the surface. Plankton-enriched water was then passed through a series of filters, which further concentrated food organisms, and the highly enriched filtrate was piped to a 760 liter storage tank. Organisms with a cross- sectional diameter of 0.028 millimeter and larger were thereby collected. Before being fed to fish larvae, concentrated plankton was graded by filters to remove organisms larger than 0.1 millimeter. The portion containing large copepods, crab larvae, chaetognaths, and the like was fed to advanced fish fry and juveniles (*6, 8*).

Comparable techniques may help to achieve survival, after forced spawning in captivity, of milkfish and mullet. Inasmuch as these two species of economic importance in Asia are now raised from fry collected on the shores and as the fry are becoming scarce regionally, domestication of the two species, including manipulations ensuring high survival of fry will be an important advance for fish culture.

Although fish larvae are recognizable as fish even though they are not like the adults, invertebrates undergo more profound transformations from egg to adult. Oysters spend their first 2 weeks before they "set" as ciliated trochophores and veliger larvae needing flagellate algae for food. About 2000 cells of two or more species, for instance *Isochrysis galbana, Monochrysis lutheri,* and *Rhodomonas* and *Nannochloris* species, have to be available for each larva per day; and larger species are required to replace the smaller species as the larvae grow. Algae must be cultured en masse when oyster larvae are raised indoors, an innovation largely developed at the U.S. Bureau of Commercial Fisheries Biological Laboratory at Milford, Connecticut (*26*), and now expanded by progressive growers of Long Island oysters.

In shrimp raising, which is successful on a commercial scale only in Japan, there are problems with the larval stages before the animals can be fed chopped trashfish and shellfish. The operation, initiated by M. Fujinaga in 1934, begins in the spring with the collection of "berried" (egg carrying) females ready to release the stored spermatophores from their seminal receptacles; raising the water temperature speeds this and subsequent processes. After three distinctly different planktonic stages and 12 molts in about as many days, the postlarvae begin to crawl on the bottom; they still have to undergo some transformations and another 20 molts before they become adults. The early part of the life cycle of the cultured shrimp takes place indoors in ceramic tile-lined wooden tanks and in water heated to between 26° and 30°C. Diatom-mainly *Skeletonema costatum* – and flagellate cultures are maintained for feeding the early larvae, which are later given finely chopped mussel or clam flesh. When they have reached a length of between 15 and 20 millimeters or a weight of about 10 milligrams, they are stocked in outside ponds with arrangements for aeration and circulation.

In October or November, the shrimp, though not fully grown, are ready for market. They are about 10 centimeters long and weigh 20 grams having been fed once daily, converting 10 to 12 kilograms of food into 1 kilogram of shrimp. When the water later cools down to below 15°C, the animals no longer feed, but many of them may be retained without feeding for a later more favorable market (*6*).

The oldest shrimp-farming enterprise is now located near Takamatsu on Shikoku. It covers almost 10 hectares and has a staff of 30 men, including some in management research. Ten million shrimp were produced there in 1967, a quarter of which were raised to adult size; the rest were sold for stocking. The cost of production of cultured shrimp is certainly higher than that in any other aquaculture enterprise, but the wholesale price in Japan for tempura-sized shrimp of 6 to 10 centimeters is between \$12 and \$13 per kilogram, and the supply does not meet the demand. Shrimp farming of this type in a country whose material or labor costs are less favorable than those of Japan would not be possible (*6*). There are, however, opportunities for greater mechanization and for feeding innovations that will simplify the most laborious parts of culture operations for larval as well as postlarval shrimp. The use of the most improved shrimp-culturing methods with fast-growing species may hold some promise for a number of regions in the world (*27*).

Making Full Use of the Water

Aside from selecting the best suited strains, a practice not yet widely followed in aquatic husbandry, aquaculture should make use of the entire water column where possible and be three-dimensional, as it is in China and other Asian countries, where common carp is stocked with other species of the minnow family (Cyprinidae) such as the grass carp (*Ctenopharyngodon idella*), the silver carp (*Hypothalmichthys molitrix*), and the bighead carp (*Aristichthys nobilis) (28)*. The success of this method is based on the different food habits of the respective species; the carp is a bottom feeder; the grass carp and the silver carp feed on plants (banana leaves, even) and beanmeal or rice bran supplied to them from outside the ponds; and the bighead carp uses the plankton surplus in the well-fertilized water. Thus, the various water layers and all potential food sources are used (*29*).

The culture of oysters in Japan's best oyster-growing district, Hiroshima Bay, also is an illustration of the use of the entire water column (*6*). Seed oysters are collected on scallop shells suspended on wires from a bamboo framework driven into the bottom. Biologists from the prefectural and municipal laboratories monitor the plankton during the spawning period and advise the growers on the best time for spat collection. It is not uncommon to collect several thousand spat per scallop shell, although the average is about 200. The shells are removed from the collecting frames after 1 month when the surviving oysters have reached a size of about 12 millimeters. They are then cleaned, culled, and restrung on heavier wires separated by bamboo (and more recently by plastic) spacers. These wire rens are suspended from bamboo rafts, buoyed by floats of various kinds, and extend to a depth of 10 to 15 meters. Floats are added as the oysters grow, and before harvest require several times the support they needed at the beginning of the growing season. Long lines instead of rafts are an innovation in the method of suspension, but they are still only a variant on the hanging-culture technique, which uses the water column efficiently and which protects the oyster from its bottom-living predators, such as starfish and oyster drills.

A typical raft, about 20 by 25 meters carries 600 rens and produces more than 4 metric tons of shucked oyster meat per year. On a perhectare basis, this harvest amounts to 58 metric tons, if it is assumed that only one-fourth of a certain area of intensive cultivation is covered by rafts, as is the current practice. Such yields result from intensive care and high primary productivity in the water that is dependent on tidal exchange and fertile terrestrial runoff. By comprarison, the average is 5 metric tons per hectare of well-managed, leased oyster ground in the United States and the peak harvest of 300 metric tons per hectare of mussels (*Mytilus edulis*) also grown with hanging culture in the bays of Galicia in Spain. On public oyster grounds in the United States, where the mollusks are a minimally managed common property resource, the average per hectare is only 10 kilograms (0.001 metric ton) or less (*6*).

Fertilization and Feeding

Fertilization of bays, fjords, or enclosures has led to increases in phytoand zooplanktons, but favorable cost-benefit ratios for use with fish have not been proved (*30*). In ponds (including brackish ones) organic and inorganic fertilization has been efficacious. In Israel, fertilized carp ponds, some with admixtures of *Tilapia* and mullet, produce twice the tonnage per hectare of unfertilized ponds, and fertilization and additional feeding doubles the yield again (*31*). Fertilized *Tilapia* ponds in which the fish were also fed have yielded as much as ten times the crop of unenriched ones (*32*).

Many kinds of inorganic or organic fertilizers can be used, but sewage which produces dense invertebrate populations certainly works well. Munich sewage ponds with a slow exchange of water produce 500 kilograms of carp per hectare per year and a profit for their operator, the Bavarian Hydropower Company; the method requires large tracts of land, however; under temperate conditions rising land values threaten to make it obsolete (*6*).

In a much warmer, rapidly flowing stream in West Java, with a high sewage content, carp, confined in bamboo cages to graze on the dense carpet of worms and insect larvae in the sandy substrate, grow rapidly to yield 50 or more kilograms of fish per square meter of cage surface, or 500 metric tons per hectare (*33*). Even with allowances made for only partial use of the stream surfaces, this practice clearly represents an extremely efficient and ecologically sound use of sewage, especially in warm waters. The main drawbacks to this practice arise because the fish are not always well cooked before they are eaten.

In addition to the fertility of the water, its temperature, especially in a colder climate, is also very important. The most spectacular use of naturally warmed water for fish culture is in Idaho's Snake River valley trout-farming district, where springs of an even 16°C (optimum temperature for trout) gush forth from the canyon wall year in and year out. A thousand tons of troup can be raised in a year on every 2830 liters per second (100 cubic feet per second) that flow from these springs. Such unprecedented results in fish husbandry depend on high-density stocking, fast growth, mechanization, and cheap feed – the latter being locally procurable since the Snake River valley is also a stock feed-growing area (*34*). Most of all, however, the high yield depends on the flushing of growth-inhibiting wastes from the trout raceways. Hence, it is more appropriate to relate weight gain to water flow rather than to water surface or volume. By such a measurement, production would be around 170 kilograms per liter per second.

Naturally warmed water is not prevalent, but man-made heated effluents occur with increasing frequency. In fact, thermal pollution may become a threat to some natural waters because it hastens eutrophication. Heated power plant effluents, however, can also be used to the advantage of the aquaculturist. At the atomic energy plant at Hunterston, Scotland, cooling water, ascertained to be nontoxic to fish, was fed into cement troughs for sole and plaice raising. Both species were grown to marketable sizes in 6 to 8 months at between 15° and 20°C, as compared with the 3 or 4 years needed for the same growth under natural conditions. (*35*).

A progressive grower of Long Island shellfish used about 57,000 liters per minute of cooling water discharge of the Long Island Lighting Company. The cooling water is taken from a deep section of the bay and has a high nutrient content, which favors oyster growth as does its warmth. Year-round production in a near 3-hectare lagoon of both oysters and hard clams (*Mercenaria mercenaria*) has been achieved, and seed oyster production in the heated lagoon promises to be highly successful. Summer water temperatures above 30°C, first feared to arrest growth or to be lethal, in fact, promoted exceptionally rapid growth (*6, 25*). At the atomic plant at Turkey Point (Florida), replicate feeding trials by the University of Miami with shrimp (*Peneus duorarum*) and pompano (*Trachinotus carolinus*) are in progress to compare the effects of different levels of water temperature and consequently of different levels of heated water admixture (*9*). Heated waste water is also used for freshwater fish culture in the Soviet Union (*36*).

Status and Potential of Aquaculture

Aquaculture, practiced with a far wider range of species than mentioned here, is found in most of the world. In many areas it occurs at a subsistence level, and its potential contribution to the food supply has not been assessed (*37*). Village ponds, once a hopeful development in Africa, for instance, are now in disrepair and their potential is not being realized (*32*). Local fish ponds can be important,, however, as has been demonstrated in Taiwan, mainland China, and Indonesia (*6*).

Husbandry of aquatic animals brings increasing financial returns as it is practiced on a larger scale. Culture intensities vary, as do the fixed and variable costs of the operations and the yields (Table 1). From a commercial point of view, the return on the investment is of most interest; in milkfish culture the annual return ranges from 10 to 20 percent or more, and increases with the intensity of cultivation. Malayan mixed pig and fish farms yield 30 percent and similar returns are noted in the oyster business (*6*)

Aquaculture can be not only a lucrative business but it may even produce yields high compared with the harvest of comparably sized land surfaces. The relative scarcity of such peaks in aquacultural production, especially in the tropics, is caused by a lack of biotechnical engineering and managerial skills, the absence of suitable credit or seed capital for even low-cost installations, and the absence of transport and marketing facilities that might encourage the development of a product for a certain market, and so forth.

This is well illustrated by a comparison of Indonesian and Taiwanese milkfish Culture in brackish water. Milkfish feed predominantly on bluegreen algae and are raised in pond complexes on land cleared of mangroves. Canals permit the control of water level and salinity by means of sluices, which regulate tidal or freshwater flow (*38*). Average Indonesian and Philippine annual harvests are 300 to 400 kilograms per hectare, whereas Taiwanese milkfish raisers attain nearly 2000 kilograms on the average in spite of a cooler climate (*39*). Cooperatives, rural reconstruction agencies, a good layout of the farms, control of predators of the fry, some fertilization, and prevention of siltation of ponds and connecting water bodies are some of the secrets of successful milkfish farming in Taiwan. For similar reasons there occur in mollusk culture the aforementioned wide range of yields, from nearly 10 kilograms per hectare on public oyster grounds in the United States to the 58,000 kilograms per hectare in Hiroshima Bay.

Filter feeding mollusks and milkfish are brackish water plankton- or algae-feeders, respectively. These hold more promise for protein-deficient regions than do the carnivores of the same environment because it is more sound to increase the fertility of the water than to produce extraneous feed, let alone to raise one aquatic animal with scrap from another, which is perhaps already being used, or could be used, directly for human consumption.

Most products of aquaculture could be called luxury foods, whether they are sold as high-priced items in a food economy with wide consumer choice (for example, shrimp in Japan, trout in the United States) or boost the scant animal protein supply of developing nations (for example, milkfish in Southeast Asia), where they also bring a good return to the producer. It might seem unrealistic, therefore, to expect aquaculture to help alleviate the world protein deficiency, but such is not necessarily the case. Luxury foods stop being a luxury when they can be mass produced, a case well exemplified by the broiler chicken industry in the United States and Western Europe.

Differences in biology between chickens

Table 1. Selected ranges of Aquaculturals Yields (6) Per Year. Results Are Given in Kilograms Per Hectare Except as Noted. The Value is in Dollars per Hectare Except as Noted.

Type of cultivation	Location	Yield	Approximate wholesale value of annual crop
Oyster			
Common property resource (public grounds)	U.S.	9	38
Intensive cultivation, heated hatchery, larval feeding	U.S.	5,000	21,000
Intensive care, hanging culture	Japan*	20,000	23,000
Mussels			
Intensive care, hanging culture	Spain*	300,000	49,000
Shrimp			
Extensive, no fertilization, no feeding	S.E. Asia	1,000	1,200
Very intensive, complete feeding	Japan	6,000	43,000
Carp			
Fertilized ponds, sewage ponds	Israel	500	600
	S. Germany	500	
Fertilized ponds, accessory feeding	Israel	2,100	
Sewage streams, fast running	Indonesia*	125,000	
Recirculating water, intensive feeding	Japan	100†	114†
Catfish			
Ponds, no fertilization or feeding	Southern U.S.	200	70
With fertilization and feeding in slowly flowing water		3,400	2,400 (net profit 300)
Milkfish			
Brackish ponds, extensive management	Indonesia	400	
With fertilization and intensive care		2,000	600
Trout			
Cement raceways, intensive feeding, rapid flow	U.S.	170†	168†

*Values for raft culture and comparable intensive practices based on 25 percent of the area being occupied.
†Per liter per second.

and aquatic animals notwithstanding, some of the latter could well become mass-produced, cheap and abundant foods at conversion rates of two parts of dry feed to one part of fish flesh. Among fresh and brackish water fish, especially trout, carp, and catfish can be raised with pellets. Chinese carp and certain tilapias eat leaves and stems of leafy plants; other fish feed on algae. In Southeast Asia well over 200,000 hectares of ponds now lie in former mangrove areas; there are in the tropics vast unused mangrove regions, some of which could be turned into pond complexes for the culture of fish. Mollusk production, though limited eventually by the suitability of grounds, could be expanded, and above all intensified in the areas where it is now prevalent. Aquaculture is only beginning to develop such practices as manipulation of the temperature regime to achieve best growth, devising simple automated feeders that fish can learn to activate themselves, and building machines that simplify harvesting. Several disciplines are expected to contribute to the development of aquaculture. Since intensive husbandry alters the conditions of nature, a knowledge of the ecology of the cultured organisms in both natural and artificial states is essential. Engineering can also make increasingly important contributions to aquaculture development as it has in the successful pilot-scale raising of plaice and sole by the British Whitefish Authority (*4*). It was the basis for the as yet theoretical calculation that "the annual British catch of plaice could be housed in shallow ponds covering 1¼ square miles in extent" (*7*).

Japanese yellowtail fish are now raised at high density, and with sequential cropping have already achieved yields of 28 kilograms per square meter (280 metric tons per hectare) and shown that it is economical to use small portions of the sea under very intensive management. The success with this oceanic schooling species and the fact that other species of similar habits had become adapted to confinement led to the speculation that still others, such as tuna, might behave similarly and that their mass culture under controlled conditions might become possible. In fact, Inoue of the Fisheries Research Laboratory, Tokay University, Japan, urged that Japan take the initative in launching a tuna-rearing project in the equatorial Pacific, where atolls and lagoons could be used as sea farms (*8*).

Such projections say nothing of the problems of translating small to modest enterprises into much vaster ones – the main one likely to be the procurement of many millions of tons of suitable food. Trashfish, in part now used for fish meal, krill and other marine organisms lower in the food chain than the highly prized fish to be cultured have been thought suitable, provided that they can be produced at a low enough cost. The theoretical potential of marine fish culture also rests on the assumption that marine fish can be induced to function sexually under artificial conditions, as have many freshwater fish. Hormone stimulation is expected to be one of the solutions to this problem along with rearing an initial breeding stock born and adjusted to life in artificial environments.

But even without further advances through research, a considerable increase of aquacultural yields appears attainable soon by consistent application of already known techniques of inefficiently managed fresh and brackish water bodies. It has been advocated (*4*) that millions of hectares of ponds be constructed in Asia, Latin America, and Africa to help satisfy the protein needs of these areas. If local economic and socio-political constraints were removed, these new waters and the upgrading of presently existing ones could yield by the year 2000 a harvest of 30 to 40 million metric tons (*3, 41*) produced near areas of need, which are still likely to lack refrigeration.

Long-term and large-scale projections of yields attainable through practicing aquaculture with marine animals, outside the brackish water zone, can hardly be attempted; true mariculture is in its infancy. However, experiments in several locations have established that it is technically feasible, and no doubt the intensive development and success of brackish water aquaculture will lead to further efforts to develop mariculture on a large scale. It is too early, however, to tell where or under what conditions such efforts could become economically sound.

Summary

The role of aquaculture in producing high-grade animal proteins for human nutrition is discussed. Raising and tending aquatic animals is mainly practiced in fresh and brackish water although there are promising pilot experiments and a few commercial applications of true mariculture. Yields vary with the organisms under culture and the intensity of the husbanding care bestowed on them. The products are now mainly luxury foods, but there are some indications that upgrading of the frequently primitive culture methods now in use could lead to increasing yields per unit of effort and to reduce production costs per unit of weight. Under favorable conditions, production of animal flesh from a unit volume of water far exceeds that attained from a unit surface of ground. With high-density stocking of aquatic animals flushing is important, and flowing water or tidal exchange is essential. Combinations of biological and engineering skills are necessary for full exploitation of aquacultural potentials; these are only partially realized because economic incentives may be lacking to tend aquatic organisms rather than to secure them from wild stocks, because of social, cultural, and political constraints. Nevertheless, a substantial development of aquaculture should occur in the next three decades and with it a severalfold increase in total yield.

References and Notes

1. W. M. Chapman, *Food Technol.* **20**, 895 (1966).
2. Editorial, *Yearb. Fish. Stat.* 20 **(1965).**
3. President's Science Advisory Council, *The World Food Problem* (U.S. Government Printing Office, Washington, D.C., 1967), vol. 2, pp. 345-361.
4. L. A. Walford, *Living Resources of the Sea* (Ronald, New York, 1958), pp. 121-132.
5. I prefer the spelling aquaculture to aquiculture because the former is etymologically more correct.
6. In 1970, 23,000 metric tons of channel catfish are expected to be harvested in the lower Mississippi River states, and this production will double again in 1972 [editorial, *Comm. Fish. Rev,* **30** (5), 18 (1968)]; J. E. Bardach and J. H. Ryther. *The Status and Potential of Aquaculture, Particularly Fish Culture,* prepared for National Council on Marine Resources and Engineering Development 1967, PB 177768 (Clearinghouse Fed. Sci. Tech. Info., Springfield, Va., 1968). Ryther and J. E. Bardach, *The Status and Potential of Aquaculture, Particularly Invertebrate and Algae Culture,* prepared for National Council on Marine Resources and Engineering Development, PB 177767 (Clearinghouse Fed. Sci. Tech. Info., Springfield, Va., 1968).
7. J. E. Shelbourne, *Advan. Mar. Biol.* **2,** 1 (1964).
8. G. O. Schumann, personal communication.
9. C. P. Idyll, personal communication.
10. The study of the status and potential of aquaculture was financed by a contract with the National Council on Marine Research and Engineering Development, Executive Office of the President.
11. W. A. Dill, *Proc. World Symp. Warm Water Pond Fish Culture, Fish. Rep.* **44** (1), i (1967).
12. H. P. Clemens and K. E. Sneed, *Bioassay and Use of Pituitary Materials to Spawn Warm-Water Fishes,* Res. Rept. 61 (U.S. Fish and Wildlife Service, Washington, D.C., (1962), 30 pp.
13. W. Steffens, *Verh. int. Ver. Limmol.* **16** (3) 1441 (1967).
14. R. Moav and G. Wohlfarth, *Bamidgeh* **12** 5 (1960).

15. Departments of Commerce and Labor, *Fisheries of the U.S. 1908; Special Report* (U.S. Government Printing Office, Washington, D.C., 1911), p. 49.
16. F. Bregnballe, *Progr. Fish. Culturist* **25** (3), 115 (1963).
17. L. R. Donaldson and D. Manasveta, *Trans. Amer. Fish. Soc.* **90,** 160 (1961).
18. K.D. Carlander. *Handbook of Freshwater Fishery Biology* (Brown, Dubuque, Iowa, 1950). pp. 30-36.
19. L. R. Donaldson, *Proc. Pac. Sci. Congr. Tokyo Sci. Counc. 11th* **7,** 4 (1966).
20. N. Fredin, personal communication.
21. *Fishery Statistics of the United States 1965,* Statistical Digest No. 59 (U.S. Department of Interior, Washington, D.C., 1967), pp. 541-547.
22. C. L. Hubbs, in *Vertebrate Speciation,* W. F. Blair, Ed. (Univ. of Texas Press, Austin, 1961), pp. 5-23.
23. L. R. Donaldson, personal communication.
24. P. S. Galtsoff, "The American oyster fish," *U.S. Dept. Interior Bull.* **64**, 297-323 (1964).
25. J. H. Ryther, personal communication.
26. V. L. Loosanoff and H. C. Davis, *Adv. Mar. Biol.* **1,** 1 (1963).
27. Research on shrimp rearing in the United States is carried on at the Laboratory of the Bureau of Commercial Fisheries in Galveston at the Bears Bluff Laboratory of the South Carolina Wildlife Resources Commission, and at the Institite of Marine Sciences of the University of Miami in Florida.
28. S. L. Hora and T. V. R. Pillay, *Handbook on Fish Culture in the Indo-Pacific Region,* FAO Fish. Biol. Tech. Paper No. 14 (Foreign Agriculture Office, Rome, 1962), pp. 124-132.
29. Yun-An Tang, personal communication.
30. F. Gross, S. M. Marshall, A. P. Orr, J. E. G. Raymont, *Proc. Roy. Soc., Edinburgh Ser. B* **63,** 1 (1947); F. Gross, S. R. Nutman, D. T. Gauld, J. E. G. Raymont. *ibid.* 64, **1 (1950).**
31. A. Yashouv, *Bamidgeh* **17** (3), 55 (1965); A. Yashouv. personal communication,
32. M. Huet, *Rech. Eaux Forêts Groenendaal- Hoeilaart Belgique, Trans. Ser. D* **22,** 1-109 (1957).
33. K. F. Vaas and S. Machlan, *Proc. Indopacif. Fish Counc. 6th* (1956), pp. 187-196.
34. Th. Rangen, personal communication.
35. Whitefish Authority, *Annual Report and Accounts 1967* (Her Majesty's Stationery Office, London, 1967).
36. L. V. Gribanov, *Use of Thermal Waters for Commercial Production of Carp in Floats in the U.S.S.R.,* Working MS 44060, World Symposium on Warm Water Pondfish Culture (Foreign Agriculture Office, Rome, 1966).
37. T. V. R. Pillay, personal communication.
38. W. H. Schuster, *Fish Culture in Salt-Water Ponds on Java* (Dept. of Agriculture and Fisheries, Div. of Inland Fisheries, publ. 2, Bandung, 1949), 277 pp.

39. Yun-An Tang, *Philippines Fish, Yearb.* (1966), p. 82.

40. The U.S. Atomic Energy Commission Laboratory at Oak Ridge studies the feasibility of agronuclear complexes as shore installations in arid regions to produce cheap power, fresh water, and fertilizer; see *New York Times,* 10 Mar. 1968, p. 74. The agronuclear complexes will furnish ideal conditions for advanced aquaculture on a large scale.

41. S. J. Holt, in *The Biological Basis of Freshwater Fish Production,* S. D. Gerking, Ed. (Wiley, New York, 1967), pp. 455-467.

42. I thank for assistance and information J. H. Ryther, G. O. Schumann, L. R. Donaldson, S. J. Holt, T. V. R. Pillay, W. Beckman, Th. Rangen, M. Fujiya, A. Yashouv, F. Bregnballe, E. Bertelsen, C. Mozzi, K. Kuronuma, S. Y. Lin, S. W. Ling, Y. A. Tang, M. Ovchynnyk, C. F. Hickling, I. Richardson, S. H. Swingle, R. V. Pantulu, F. P. Meyer, J. Donahue, M. Bohl, M. Delmendo, H. H. Reichenback-Klinke, and D. E. Thackrey.

III. Cells

Of fundamental importance to the functioning of all living systems are the control mechanisms found among the components of the cell. An understanding of their form and function relationships is essential in the development of one's concepts of organisms.

Lawrence Lessing, author of a number of science-oriented articles for *Fortune*, explores the complexities of the Operon Theory, relating to the functioning of DNA, in an unusually clear presentation of a difficult subject.

Christian De Duve first recognized lysosomes as distinctive cellular particles in 1955. He referred to them as ''suicide bags'' alluding to the autolytic enzymes found in them. The actual role of these enzymes intrigued De Duve and other investigators who later related them to cell damage and to many aspects of classical cellular reactions in disease. In ''Lysosomes in Disease'', A.C. Allison of the British National Institute of Medical Research cites investigations into their role in bacterial diseases, cell division control and hence cancer, as well as in other disturbed functions.

The third essay of this unit, ''The Chalones,'' introduces a recently discovered class of chemical agents. The author, W. S. Bullough of Birbeck College, University of London, considers them mitotic inhibitors. When synthesized by the maturing cells of a tissue such as epidermis, chalones diffuse to the basal epidermal layer to act as a feedback message and inhibit further mitotic activity. Should this method of control become unbalanced, is there a possibility of uncontrolled growth or cancer? His evidence would point in this direction.

5.

At the Controls of the Living Cell

Lawrence Lessing

. . . with the discovery that the genesis and control of all life lies in a unique, double-stranded, helical molecule known as deoxyribonucleic acid or DNA, biologists have opened the way to solving the central mystery. AUTHOR

The ultimate mystery of life is how a single living cell, dividing and subdividing again and again, in the well-ordered drama of conception, birth, and development, becomes in time a complex, sentient organism such as man. For about a century, embryologists have wrestled with this great problem, gaining many insights into its depths but arriving at no firm conclusions as to exactly how life takes shape. Now, with the discovery that the genesis and control of all life lies in a unique, double-stranded, helical molecule known as deoxyribonucleic acid or DNA, biologists have opened the way to solving the central mystery.

This code works in two elementary ways, First, in cell division, DNA transmits the full code to succeding cell generations by splitting its two strands apart and forming on the separated strands two exact replicas of itself. Thus DNA is the self-reproducing molecule responsible for the continuity of life. Second, DNA transmits its code by transcribing portions of it on a secondary, single-stranded form of ribonucleic acid called messenger RNA, which goes out into the cell to direct the formation of all the characteristic proteins making up the cell or organism. But while this DNA-RNA-protein synthesis explains how individual genes direct the production of specific proteins, it does not explain how many genes acting in concert give rise to a complex new organism.

The older genetics had long ago settled on the general laws by which new life is born. In nearly all higher forms of life, above the one-cell organisms, reproduction takes place not by simple cell division but by a sexual joining of parental chromosomes to form a new combination. The chromosomes of all higher organisms come in two complete, duplicate sets. In reproduction, one set of chromosomes from the father and one from the mother are joined together by means of special sex cells containing only half of each parent's chromosomes. Each set may come from the parent's paternal or its maternal side, and in the act of joining, the composition of the chromosome sets may be further reshuffled by a mysterious process called recombination or crossover, in which some gene groups are broken apart and exchanged between pairs of chromosomes. Thus an immense diversity of traits is assured, within species limits, in the single-cell fertiled egg.

From the moment of fertilzation an almost magical process takes place. In the explosive growth of millions of cells from a single cell an entirely new organism gradually takes shape. This development proceeds in well-marked phases, almost with the precision of a preset program. At first the embryo is just a blob of cells hardly distinguishable from one another, then the cells begin to be segregated in slightly different regions, and finally recognizable features of an organism appear.

The program directing this development is written, as we now know, in the DNA contained in the chromosomes. It is written in the endlessly variable order of four simple organic chemicals in DNA's structure, designating, in triplet sequences or code "words," the order in which some twenty amino acids are to be built by the cell into specific proteins. With these physical units in hand, it is possible to get to the bottom of the mysterious process of recombination and of conception itself. More important, the new knowledge allows the great central question of development to be asked with more point and precision. If each embryonic cell contains in its nucleus a full replica of the DNA designating the whole organism, then how is it that some cells become muscle and bone, while others become brain, nerve, liver, kidney, spleen, and all the other quite different bundles of specialized cells that make up a man?

It is this question that is currently occupying the most active frontier of molecular biology. This new frontier is concerned with nothing less than the organization and control of whole organs and the total organism, involving the interactions of many genes with the cell environment. A fuller understanding of these interactions will lead to much more fundamental knowledge of how the human body works, how diseases affect it, and how drugs act upon it. Potentially within grasp is the first basic understanding of such processes as immunization, viral infection, and aging. and of such widespread degenerative disorders as hypertension, heart disease, and cancer.

At the center of this new quest for knowledge is the cell. the basic structural unit of life. Cell theory, like genetic theory, was formulated in the last century, by two prescient German biologists, Matthias J. Schleiden and Theodor Schwann, long before most of the physical parts of the cell were known. Only in the last two decades or so, however, with the development of such instruments and techniques as the electron microscope, the ultracentrifuge, and radioactive tracers, have biologists been able to view the complex components of this fundamental unit as a functioning whole.

In the nineteenth century, under the spell of the early industrial revolution, the cell was most often likened to a factory. A more sophisticated analogy today would be a completely automatic electrochemical plant run by a central computer – the DNA of the chromosomes. Packed into this tiny plant is a maze of circuitry, energy-coversion, information-transfer, and processing devices that makes the most advanced of today's microelectronics look crude indeed. A million human cells, for instance, could be put on the head of a pin. And even a one-cell organism carries on hundreds of intricate chemical reactions, side by side, without confusion. Most of the major reactions, such as the one in which energy is stored and released in complex molecules through the respiratory oxidation of sugars and fats, are many-stage cyclic reactions. The product of one stage initiates the next, with the aid of specific enzymes, until in the final stage a product is produced that starts the cycle over again. Reactions proceed at different times and different rates, some only at special tines or intermittently. A set of exquisite controls must therefore be at work, keeping the cell in equilibrium, producing not too much of one product or too little of another, to sustain the orderly processes of growth and development.

The simplest of these controls, biochemists early found, is a form of negative "feedback" control common in chemical engineering and electronics. When an excess of end product appears, it reacts back upon the start of the process to inhibit production. Extensive experiments have now shown that in the cell this may work in one of two ways: the excess product may lock molecularly into control sites on the activating enzymes, making them inactive; or the product may join somehow with the messenger RNA responsible for synthesizing the enzymes, cutting off their further production. While this feedback system may account for much of the local control of individual reactions, it does not begin to answer the major question. How, in a basic way, are a great number of processes turned on and off and coordinated in the variety of cells making up a complex organism? Ultimately, a great loop of feedback controls must extend back to the DNA in the cell nucleus.

The first glimmer of a master control system appeared in a notable review paper in 1961 titled "Genetic Regulatory Mechanisms in the Synthesis of Proteins," by Francois Jacob and Jacques Monod of the Pasteur Institute of Paris. Jacob and Monod, who had been Free French and Resistance fighters throughout World War II, came to the Pasteur Institute after the war, and in the laboratory of André Lwoff, a noted microbiologist, took up the new molecular biology, particularly the study of enzyme synthesis in bacteria. By 1960 their own and others' experiments had piled up many odd bits and pieces of evidence of how enzyme synthesis is induced and repressed in these one-cell creatures. So they sat down to try to bring some theoretical order out of the then abounding confusion.

Jacob and Monod first proposed that DNA transcribes a section of its code specifying a particular enzyme on a short-lived, mobile molecule that then goes out to act as a template upon which the protein is formed. Such a molecule, in fact, had been identified by U.S. biologists even before their paper appeared, and it became known as messenger RNA. Jacob and Monod's thesis was that DNA production of such a molecule sharply controls enzyme production. They then went on to suggest in a more far-reaching way how this control works. DNA, they postulated, contains two kinds of genes: structural genes that specify the formation of specific proteins and, some distance away from these on the DNA chain, regulator genes that act upon the structural genes, controlling their produuction of messenger RNA. The molecular make-up of each structural gene had at its starting point a short DNA-coded sequence that triggered gene action. Jacob and Monod called this coded sequence the operator, and the combination of operator and structural gene, an operon. Since information or orders can be transmitted in a chemical system only through the physical shape of specific molecules, they proposed two additional entities to make the control system work: a repressor and an inducer substance. The regulator gene, they theorized, produced a specific repressor that bound itself to the operator, preventing it from triggering the operon into synthesizing messenger RNA. The operator was released, and RNA synthesis proceeded, when an inducer – a small specialized molecule or metabolite – disengaged the repressor and allowed the operon to function.

For the promulgation of this so-called operon theory, which presented microbiologists with a new vocabulary and a range of new ideas with which to experiment, Jacob and Monod received the Nobel Prize, along with Lwoff, who had made many basic contributions to the understanding of gene operation in viruses and bacteria. It was France's first Nobel Prize in science in thirty years. Monod seized the occasion to launch a blistering attack upon a rigid, minister-dominated educational system, laid down by Napoleon, which had sapped innovation and was experiencing no improvement under another rigid military man, de Gaulle. Monod's group had waged a long struggle to conduct its research, which was unrecognized in the classical curriculum of the universities, given only limited recognition by the French Academy of Sciences, and even hampered by the administration of the Pasteur Institute. Monod's attack sparked an upheaval in the institute second only to the commotion loosed in biological circles by the operon theory itself.

In a fever of experimentation, parts of the theory have been confirmed, parts remain to be proved, and the whole is still a subject of some controversy. The validity of the operon concept has been demonstrated, through experiments involving controlled mutations, in many types of bacteria. Its operation has been fitted to the action of many well-known inducers, mainly key substances or substrates of the reactions to be induced. For instance, when a molecule of the milk sugar, lactose, is presented to a specific repressed operon, the operon is immediately unlocked and begins making messenger RNA to produce the enzyme needed to break down lactose. In the absence of lactose, the operon is again repressed. The location of the regulator genes that effect this repression also has been confirmed, through genetic mapping techniques, but the nature of the repressor substance itself is still unknown, and the whole regulatory mechanism remains obscure.

Because there are these gaps, many biologists believe that the operon theory will be found to apply only to certain genetic sites, and only to bacteria, whose cells are all alike and simple. It can only begin to suggest the much more definite and elaborate control apparatus required for multicell higher organisms.

The chromosomes of higher plants, insects, animals, and man were early found to differ in one important respect from those of bacteria and other one-cell organisms. Besides being bigger, more numerous, and more complex, the higher chromosomes had an ingredient never found in bacteria – a complex of nucleoproteins called histones. Indeed, these proteins were discovered in close association with DNA by the Swiss chemist J. Friedrich Miescher nearly a century ago. But since no one had any idea what purpose they served, both the histones and DNA remained merely curious chemical compounds. About 1950, however, with DNA finally identified as the genetic material, two British biochemists, Edgar Stedman and his wife Ellen, took another look at the histones and suggested they might somehow act as gene inhibitors. The suggestion was ignored for several years. Then the rapid unraveling of the DNA code and the rise of the operon theory impelled a number of laboratories to take off after the histones.

A group at the California Institute of Technology, led by James Bonner and Ruchih Huang, conducted a series of elegant experiments on pea-plant chromosomes – the classic subject of Mendel's original genetic observations a century ago. By 1963 the Caltech team had demonstrated that histones appeared to be the main repressors of gene action in multicell organisms. They isolated the chromosomes and showed that in their natural state they synthesized only 5 to 20 percent as much messenger RNA as chromosomes from which all histones had been removed. Some 80 to 95 percent of the DNA in the chromosomes, they found, was covered by histones, blocking RNA synthesis. They then focused on one gene responsible for storing a specific protein in pea seeds. This gene operated only in pea seeds during growth, and in no other part of the plant at any time, though the gene was present but dormant in all the other plant cells. This they demostrated by removing histones from the chromosomes of bud cells, whereupon the bud-cell chromosomes were induced to produce seed protein.

At about the same time a group at Rockefeller University in New York, led by Vincent G. Allfrey and Alfred E. Mirsky, reported an extensive series of related experiments on the nuclear chromosomes of calf thymus-gland cells. They showed that adding free histones to the isolated nuclei substantially cut down their synthesis of RNA, and, conversely, that selectively stripping histones from the nuclear chromosomes greatly increased RNA production. Later Allfrey showed that histones varied in their repressing activity according to how strongly they bonded themselves to the DNA, and this depended in turn on the degree of acetylation in the histone molecules.

Meanwhile, Bonner and his group went on to make an ever more detailed physicochemical study of histones. They found that histones worked in much the same way in cows, rats, insects, and humans as they did in peas. In all these organisms they found that histone complexes blanketed 80 to 95 percent of the chromosomal DNA, but in different regions on the chromosomes of different cells, as was to be expected in cells having various specialized structures and purposes. Only the genes needed for a particular cell or organ were left free to operate. They also concluded that there are only five types of histones, common to all higher organisms, and that these wrap themselves around the DNA and adhere to it with various degrees of bond strength. Finally, they came upon what may be an answer to the most puzzling of all questions concerning histones: how do they know which genes to sit on and which to leave alone? The histones do not come in enough structural varieties to be selective, and no one has been able to discover that they bond themselves to DNA by anything but simple electrostatic attraction. Last summer, however, Bonner and Huang discovered what they believe is a new form of RNA, attached to long chains of histones of one type. Presumably made by a regulator gene and therefore specifically coded, this RNA tag might act as a guide – Bonner whimsically calls it "seeing-eye-dog RNA" – drawing its chain of histones to a specific site on a structural gene, where it clamps on by simple attraction.

At the same time that the histones were being run down, a hunt was also on for the inducers that release these repressors. A major candidate soon turned up in another old but more familiar class of compounds – the hormones. These complex protein and hybrid molecules, secreted by special glands, long had been known for their ability to move about in the body and selectively influence distant cells and tissues. Their powerfully stimulating, multiple effects on cell activity, as seen in the performance of insulin and the sex hormones had already made them useful in modern medicine. But little was known about how they really worked until molecular biologists began to look at them in the new context. The nature of the hormonal action has come into focus only in the last two or three years, under the scrutiny of many laboratories, from the Max Planck Institute for Biology in West Germany to the University of California School of Medicine at Los Angeles. It is now generally evident that the hormones act as the main releasers or inducers of genetic action in higher organisms.

The general pattern of experiments by which this was proved involved administering a specific hormone to test animals, and then measuring at intervals the rate of protein synthesis it induced in certain target cells, hence the rate at which the DNA in these cells was producing RNA. In one instance, protein production started to rise within thirty minutes after hormone injection and soon jumped as much as 300 percent. Then, as a clincher, a further experiment was performed. It was based on the discovery, early in the research on gene activity, that some antibiotics temporarily block all DNA manufacture of RNA. Indeed, this turned out to be one of the basic ways in which antibiotics work; they deny to an infecting organism the supply of cellular enzymes and other proteins it needs in order to grow. So, as a final check on how the hormone acts, test animals were given an antibiotic known to block all RNA production. When the hormone was administered to these animals, it had no effect at all. This proved that the hormone acts upon DNA at the gene level.

The exact mode of hormone action is not yet quite clear. One general theory is that a hormone, entering a cell, combines with or activates a waiting target protein, which then goes to a specific site on the DNA to disengage the repressive histones, thus uncovering the genes so they can make RNA. But this is not the only way hormones work. Some hormones have been found to activate enzymes directly to make more protein, while others have been discovered attached to cell walls, regulating the passage of such elements as salt across cell membranes. Nor are hormones the only inducers of gene activity. A host of metabolites and other agents, such as the sunlight that wakes a green leaf's chlorophyll to action, are known to unlock the genes in many metabolic systeme. Level upon level of controls are needed to ensure that the right proteins are made in the right amounts at the right time. But in the discovery of how hormones and histones act, biologists at last have their hands upon the main throttles that turn genes on and off.

Still to be explained, however, is the great central process by which DNA controls the development of many different types of cells from a single precursor to form a complete, complex organism. Biologists call this process differentiation. It is so delicately tuned that a plant produces its flower and fruit precisely in season, and each feature of the human form takes shape and grows only so big and no bigger. The basic mechanism, as we have seen, is the switching off of some genes and the turning on of others in sequence, so as to produce different types of cells for different purposes. But what is it that selects the right genes to repress and the right ones to activate in these differentiated cells – each of which contains, be it remembered, the full DNA code of the organism?

To begin to think about the problem, Bonner and his group at Caltech have set up a computer program to try to simulate, in a simplified way, the logical stages by which a single apical cell at the tip of a growing plant continuously divides and gives rise to a bud, stem tissue, and finally a flower. At each cell division, Bonner reasoned, the growing cells must test themselves for position and proximity to neighboring cells. If a cell finds itself all alone at the very tip of a stem, it concludes it is the apical cell and continues to divide. (Frederick C. Steward of Cornell has shown that if any one cell of a carrot plant is isolated and nurtured, it proceeds to act like an apical cell and grows into a whole new plant.) If a cell finds itself surrounded by neighbors, then it concludes it is a bud cell, some of its genes are turned off, and further growth is inhibited. (Cells have been shown to have a means, through cell-wall contacts, of identifying and influencing one another in development; in the course of differentiation they manufacture a special mucilage that causes like cells to stick together, inhibiting growth.)

As the bud swells under the ever dividing apical cell, older cells near the bud's surface and base enter a second stage of differentiation and become leaf and stem tissue. At this point another round of self-testing and turning on and off of genes takes place. If, for instance, a bud cell near the base finds one surface exposed to air, it develops into an epidermal cell; if it finds itself inside a cell mass, it becomes xylem, the pithy inner structure of the stem.

Bonner translated all these events into computer language so that he could test out his logical assumptions of what happened at each stage of the plant's development. He has still to work out a computer program for flowering. Experiments have shown that this takes place when the leaves of the mature plant "notice" that daylight is lasting a certain length of time. They thereupon release a hormone that goes to the bud, where it unlocks the genes that control flowering.

"The development process behaves as though it were a pre-programed routine, a pre-programed sequence written down in the genetic DNA," says Bonner. It is a process that, once started on a pathway or on a subroutine such as flowering, flows on almost automatically. Yet it is also a process in which there is a constant feedback of information to the nuclear DNA from the cell's cytoplasm, cell walls, and the environment that triggers, paces, and even modifies the stages of development. A steady flow of information and stimulation, particularly in the initial stages of differentiation, is required to evoke the genes to proper development. Bonner believes that the computer may help in understanding this complicated interplay, if only because it forces biologists to think through the development process more rigorously in order to program it.

Many laboratories are now beginning to get at these dual aspects of differentiation – the action of the genetic code and the reaction of the environment – by working on living embryos. One of the leading investigators is Eric H. Davidson, a young assistant in Alfred Mirsky's laboratory at Rockefeller University, who is working on embryos of the curious marine snail, *Ilyanassa.* Over half a century ago embryologists discovered through the study of various creatures, including *Ilyanassa*, that egg cells contain in their cytoplasm mysterious, unevenly distributed, localized "factors" that seem to determine how various regions of the cell will develop. The snail *Ilyanassa* was selected for study because, before cell division, its eggs take on a convenient trefoil form. One of the three lobes of the trefoil contains none of the egg cell's genetic material, and hence it could easily be lopped off to study how the organism develops without part of its original cytoplasm. The dismembered cell grew into a wriggling, free-swimming creature, but with some of its most important tissues missing. No one knew why.

Davidson returned to the study of *Ilyanassa* with all the new found knowledge of how DNA functions. He showed that removal of the lobe eventually causes the embryo's rate of RNA synthesis, and hence its protein production, to drop below normal, precisely at the stage of cell differentiation when gene activity usually rises dramatically. The inescapable conclusion is that some important genes are left repressed and inactive in the absence of the cytoplasmic factors contained in the missing lobe. This conclusion is being fortified by similar microsurgical experiments on the eggs of the toad *Xenopus.* In addition, the work on *Xenopus* is demonstrating that the genes left inactive by cytoplasmic surgery are fully capable of activation.

Thus the cytoplasm of the egg cell is being viewed more and more as the crux of development. Laid down in this jellylike fluid during the cell's long maturation in the maternal body is a rich store of special factors without which the orderly sequences of development and differentiation cannot take place. These factors are still unknown, but are probably traceable to specialized RNA and protein inducers specified by the maternal DNA. As some scientists now see it, the cell is a kind of federation, in which the nuclear DNA of the chromosomes acts as the central government, while certain elements of the cytoplasm operate as semi-autonomous states. Recent studies have shown, for instance, that such particles as the mitochondria, the power plants of the cell, have a small, non-nuclear DNA component in their own makeup. The DNA of the chromosomes remains in strong, central control, but the rest of the cell, the whole organism, and the total environment all have a critical say in development. The task of unraveling these subtle interrelationships in the human egg cell will be a vast one for many investigators.

But the basic principles are at hand, a start has been made, and progress should be rapid.

Out of this rapidly broadening base of new knowledge may come developments as varied and spectacular as life itself. One of the earliest and most predictable will be a great expansion in the synthesis and use of hormones or other metabolic agents to intervene more precisely in the activation of specific genes controlling body processes. Great developments have already taken place in this field, notably the introduction of such hormonal agents as the birth-control pills, but the surface has only been scratched.

Only last year a successful experiment was reported by M. James Whitelaw of San Jose, California, who used a cortisone-like hormone to reduce a hereditary tendency to excessive tallness in fifty teen-age girls. The hormone accelerated puberty so as to shorten the growth period, keeping the girls' heights an estimated two to six inches below their ultimate theoretical growth. Not only height but body build may someday be regulated in the same way. And only this year a team at London's Postgraduate Medical School reported that a newly discovered thyroid hormone, named thyrocalcitonin, sharply controls excess calcium in the blood, seemingly by depositing it in bone. The behavior of this hormone may go far to explain and treat certain bone diseases, including bone cancer, in which calcium from bone is lost to the blood. Many more hormones and their intricate pathways in the body remain to be discovered.

There is also likely to be sharply rising interest in the synthesis of inhibitors or blocking agents to curb gene activities that have got out of control. For instance, man has inherited from primitive ancestors a genetic "fight or flight" mechanism that swiftly pours adrenal hormones into the bloodstream to meet any sudden danger. These agents, chiefly epinephrine, secreted in the inner part of the adrenal glands, powerfully step up the heart beat, dilate some arteries, and constrict others, to send a massive flow of blood to the muscles for instant physical action. This mechanism is no longer widely needed in civilized life, but it continues to be triggered in some individuals by less tangible threats, fears, and frustrations, leading in some cases to chronic high blood pressure and heart attacks. Two classes of chemical compounds have now been found to short-circuit this mechanism. One is an agent that blocks epinephrine action on the arteries; the other, more recently discovered by scientists of the Lily research laboratories, blocks epinephrine excitation of the heart muscle. At the same time, the National Heart Institute is pressing a search further back in the genetic system to find a means of curbing the production of adrenal hormones themselves by inhibiting a key enzyme in their synthesis. Most of the solutions to heart and arterial diseases seem to lie in these devious pathways of genetic control.

The most striking example of the genetic system's getting out of control is, of course, cancer. In this malady the cells of an organ or tissue become unstuck, so that their growth is no longer regulated by contact with other cells. They then wildly multiply and degenerate back toward an undifferentiated state. Much of the basic research on DNA and molecular biology is being supported by cancer research funds, for cancer involves in a basic way the derangement of the central repressor and inducer mechanism of DNA. Any deeper understanding of this basic mechanism will aid in tracking down the causes not only of cancer but of a host of other degenerative diseases – and aid in developing the drugs to control them. The implication of viruses, those infectious threads of DNA or RNA, in the stealthy triggering of certain types of cancer grows stronger by the day. And drug research is piling up so many promising leads that the National Cancer Institute, for the first time, is entering into industrial research contracts to run them down more quickly.

Meanwhile, stirring progress is being made in clarifying the body's basic defense system itself. This work focuses upon the antibodies, those long mysterious protein agents that are manufactured by special plasma cells and arise in magical numbers to repel a disease invader or any foreign substance, providing future immunity against it. Only in the past few years, through work carried on in many laboratories, has the detailed structure of the antibodies become known. The antibody is a large, elongated molecule, composed of four subunit chains, joined in such a way that at both ends of the molecule there are identically shaped "slots" that fit and lock upon a specific invader, or antigen. Laborious step-by-step analysis revealed that the main sections of all antibodies of a particular class have a common amino-acid sequence, and hence are controlled by a single gene. But the end sections, bearing the specifically shaped slots, differ from antibody to antibody. When all the various end-section sequences are translated back into the DNA code, it looks strongly as though each distinctive shape was made by a different gene. As many as 10,000 different genes might be needed, if this is so, to arm a sufficient variety of antibodies to meet the range of possible disease invaders.

Several theories have been put forward to explain the unique composite structure of the antibody, which seems to violate the dictum that one gene controls the production of one protein chain. One such theory was proposed last year by William J. Dreyer and J. Claude Bennett of Caltech, who focused on the primitive plasma cell that makes antibodies. They maintained that as each plasma cell goes through the process of differentiation it combines one common gene with one of the genes that specifies a particular end shape; it leaves all the rest of the end-shape genes turned off. That cell can then make only one kind of antibody. Each primitive plasma cell is preset to make a different antibody and it then lies dormant until it is invaded by a foreign agent that fits – e.g., a specific virus – whereupon the cell jumps into action, vigorously dividing and pouring out large amounts of the specific antibody to overcome the invasion. Thus plasma cells may be the most widely differentiated cells in the body, genetically pooling all the antibody shapes devised by living organisms over eons of time and evolution to protect themselves.

With this growing knowledge of antibody structure and how it is put together, scientists may soon be making antibodies in the laboratory and thus mobilizing a powerful new agent against disease. Dreyer thinks such a step may be only five years away. Ultimately it may be possible to build up resistance to disease in a more basic way. The resistance to disease varies widely in human populations. Some people, for instance, seem to have a hereditary resistance to tuberculosis or cancer, while others do not, because they have a deficiency in antibody production. It may be possible to correct this deficiency by implantation of specific plasma cells or by correction of the genes themselves.

When scientists let their imaginations reach out from this new body of knowledge, they go much further than this. They believe it may be possible to intervene genetically in the development of tissues and organs themselves. In highly complex animals differentiation is normally irreversible. Once a man's cells are set in a specialized way, he cannot regrow lost parts or generate new ones, as less highly differentiated creatures do; a lizard, for example, can regrow a lost tail, or a plant may generate new roots or branches from most of its parts. But James Bonner, for one, thinks that now that biologists have the key to turning genes on and off, they should be able to take any human cell and reset its program at will to any stage in the developmental process. It may become possible, for instance, to cure diabetics by resetting and reactivating the genes responsible for making insulin, which have become repressed through degenerative disease. It is even conceivable that whole genetic programs can be reset to replace tired or damaged organs, or to generate entirely new or additional ones.

"I have tried to think about what further organs I would like to have," says Bonner, only half jokingly, "and I have decided that I would like to have four hands, since there is so much for biologists to do. Recently, as I was trying to light my pipe in the laboratory, my colleague, Professor Huang, said to me, 'If you're going to smoke a pipe in the laboratory you'll need five.' "

An even more immediately promising prospect is the transplantation of whole organs. The basic obstacle to such transplantation is that antibody action automatically rejects all grafts of foreign tissue. But scientists are very close to finding a way around this barrier through increasing knowledge of the immunological process. Another line of research is exploring ways to make damaged organs repair themselves. In 1960, Ruth Hill of Columbia University, studying the effects of ultraviolet radiation on viruses and bacteria, found a clue to the existence of a self-repair mechanism. It cut out radiation-damaged genes and replaced them with new ones in the living DNA. Groups working under Paul Howard-Flanders at Yale University and Richard B. Setlow at Oak Ridge National Laboratory have now worked out the essential details of this repair mechanism, which is controlled by three genes and their resulting enzymes. A similar mechanism, operating in man, may help him to survive in a world of increasing radioactivity and space travel. Howard-Flanders and others have gone on to suggest that this repair mechanism may also be the means by which recombination takes place in the fertilization of egg cells in higher organisms. Thus a fuller understanding of the mechanism may not only show how it might be used to make repairs in man's chromosomes, but it could also reveal the great secret of how chromosomes recombine in human conception.

A number of bold experimenters in the U.S. and Europe are pressing into this mysterious realm. Robert G. Edwards, who works both at Johns Hopkins Hospital in Baltimore and at Cambridge University, England, has succeeded for the first time in growing human eggs outside the body, up through what appears to be the stage of fertilization. At birth the female ovaries contain some 500,000 potential ova, one of which normally matures each month during the childbearing period. Edwards obtains his eggs from ovaries that have been removed from women for medical reasons. There is no inherent reason why such eggs, given the right laboratory environment – a big order – cannot be grown up to the stage of a full-term fetus.

At the same time, Teh Ping Lin of the University of California School of Medicine in San Francisco has developed a technique for removing and delicately injecting minute amounts of material into fertilized mouse eggs – which are only ten times larger than a human red blood cell – and then reimplanting them in foster mothers, where some continue to grow into normal embryos. Together these two techniques offer the possibility of manipulating the human egg outside the body and of introducing specific materials to influence its genetic development, for instance, altering sex characteristics.

No complete account can be given of the numerous paths and bypaths down which the revolution in molecular biology is moving. If, as has been amply demonstrated, all genetic information is stored in the four simple organic chemical bases of DNA, then it follows that all of life's complexity and variety, all its patterns of behavior, and all mental traits and abilities are basically open to study and manipulation. There is no ironbound tyranny in the genes. A new light has been thrown on the old controversy of genetic inheritance versus environment. For it is now reaffirmed on the deepest molecular level that the individual genes cannot develop fully without the right physical environment and stimulation. In addition to learning how to manipulate the genes, therefore, we may also learn how to awaken the great pool of dormant genes that lies untapped in the human race.

6.

Lysosomes in Disease

A. C. Allison

Increasing attention is now being given them by pathologists because it is already clear that lysosomes play a part in many of the classical *cellular* reactions in disease. AUTHOR

The study of pathology – the manifestations of disease in organs and tissues – has had several phases of development. Early studies initiated by the Italians H. Fracastoro, Marcello Malpighi and Giovanni Battista Morgagni and continued by John Hunter and the Frenchman Marie Francois Xavier Bichat in the 18th century, provided a great deal of information about the appearances of diseased organs to the naked eye. About the middle of the 19th century improved microscopical techniques showed that all tissues are composed of cells, and that changes in healthy and diseased tissues depend on cellular reactions. Thus normal growth follows orderly and controlled multiplication of cells, and malignant growth follows uncontrolled multiplication and spread of cancerous cells. The reactions in inflammation are associated with invasion of tissues by white blood cells, fibrosis depends on the mobilization and activity of fibroblasts, and so on. The science of cellular pathology, pioneered by Rudolf Virchow and others a century ago, has dominated thought to the present day.

But now the next phase of development is in progress – that of analysing reactions at the subcellular level. This means identifying which constituents of cells are primarily involved in different reactions and what is the biochemical basis of the disturbed functions. All cells, except mammalian red blood corpuscles and blood platelets, have a nucleus surrounded by cytoplasm in which various organelles move around in a fluid matrix. The organelles include the ribosomes, which are sites of protein synthesis, and mitochondria, in which the oxidation of food-stuffs with release of energy takes place.

There is no doubt that mitochondria are the main sites of action of certain poisons which lead to disturbed functions. Thus, carbon monoxide combines quite specifically with the red pigment – haemoglobin – in the blood, reducing the supply of oxygen to the tissues, and it also combines with cytochrome enzymes in mitochondria. In carbon monoxide poisoning the primary damage is to the oxidative energy-producing reactions in mitochondria. Cyanide is another poison acting in a similar way, but less specifically. A third example is provided by the work of Sir Rudolph Peters and his colleagues on poisoning by fluoroacetate, which occurs in the plant *Dichapetalum*. If herbivorous animals ingest the plant, the fluoroacetate is converted to fluorocitrate which inhibits the enzyme aconitase, thus blocking the main pathway of oxidative metabolism in mitochondria.

It is also known that abnormal protein synthesis in ribosomes is the cause of certain diseases. One of the most striking

---

Reprinted from the November 1965 issue of *Science Journal* by courtesy of *Science Journal*, London, England, and the author.

examples is the formation of an abnormal type of haemoglobin in red blood cells in the inherited disorder known as sickle cell disease. The abnormal pigment, sickle cell haemoglobin, is very insoluble when it loses oxygen, and crystallizes in the red blood cells, distorting them into a sickle shape. Anaemia and obstruction of blood flow produce a severe and often fatal disorder. Many inherited diseases are now known to be due to synthesis of abnormal enzymes or other proteins in ribosomes, and a great deal of information about the enzymes involved and their location in cells has accumulated.

Thus there is no doubt that primary disorders can affect the functioning of mitochondria or ribosomes. But another organelle, the lysosome, seems to be of particular interest in pathology. Lysosomes were first recognized as a separate group of particles in 1955 by Professor Christian de Duve and his colleagues in Louvain. They carefully homogenized liver cells and then separated their constituents, according to size and density, in a centrifuge. This revealed organelles which were distinct from mitochondria and ribosomes and had their own groups of enzymes, including those which can break down proteins, the hereditary materials deoxyribonucleic acid (DNA) and ribonucleic acid (RNA) and also certain carbohydrates, under slightly acidic conditions.

In healthy cells the lysosomal enzymes are safely contained in small sacs surrounded by membranes or lipoprotein. However, under certain conditions the enzymes can escape and dissolve, or lyse, the cells and from this property the name lysosome was derived. At first the lysosomes were investigated mainly by biochemists and were found in many different cell types. Increasing attention is now being given to them by pathologists because it is already clear that lysosomes play a part in many of the classical cellular reactions in disease.

The simplest and most obvious involvement of lysosomes is in necrosis or autolysis, the death of tissues inside or outside the body. If a tissue or organ is isolated from its supply of oxygen and nutrients under sterile conditions, it quite rapidly breaks down. All the macromolecules – protein, nucleic acids and complex carbohydrates such as glycogen – are digested, and evidence has accumulated that enzymes liberated from lysosomes play an important part in this process.

The second obvious function of lysosomes is in phagocytosis, the uptake and destruction of bacteria by white blood corpuscles, phagocytes, to which Elie Metchnikoff drew attention almost a century ago. This process has been studied in detail by Z. A. Cohn and J. G. Hirsch of the Rockefeller Institute, New York. The bacteria are engulfed into food vacuoles or phagosomes in the phagocytes. Numerous lysosomes (neutrophil granules) are present in these cells; the lysosomes adhere to the membrane surrounding the phagosome and discharge their enzymes into it. The enzymes then kill and digest the bacteria.

This is an important defence mechanism of the body, and it is not surprising that the bacteria have evolved ways of resisting such attack. The mycobacteria – which cause tuberculosis and leprosy, and many of which normally grow inside cells – have a thick waxy coat which resists enzymic digestion. Other bacteria have toxins which when taken up by white blood cells make the membranes of their lysosomes permeable. Hence the lysosomal enzymes escape into the cytoplasm and the cells are rapidly killed, leaving the bacteria free to multiply. Such toxins include streptolysin S and O which are produced by streptococci, the *a*-toxin of staphylococci and many others. These all increase the permeability of various biological membranes. They liberate haemoglobin from red blood cells, for instance, and so have been known as haemolysins.

White blood cells ingest not only bacteria but also inert particles such as carbon from polluted air or silica or asbestos dust inhaled by people mining or working with these materials. Ingested carbon is harmless but ingested particles of silica or asbestos bring about liberations of lysosomal enzymes into the cytoplasm (apparently by reaction with the phagosomal membrane) and this kills the phagocytic cells. Eventually fibrous tissue is laid down and the typical pathological pictures of silicosis or asbestosis develop.

Damaged tissues show a characteristic reponse termed inflammation, the cardinal signs of which were known to the ancients: tumor, calor, rubor, dolor – swelling, heat, redness, pain. This can occur as a result of infection, for instance by bacteria in a finger, but follows other types of injury too, for instance burns or injection of a sterile solution of a noxious substance such as turpentine. The swelling, redness and warmth are due to opening up of the blood vessels supplying the inflamed area and passage of fluid containing protein (which is normally confined within blood vessels) through the walls of blood vessels into the spaces between them. One of the proteins, fibrin (which provides the structural basis of blood clots) is deposited in the tissues. There is also emigration, from the vessels into the tissues, of white blood corpuscles – characteristically polymorphonuclear leucocytes in acute inflammation of short duration and mononuclear leucocytes in chronic or long lasting inflammation.

These and other reactions in inflammation are complex and it is clear that many different factors are involved. A number of chemical substances are liberated from damaged cells or formed in inflamed tissues. Among these are permeability factors such as histamine, serotonin (5-hydroxytryptamine) and bradykinin, a peptide which is produced from serum proteins by enzyme action. There are also pain producing factors, which seem to be produced in a way somewhat analogous to bradykinin and, in certain cases, pyrogens which induce a general increase in body temperature. Another substance known as leucotaxin attracts white blood corpuscles.

In this complex situation it is difficult to define which reactions are primary and which take place secondarily or later. However, it is clear that lysosomes can be implicated in some of the reactions. Thus histamine and serotonin are liberated from granules in mast cells which show many properties of lysosomes. There is also evidence that another permeability factor can be released from lysosomes and that a basic protein from lysosomes causes adhesion of white blood corpuscles to the walls of blood vessels. Moreover, an enzyme from lysosomes starts off a chain of reactions leading to breakdown of fibrin, and it is possible that other such chain reactions are initiated in a similar way. The role of lysosomes in inflammation is now under intensive study and much more information is likely to emerge in the next few years.

Cytologists have long known that living cells take up dyes and concentrate them in specific organelles. Janus green, for instance, has been used to show mitochondria and neutral red to stain other cytoplasmic bodies which we now know to be lysosomes. Fluorescence microscopy is very useful for this purpose because a number of dyes and drugs are fluorescent and even low concentrations can be shown inside living cells by this technique.

At the National Institute for Medical Research, M.R. Young and I have studied mammalian cells exposed to a variety of dyes and drugs in this way. The most brilliant picture is obtained after exposure of living cells to low concentrations of acridine orange. Nuclei and nucleoli fluoresce green whereas lysosomes (with a higher concentration of dye and molecular stacking) fluoresce orange. In this way useful information about the number and characteristic distribution of lysosomes in different cell types is obtained. Not all substances are taken up by lysosomes. Many acid dyes, such as eosin and primulin, do not enter living cells at all. Some drugs such as tetracyclines show fluorescence in mitochondria. But the majority of the substances that are taken up by living cells seem to be concentrated in lysosomes. These include porphyrins, vitamin A, the anti-malarial drugs quinine, mepacrine and chloroquine, the anti-inflammatory drug indomethacin, and drugs used against trypanosomiasis such as stilbamidine and antrycide.

Polybenzenoid hydrocarbons are likewise concentrated in lysosomes and since these are uncharged it is clear that not only basic substances are concentrated in lysosomes. Why lysosomes should avidly take up so many different substances is at present unknown. This seems to be more than just lipid-water partitioning, and it may help to explain some actions of hormones and drugs. Thus cortisone and chloroquine stablize lysosomal membranes, and this may contribute to their anti-inflammatory activity, whereas vitamin A makes the membranes of lysosomes more permeable, and this accounts for the effects of overdosage of this vitamin on cartilage and bone.

Viruses contain genetic material, which can be either DNA or RNA, enclosed in a protein coat; in some viruses the coat has lipid as well. They become attached to the walls of host cells and are taken up by phagocytosis in the same way as other foreign particles. In phagosomes the viruses are exposed to lysosomal enzymes and evidence is accumulating that these play an important part in digesting away the outer coat of the virus so as to liberate the nucleic acid for replication. Eventually many copies of the virus nucleic acid and large amounts of coat materials are made and these are finally reassembled to form daughter virus particles which can start the cycle of infection again in other cells.

A cell in which virus reproduces is often killed and it seems that massive liberation of lysosomal enzymes is one of the factors that has to be taken into account in the production of cell damaging effects. Lysosomal changes are readily shown by neutral red staining of layers of cultured cells in which virus is replicating because the stain is taken up into the lysosomes of uninfected cells, giving them a uniform red appearance, whereas areas of infected cells with disrupted lysosomes stand out as white 'plaques' against this background. This technique allows the number of infective virus particles in any preparation to be counted quickly and accurately, and gives valuable information about the types of virus involved.

Lysosomal changes in virus infected cells are also demonstrable by fluorescence microscopy and histochemistry. Among the effects of virus infection is the formation of giant cells with many nuclei. When the unfixed cells are stained for the lysosomal enzyme acid phosphatase, intense enzyme activity is seen in the infected foci, showing that the lysosomal enzymes have been activated. Fluorescence microscopy after acridine orange staining shows orange fluorescence of intact lysosomes in uninfected cells but diffuse green fluorescence of cytoplasm in infected foci – which implies that disruption of lysosomes has occurred.

Cells infected with viruses have three possible fates. Often the cells are killed; sometimes the virus multiplies without apparently damaging the cells; or, after infection with tumour producing viruses, the cells can be transformed into malignant cells. The latter changes can be achieved in cultures, in which case the transformed cells show altered growth patterns and when inoculated into animals produce cancerous growths.

One of the things that is noticeable in cultures of cells vitally stained with acridine orange is that mitotic or dividing cells have relatively few lysosomes and these lie in a peripheral position instead of alongside or around the nuclei. Moreover, populations of rapidly growing cells have fewer lysosomes than cells which are not dividing. One possible interpretation of these observations is that breakdown of lysosomes acts as trigger for mitosis in cells that are prepared for it. The relationship might not be causal, however, and to obtain more direct evidence for such a mechanism it was necessary to show that agents increasing permeability of lysosomes could induce mitosis in cells that would not otherwise divide, whereas agents stabilizing lysosomal membranes could prevent this mitogenic effect.

Two systems that are convenient for study are peripheral blood lymphocytes and unfertilized animal egg cells. It has been known for some years that human lymphocytes in culture seldom divide but, if treated with phytohaemagglutinin, an extract of the red kidney bean *Phaseolus*, a high proportion enlarge and divide. A similar mitogenic effect is produced by bacterial toxins such as streptolysin S and staphylococcal a-toxin, which – as I have mentioned – are known to induce permeability changes in lysosomes. It has also been shown that treating human lymphocytes with antibody, which has been produced in rabbits by injecting them with human lymphocytes, stimulates them to divide. This must be done in the absence of complement – a factor found is normal serum – otherwise the cells are killed. It is known that anti-cellular antibodies in the absence of complement increase the permeability of lysosomal membranes. The mitogenic effects of phytohaemagglutinin and other agents can be prevented by cortisone or chloroquine which stabilize lysosomal membranes.

The animal egg is another cell that does not normally divide until stimulated to do so by fertilization. The spermatozoon has a special organelle, the acrosome, which becomes detached soon after fertilzation. The acrosome contains several enzymes characteristic of lysosomes and also hyaluronidase which breaks down the gelatinous material around the egg. The acrosome also shows the same uptake of dyes as lysosomes and must presumably be regarded as specialized lysosome (*cover picture*.) The egg contains cortical granules which also have the same enzymes and staining reactions as lysosomes. It seems likely that enzymes released from the acrosome at fertilzation initiate disruption of the cortical granules in a chain reaction which rapidly spreads around the egg. The outer layers of the egg are broken down, a new membrane resistant to enzymic attack is formed underneath and the cell prepares to divide. Professor J. Brachet of Brussels has also concluded that cortical granules belong to the lysosome family. The cortical granules in frogs' eggs can be disrupted by injury such as the prick of a needle and this stimulates parthenogenetic cleavage in the absence of a sperm. In sea urchin eggs partheonogenetic cleavage can be induced by anti-bodies which disrupt cortical granules but antibodies against other constituents of the egg cell do not have the effect.

Hence in both lymphocytes and egg cells there is evidence that material escaping from lysosomes can release cells from mitotic inhibition. Evidence from other cell systems points to a similar conclusion so that the mechanism may be quite general.

It is well known that in the presence of certain dyes cells become sensitive to light. Experiments carried out in collaboration with I. A. Magnus and M. R. Young have shown that there are two main types of photo-sensitization. First, some dyes such as eosin and rose bengal do not actually enter living cells but, in the presence of light, damage the cell membrane. This very rapidly kills the cells which then show cytoplasmic extrusion. The second type of photo-sensitization is due to lysosomal damage, and occurs with substances such as neutral red, porphyrin or anthracene which are concentrated in lysosomes. Enzymes released from lysosomes kill cells after some delay and without cytoplasmic extrusion. Studies with high intensity illumination from a quartz prism monochromator have shown that the wavelengths which are most efficient for cell killing correspond closely to those absorbed by the photosensitizer.

Some compounds which produce photosensitization through lysosomes lead to light sensitivity in the human skin and of these a notable example is porphyrin, varieties of which accumulate in diseases known as porphyria. Under these circumstances, exposure to light produces dilation of the blood vessels, exudation of fluid to form weals and pain within a few minutes. The photosensitizing agents are concentrated in the lysosomes of living cells and it seems likely that lysosomal damage in the deeper layers of the skin is involved in this process, which resembles an inflammatory reaction. However, damage to lysosomes in the outer layers of the skin appears to play an important part in the short wavelength sunburn reactions. But because the factors released have to diffuse down to the deeper layers of the skin, it takes some hours for the burning sensation, dilation of blood vessels and blister formation to develop.

These observations on photosensitization suggested a method of selectively damaging lysosomes which we have used to study the effects of lysosomal enzyme release on cells. We know that if enough dye is taken up in lysosomes and enough light is given, cells are killed, but reducing the amount of dye or time of illumination causes lesser degrees of damage. Miss G. Paton and I have found that this treatment also produces many breaks in chromatids – the twin constituents of each chromosome – in human cells. We believe that the lysosomal enzyme deoxyribonuclease (DNAse) enters the nucleus and breaks the double helical strand of the genetic material, deoxyribonucleic acid. The chromosomes in the resting mammalian cell cannot be seen but it is known that visible chromosomes, for example the lampbrush chromosomes in the developing egg cells of amphibians, are readily broken by DNAse but not by other enzymes, The lysosomal DNAse has the interesting property of breaking both strands of a double helix of DNA at the same time. Such a breakage would be much more difficult to repair than breakage of one of the two strands. Hence the break persists and is visble in the coiled chromosomes seen at the metaphase stage of mitosis.

We think that a similar mechanism plays a part in chromosome breakage produced by other agents including ultraviolet and ionizing radiation, various chemicals and certain virus infections.

Many different treatments can produce cancer in experimental animals and can probably also do so in man. These included exposure to chemicals of several different classes, ionizing or ultraviolet irradiation and certain virus infections. Even implanting films of inert plastics induces tumours in rats. It seems likely that these act through a common cellular mechanism, and in looking for such a mechanism our attention has been directed to lysosomes. In the first place, we found that the polybenzoid hydrocarbons dimethylbenzenthracene, methylcholanthrene and benzpyrene – the most potent carcinogenic or cancer-producing chemicals known – are taken up by cells and concentrated in lysosomes.

Under certain conditions massive release of lysosomal enzymes occurs which accounts for the cytotoxic effects of carcinogens. The selective damage caused to the adrenal glands of rats by a metabolite of dimethylbenzanthracene is a case in point – this same metabolite also releases enzymes from isolated rat adrenal lysosomes. Under other circumstances these carcinogens could bring about gradual release of material from lysosomes – insufficient to kill cells but enough to transform them into malignant cells. Secondly, all the classes of agents known to be carcinogenic or co–carcinogenic-that is enhancing the cancer-producing effect of small doses of chemicals – can bring about permeability changes in lysosomes.

Thirdly, we have evidence, as I have mentioned above, that activation of lysosomal constituents can release cells from mitotic inhibition. Fourthly, activated lysosomal enzymes can produce changes in genetic material, as shown by the induction of chromosome breaks. This might make permanent the release from mitotic inhibition of later generations of progeny cells, and the hallmark of cancer cells is uncontrolled growth. We have already shown that cultured normal mammalian cells in which selective damage to lysosomes has been produced, by the photosensitization technique, have an altered pattern of growth in culture similar to that of malignant cells. From all these experiments we think it is highly probable that the initial event in the transformation from a normal to a malignant cell is release of lysosomal constituents.

Our view is admittedly speculative, but no more so than alternatives which have been widely canvassed. One of these is the hypothesis that chemical carcinogens become bound to and delete a specific protein. There are two main difficulties about this hypothesis. The first is that some compounds which are highly carcinogenic – such as the heterocyclic nitrogen compound, tricycloquinazoline – do not seem to bind proteins at all. The second difficulty is in explaining how such an effect can be perpetuated among progeny cells. This can be done by some mental gymnastics and assumptions that genetic control mechanisms in mammalian cells are like those in bacteria, but such an assumption is quite unwarranted in the present state of knowledge.

The other main theory that has been put forward is that chemical carcinogens are bound to DNA and produce mutations in it. Limited binding of carcinogens to DNA has been demonstrated experimentally, but there is no evidence that this happens in cells that are still living. There is, in fact, a very poor correlation between the effectiveness of a chemical in producing classical mutations and producing cancer. On this hypothesis it is difficult to account for certain observations – such as the effect of high oxygen concentrations or cytoplasmic viruses in increasing hydrocarbon carcinogenicity or the carcinogenic effects of plastics films. All these can be readily accommodated in the lysosome hypothesis which has more generality and more plausibility.

FURTHER READING

CIBA Foundation Symposium on Lysosomes (Churchill, London, 1963)

"Lysosomes," by C. de Duve (in *Scientific American*), 208, 5, 1963)

"Histochemical Studies of Lysosomes," by A. C. Allison and L. Mallucci (in *Journal of the Expermental Medicine,* 121, 463, 1965)

ACKNOWLEDGMENTS

M. R. Young, National Institute for Medical Research, London.

7.

The Chalones

W. S. Bullough

In contrast to the hormones we have proposed that these substances shall be called chalones, . . . a name which implies that they act as inhibitors. . . . when a cell emerges from mitosis its future activity is determined by the concentration of the tissue specific chalone in its immediate neighborhood. AUTHOR

In the past 15 years molecular biologists have discovered the structure of DNA, developed a model to explain how genes are controlled in bacteria and elucidated the genetic code. What are the next major biological problems likely to be? One such problem is surely that of the application of this new knowledge to the infinitely more complex question of gene control in the multicellular organisms, especially the mammals. In such organisms the most important single biological problem at this moment is that posed by the normal processes of cellular differentiation, while one of the most important medical problems is that posed by the abnormal distortion of cellular differentiation, which may sometimes lead to cancer.

The life of a mammal begins with a fertilized egg containing a double set of all those genes which are needed to direct the full development of a sexually mature adult. The process of development in an embryo depends, first; on cell division whereby a mass of cells is built up; second, on cell movement whereby groups of cells migrate to their proper positions in the embryo; and, third, on the progressive differentiation of these groups of cells to form the tissues and organs and so the general body form. Although the details of the controls that lie behind these vastly complex processes are unknown, the embyronic differentiation of tissues is believed to be achieved in response to an array of regulatory substances. These substances, called inducing agents, operate in the correct positions and in the correct sequences to dictate to the responding cells which genes shall be repressed and which genes shall be activated. At the end of the process the cells of each differentiated tissue still contain the full genetic instructions for making the whole organism but only those instructions relevant to the function of that particular tissue remain either potentially or actually active. This article deals mainly with these differentiated tissues.

The first question to be settled is whether any alternative genetic programmes, that is, any alternative states of differentiation, remain available to the cells in differentiated tissues. The answer is that with a few important exceptions, they are commonly capable of at least two distinct gene directed activities: the cells may either differentiate for cell division by the complex mitotic process, or they may differentiate for tissue function, which also automatically leads to the death of the cells involved.

In attempting to analyse the ways in

Reprinted from the April 1969 issue of *Science Journal* by courtesy of the author and, *Science Journal*, London, England.

which these alternative cellular activities are controlled, the simplest and most convenient experimental tissue is the epidermis. As is well known, new cells are constantly produced in the basal epidermal layers and constantly lost from the outer surface. In the basal layers the cells synthesize those enzymes that direct the process of cell division, which indicates that the "mitotic genes" are active; in the more superficial layers the cells, which are moving towards death, synthesize instead those enzymes that direct keratin synthesis, which indicates that the "tissue function genes" are active.

An impressively perfect balance is constantly maintained between the number of new cells produced basally and the number of dead keratinized cells lost distally, and furthermore this balance is a flexible one. When epidermal cells are destroyed by any form of tissue damage the rate of cell production in the basal layer increases dramatically to produce unusually large numbers of new cells. When sufficient cells have been produced to restore the epidermis to its normal thickness production falls back to normal. A balance of this kind could probably not be maintained unless controlled by some type of feedback mechansism.

The critical moment in the life of an epidermal cell is when it emerges from mitosis and enters what is called the dichophase. It then faces the decision whether to prepare for division once again or whether instead to prepare for keratin synthesis and death. This decision is normally taken in terms of the position of the cell: if it remains basal the cell prepares for mitosis but if it is squeezed out towards the surface the cell prepares for function and death. This in turn suggests that the decision is normally made in terms of the micro-environment in which the cell finds itself.

In analysing this problem in my laboratory, Dr Edna Laurence and I first focused our attention on the way in which the mitotic activity of the basal cells may vary. In epidermis, as in many other tissues, mitotic activity is high during sleep and low at other times, especially during heavy exercise. This is called the diurnal mitotic cycle and although the analysis of this cycle was long and difficult, the explanation is short and simple. Mitotic activity is high whenever, as in sleep, the concentration of the stress hormones – adrenalin and the glucocorticoid hormone – is low. These two stress hormones together act to inhibit mitotis. Therefore, whenever the adrenal glands are active, tissue replacement takes place more slowly and during severe and prolonged stress may stop almost completely.

However, mitotic control in the epidermis, or in any other tissue, is extremely unlikely to be excercised in terms of the stress hormones, which merely exert a general action throughout the body. Attention was therefore turned to the second way in which mitotic activity can vary.

When epidermis is damaged, the mitotic rate in those cells lying in the close vicinity of the damaged area increases sharply. For more than 50 years it has been held, without any adequate evidence, that this reaction occurs in response to the stimulating action of a "wound hormone" coming from the damaged cells. In recent years, however, we have explored the opposite point of view, that the epidermis normally contains a mitotic inhibitor, and that it is the loss of this inhibitor from the damaged cells in and around a wound which permits the increase in mitotic activity. A series of critical experiments confirmed this and also showed that the mitotic inhibitor wiithin the epidermis must be strictly tissue specific in its action.

Attempts to extract the mitotic inhibitor then began. In the first experiments the epidermis was scraped from mouse skin and macerated in water in a tube immersed in crushed ice. When the insoluble material was centrifuged and the supernatant was tested for the inhibitory activity by subcutaneous injections into adult male mice, the epidermal mitotic rate immediately fell. It was also found that similar extracts made from other tissues had no antimitotic action in the epidermis.

A more refined approach to the problem was then made by studying the reactions of mouse ear epidermis in a simple culture medium. Again the results were clear cut. The addition of the epidermal extract caused the epidermal mitotic rate to fall to a low level, but then an unexpected reaction was noted. After about five hours *in vitro* the skin extracts lost their power to depress mitosis, although they were found to be still active when fresh pieces of epidermis were added. The explanation proved to be that during the first five hours adrenalin, present in the epidermis when it was taken from the body, was all lost. To restore the inhibiting activity of the epidermal extract it was necessary only to add small amounts of adrenalin, so small that by themselves they had no detectable antimitotic activity.

The explanation of the diurnal mitotic cycle was thus obvious: the epidermal mitotic inhibitor has a stronger action whenever, during waking hours, the stress hormones are abundant in the blood. The explanation of the wound reaction is also clear: the cells adjacent to a wound lose a large part of their endogenous mitotic inhibitor and this leads to the weakening or even disappearance of the diurnal mitotic cycle in the affected cell, since there is too little mitotic inhibitor left with which the stress hormones can cooperate.

In conjunction with Dr Hondius Boldingh in Holland, attempts were made by ethanol fractionation, electrophoresis and dialysis to purify this inhibitor. After a purification of about 2000 times, it now appears to be a small protein or glycoprotein.

At the moment the conclusion is that this mitotic inhibitor is the feedback message synthesized by the maturing cells of the epidermis to diffuse to the basal epidermal layer to inhibit mitosis. It is interesting that tests, made jointly with Professor O.H. Iversen and his colleagues in Norway, have also shown that when this substance is extracted from the skins of any of a wide range of mammals, or even from the skin of the cod, it is equally active when added to mouse epidermis or to human epidermis. The inhibitor is therefore tissue specific but not species specific.

The great question arising from these studies of the epidermis was whether similar feed-back systems also operate in other mitotic tissues. A partial answer is given by the fact that a diurnal mitotic rhythm has been described in about 20 different mammalian tissues and if all these depend, as in epidermis, on the interaction of the stress hormones with tissue specific inhibitors of mitosis, such systems must indeed be widespread. More direct and conclusive evidence also comes from three dissimilar tissues, the granulocytes and the erythrocytes of the blood and the melanocytes of the skin.

Extensive investigations made by Dr T. Rytömaa in Finland have shown that the mature post-mitotic granulocytes synthesize a tissue specific inhibitor of mitosis which escapes into the blood and so reaches the mitotically active progranulocytes in the bone marrow. When the stock of mature granulocytes is large enough, the concentration of the inhibitor is sufficiently high to inhibit the mitotic activity of the progranulocytes. When after infection or haemorrhage the stock of granulocytes is depleted, the concentration of the inhibitor falls and mitotic activity in the bone marrow increases.

Recently Rytömaa has also published information showing that serum contains an inhibitor which comes for the mature erythrocytes and which selectively inhibits the mitotic activity of the proerythrocytes in the bone marrow.

Regarding the pigment cells, or melanocytes, which lie within the mammalian epidermis, our own work has shown that they too produce a tissue specific mitotic inhibitor.

Negative feedback mechanisms for the control of tissue mitosis are therefore apparently widespread and what has been discovered is a new class of physiologically active substances which act as the chemical messengers within these feedback systems. The characteristics of these substances are as follows: they are produced within the tissue on which they act; they are tissue specific in their antimitotic action; their action is readily reversible; and. finally, the stress hormones commonly act as cofactors. In contrast to the hormones, we have proposed that these substances shall be called chalones (from the Greek: *chalan,* to slacken), a name which implies that they act as inhibitors. The suggestion is that when a cell emerges from mitosis its future activity is determined by the concentration of the tissue specific chalone in its immediate neighbourhood: the mitotic genes are activated if the chalone action is weak, while the genes for tissue function (which also ultimately involves the death of the cell) are activated if the chalone action is strong.

One further point must be emphasized: the post-mitotic and functional cells have only a very short life in those tissues with a naturally high mitotic rate. Thus in the gastric and duodenal mucosae the postmitotic cells have a life expectancy of only about two days. Conversely, in those tissues with a naturally low mitotic rate the post-mitotic and functional cells have a long life expectancy. Thus in liver the post-mitotic cells may survive for years.

Even more important is the discovery that when in any one tissue the mitotic rate changes dramatically (as during stress, after tissue damage, or even in certain tissues after hormone stimulation), so also does the life expectancy of the post-mitotic cells. Thus when in wounded epidermis the mitotic rate increases, the cells pass to their death in only some four to five days instead of the usual 14 to 20 days; similarly in the uterine lining when the mitotic rate rises due to oestrogen action the cells pass to their death in only about 36 hours instead of surviving for months in the unstimulated condition. Conversely, when in extreme stress, as for example hunger in winter, the epidermal mitotic rate falls almost to zero, the life expectancy of the stock of post-mitotic cells increases from between 14 and 20 days to several months. This close interrelationship between mitotic rate and the life expectancy of the post-mitotic cells obviously introduces a critically important safety factor into tissue homeostasis. With constant irritaion or hormone stimulation a tissue may enlarge but does not grow into a tumour; with a greatly reduced mitotic rate a tissue may shrink but does not disappear.

The longest lived post-mitotic cells are those found in the non-mitotic tissues, such as the striped and cardiac muscles and the neurones. In these tissues the stock of cells is built up by active mitosis during embyronic and foetal life and soon afterwards the mitotic genes are permanently silenced. The post-mitotic cells must then last throughout life. In fact, however, they continue to pass extremely slowly to their death. Dr H. J. Curtis has calculated that in man some 10,000 ageing neurones are lost without replacement every day so that at the age of 70 about 20 per cent of the brain has permanently disappeared. The shrinkage of both skeletal and heart muscles in old age is even more obvious.

The next question to arise is whether within the mitotic tissues this system of control, whereby cell birth is balanced neatly against cell death, is ever disrupted or destroyed. It is this question which leads to a consideration of the problem of cancer. Whenever a growing tumour appears its rate of cell production clearly exceeds its rate of cell loss, or in other words the homeostatic mechanism, of which the chalone forms a part, has for some reason been distorted. This distortion is permanent. All the daughter cells inherit it in the manner of a mutation. In considering this problem in terms of chalones it must be remembered that a chalone mechanism is presumably complex: there is the synthetic mechanism that produces the chalone at a particular rate; there is the manner of chalone passage within and between cells; there is the responding mechanism which evidently leads to gene activation and repression; and there is the part played by the stress hormones. Clearly permanent damage could be inflicted at many points.

In preliminary investigations, carried out over the past two years, a series of studies have been made of three different types of highly malignant tumours: an epidermal carcinoma, a granulocytic leukaemia and, most recently, two melanomata. These were all transplantable tumours, which are necessary in order that proper comparisons could be made between the various experimental groups.

The first experiments were those done with a rabbit epidermal squamous cell carcinoma. Known as the Vx2 tumour, it grows and spreads so rapidly that one small inoculation can kill a full grown rabbit in three to four weeks. The first question was whether this tumour, the cells of which showed no obvious trace of their epidermal origin, was still continuing to synthesize the epidermal chalone. Extracts were made as in the case of the normal epidermis and it was immediately found that the epidermal chalone was present. However, quantitative studies suggested that each tumour cell must contain less than one tenth of the chalone concentration found in a normal epidermal cell. It was also found that when a large tumour is present there is so much extra epidermal chalone in the body that the mitotic rate of the normal epidermis is deeply depressed.

The next question was whether these tumour cells are still capable of reacting to the epidermal chalone. To test this the body space of the rabbit was flooded with a high dose of purified epidermal chalone taken from pig skin; within four hours the mitotic rate of the tumour cells fell by more than 80 per cent.

A similar study of a rat granulocytic leukaemia by Rytömaa led to similar conclusions. The leukaemic cells proved to contain less than one tenth of the chalone content of the normal mature granulocytes while the serum was so abnormally rich in chalone that the mitotic activity of the progranulocytes in the bone marrow was almost entirely inhibited and the numbers of normal granulocytes fell to a low level. Indeed this tumour is fatal when, with inadequate numbers of normal granulocytes in the circulation, the animal contracts a serious infection. Finally, Rytömaa showed that when a leukaemic rat was injected repeatedly for two days with the granulocytic chalone, the number of leukaemic cells in the circulation fell to about half its previous level.

At this point Dr U. Mohr, then of the Deutsches Krebsforschungszentrum in Heidelberg, suggested that two types of melanoma, one in the mouse and the other in the hamster, might be treated with extracts of pig skin which were known to contain the melanocyte chalone. A study of the chalone mechanisms of these two melanomata, both of which are fast growing and fast spreading, was therefore made in London. Again the results were similar: the cells in both tumours contained the melanocyte chalone, evidently in reduced concentrations, and in both cases the cells were able to respond by mitotic inhibition to the addition of extra melanocyte chalone, whether, *in vivo* or, *in vitro*.

An attempt was then made, in Heidelberg, to prevent the growth of these two melanomata by injections, repeated for five days, of skin extracts containing the melanocyte chalone. The large amounts needed were made from pig skins by N. V. Organon at Oss in the Netherlands, who have also supported a great part of the work reported here.

To deal first with the Harding-Passey melanoma of mice, the implanted tumours were allowed to grow to the size of a cherry before the experiments began. With tumours of this size no spontaneous regressions occurred in more than 1000 untreated control animals; in contrast, with 75 mice treated with adequate amounts of the skin extract all the melanomata regressed completely. During their regression the tumours first softened, then ulcerated through the skin, and finally healed completely. However, in all except eight mice the tumours recurred later; at the moment these eight mice, now more than two years old, are beginning to die of old age.

The results with the hamster amelanotic melanoma were similar. Again the implanted tumours were allowed to grow to the size of a cherry and in more than 3000 untreated control animals there were no spontaneous regressions. In contrast, in 200 hamsters injected with adequate doses of the skin extract, all the melanomata regressed completely. As in the mice, the tumours softened, ulcerated through the skin and, finally, healed. In these animals all the tumours later recurred. However, in both the mice and the hamsters, the question remains as to whether such recurrences can be prevented if the treatment is continued for longer than five days.

Of course, since the pig skin extracts contained much contaminating material, the possibility did remain that this dramatic tumour destruction was, in fact, caused by something other than the melanocyte chalone. The action of the extracts was therefore tested against three tumours derived from quite different tissues (Walker carcinoma, plasmocytoma and mastocytoma) but without effect. Conversely, extracts of other tissues made in the same way were tested against the two melanomata, again without any effect.

As a final test, it was suggested that since both these melanomata were known to contain the melanocyte chalone it was possible that extracts of melanomata could destroy melanomata. Extracts of both the melanomata were therefore tested against the hamster amelanotic melanoma and in both cases they proved to be active. The tumours were destroyed by five days of treatment although in all cases they later recurred.

Thus it seems probable that melanocyte chalone had indeed destroyed these two melanomata. The next question was why this should have happened at all. There is no evidence to suggest that chalones are in any way toxic substances, even in unnaturally high concentrations, and therefore their influences cannot be compared with the destructive actions of cytotoxic agents or of radiation. Rather the disappearance of these tumours appears to have been due, first, to the diversion of the mitotic cells towards their death – evidently the natural function of a chalone – and second, to the fact that this inhibition of mitosis, unlike that in a normal tissue, was not accompanied by any significant delay in the time of death. Indeed, in the case of the Harding-Passey melanoma, there was clear evidence that tumour regression does involve the diversion of the mitotic cells to melanin synthesis and to an early death.

Exciting as this discovery may be, it is essential to discourage any great mood of optimism. It is not only possible but probable that other tumours will prove to be quite different in their characteristics and in their reactions. However, if chalones can sometimes act to arrest or even to destroy just a few types of tumours, then they are substances which, like many hormones, have the great and rare advantage of strict tissue specificity. Furthermore in most cases they will probably do no serious damage to the normal tissues from which the tumour came.

Our main interest in this work, however, rests in its contribution towards understanding normal tissue organization. A great deal has now been learnt of the manner in which cells are produced within a tissue and of the way they pass through function to death. This has added significantly to our understanding of cellular differentiation. But it is already obvious that as the behavior of normal tissue cells comes to be better understood the information obtained cannot fail to contribute to a better understanding of cancer.

This article began with references to the genetic code and to the manner in which genes are controlled in bacteria; a fascinating uniformity of pattern is now emerging. Not only is it becoming clearer that the genetic code is common to all living organisms, but it is also becoming apparent that the manner in which the alternative available gene programmes are controlled may depend on a common principle. In a bacterium, groups of related genes are activated or repressed in terms of the concentrations of what are known as effector substances. It now seems that closely parallel mechanisms, depending on an array of specialized effector substances, must operate in a mammal, both in the undifferentiated embryonic cells and in the differentiated adult tissue cells.

Footnote. When this paper was in proof news was received from Dr T. Rytömaa in Helsinki that he had succeeded in destroying a solid granulocytic tumour in rats by means of the granulocytic chalone. The regression of this type of tumour followed an exactly similar course to that described by Professor Mohr in the case of the two melanomata.

FURTHER READING

"Mitotic and Functional Homeostasis," by W.S. Bullough (in *Cancer Research,* 25, 1683-1727,1965)

"Tissue Homeostasis in Adult Mammals," by W.S. Bullough and E.B. Laurence (in *Advances in Biology of Skin, VII, Carcinogenesis, edited by W. Montagna and R.L. Dobson, Pergamon Press, 1966)*

"The Evolution of Differentiation," by W.S. Bullough *(Academic Press, 1967)*

"The Control of Tissue Growth," by W.S. Bullough (in *The Biological Basis of Medicine, Vol 1, edited by E. and N. Bittar, Academic Press, 1968)*

"Chalones and Cancer," by W.S. Bullough *et al* (in *Nature 220, 134-139, 1968)*

IV. Nervous System

One of the research frontiers of modern biology, and perhaps the most important, is that of the functioning of the human brain. Many of man's present limitations reflect his lack of understanding of such neural phenomena as memory, learning, and intelligence. In the first essay of this unit, "The Search for the Memory Molecule," David Perlman, Science Editor of the *San Francisco Chronicle*, ably relates the work of David Krech and colleagues at Berkeley, Prof. Holger Hyden of Goteborg, James McGaugh at Irvine, and others.

Any discussion of neural functioning in man today usually turns to a consideration of drugs and their effects through use and abuse. We selected three papers which examine quite different aspects. "Hallucinogens of Plant Origin," a technical paper written in the classical style by Richard Schultes, director and curator of Economic Botany, Botanical Museum of Harvard, presents an informative historical viewpoint of man's use of the wide variety of plants with hallucinogenic qualities. The essay "*Cannabis*," updates man's primitive knowledge about the (in) famous hemp and points to the need of complete objectivity in all aspects of investigations relating to it. The author, Andrew T. Weil of the National Institute of Mental Health, with Norman Zinberg and Judith Nelson, in 1968 reported in *Science* the first significant results of a human study on marihuana in recent years. In "The Speed That Kills," Jonathan Black, staffer for the *Village Voice*, New York City, clearly explores the abuse of amphetamines which is widespread in our society at this time.

8.

The Search For the Memory Molecule

David Perlman

In the explosive quest for the "memory molecule" there is scarcely a major psychology laboratory anywhere today that is not populated by colonies of animals industriously running mazes, scooting through shuttle boxes, dancing away from electrical shorting machines, or frisking about with microelectrodes implanted in their skulls to radio out brain-wave signals awake or asleep. **AUTHOR**

In the long parade of fish and flatworms, rats, cats and mice that have given their all for science, it is just possible that Rat No. 895 – with the help of a bearded psychologist named David Krech – will hold a special significance for the education of our children's children. The rat is one of a group of animals whose life experience may eventually spell out the structural and biochemical nature of memory; he may help point the way to providing education meaningfully enriched – perhaps even chemically enhanced – for every child's unique individual mind.

Rat No. 895 is a gentle, gray-brown creature with the sleek coat and well-fed look of an animal reared in privilege. He lives in the basement of Tolman Hall at the University of California at Berkeley, where Dr. Krech is a professor of psychology.

One recent afternoon, the professor looked on as the rat sat quietly in a white-walled, smooth-floored chamber that was part of a tiny two-room suite in the Tolman Hall laboratories. It was a comfortable room for a rat, perhaps 8 inches square, with a light shining brightly down. In one white wall a small hole opened into the second room of the suite, and for the rat the hole was tempting: it led to an even more attractive room of equal size but black-walled and dimly lighted. To a rat, darkness is far more inviting than light; dark rooms are where the action is.

Krech could not know it, because the experiment was being conducted in scientifically "blind" fashion, but Rat No. 895 was carrying inside his body an injection of homogenized fluid extracted from the minced-up brain of a late-lamented cousin who, just before death, had been trained to fear a black chamber as a source of a highly unpleasant electric shock from a wire-grid floor. Krech had taught No. 895 to fear the shock in the black chamber, too, but that had been four weeks earlier; by now, any normal rat should have largely forgotten the shock and its lesson.

Inside the white chamber, No. 895 now hesitated, He poked his nose through the hole, whiskers twitching, and withdrew abruptly from the black room as his whiskers touched its floor. It was a different floor, not smooth like the white room's, but a grid of slender metal bars. The rat backed to a corner of its white room, licked a fore-paw – a layman might have said reflectively – and stared again at the inviting entrance to the dark room. For 10 minutes the rat considered the matter, then tentatively slipped into the darkness, nose first, then rump, then tail. Nothing happened. The floor did not shock. The rat explored the room cautiously, then more actively and, finally, scampered back and forth between the two rooms 30 times within the next 20 minutes.

Now the rat's half-hour of testing was over; a white-coated lab technician, who had sat quietly by, recording every movement, lifted him from the suite and replaced him with another one. This rat behaved quite differently: he frisked about the white room for barely two minutes, poked his head into the dark chamber – an observer might have said insouciantly – then scampered through. For 28 minutes, he explored the dark half of his suite happily, just as if he had never received a shock in his life. Krech could not know it in this case, either, but this carefree animal also carried an injection of brain extract in his body – an extract from a wholly untrained rat, which had never been taught to fear the darkness that rats love.

These two rats in their two-room suite – displaying responses far less ambiguous than most of their rodent colleagues – are part of only a single experiment in a hotly debated new research area known as interanimal memory transfer. Psychologists in scores of laboratories are trying to prove that precise mechanisms of memory and learning lie within the chemistry and structure of the brain. Some hunt the chemicals by testing new compounds in animal brains; others analyze brain tissue from "educated" animals to look for new-formed molecules. In the most controversial current research scientists like Krech are hoping to establish that memory itself can be transferred, at least partly in chemical form, from one animal to another.

Did Rat No. 895 really pick up molecules of memory in his injection of trained brain extract? Krech himself believes the suggestion is there, but by no means established. Yet if the transfer phenomenon does in truth exist, it will be the most exciting discovery in the history of psychology. It will prove the molecular basis of learning, open the way for its identification, and mark the first step toward its precise control.

Krech has already pioneered in experiments showing how memory and learning can change the brain's own chemistry and physical form. Now he is deeply immersed in the jungle of memory transfer, where the evidence is conflicting, the pitfalls abound, but the possibilities are enormously exciting. He is cautious even on the tentatively encouraging results in his own lab. But if he and other workers succeed, Krech says with mock exuberance: "The world of the mind is ours!".'

His most recent decade of laboratory work, most of it with a larger team of psychologists, biochemists and anatomists, has already helped develop remarkably original insights into the chemical workings of memory and learning. The day is very near, Krech believes, when the intellectual capacity of human beings will be subject to the products of the chemical as well as the teaching laboratory – even if full theoretical understanding lags behind. Like Lewis Carroll, the unorthodox Krech likes to use portmanteau words: "psychoneurobiochemistry," he calls the present effort in brain research. He offers another, for the day when psychology, brain chemistry and the classroom all merge: "psychoneurobiochemeducation." Krech can even pronounce it.

In the explosive quest for the "memory molecule' there is scarcely a major psychology laboratory anywhere today that is not populated by colonies of animals industriously running mazes, scooting through shuttle boxes, dancing away from electrical shocking machines, or frisking about with microelectrodes implanted in their skulls to radio out brain-wave signals awake or asleep. In one lab an educated goldfish may have his memory cut short by an antibiotic; in another a dull-witted mouse may smarten up on pep drugs; a soup of brain tissue may or may not alter a cat's intelligence. Even the questions are still uncertain: Just what is learning? How and where do we store the remembrance of things past?

In human beings and rats alike each new day of fresh experience adds new knowledge somewhere within the brain. Much of that knowledge even the most intelligent person forgets almost instantly: the name of the dull guest just introduced at a cocktail party, for example. Some knowledge we remember for a few hours, no longer: who, since cramming for a geometry final, can recall pi beyond the first few decimal places? And then there are the memories that seem to last forever: the streetcar to the zoo in a city long since left behind; the facts and ideas transmitted by an inspired teacher,

The human brain, where learning and memory certainly reside, contains some 10 billion large, complex nerve cells called neurons, plus a vast number of smaller cells known as glias, which help nourish the neurons and may also synthesize some of the chemicals of memory. The neurons receive and transmit all the impulses that make up sensory inputs and responses, thoughts and emotions and memories. Each neuronal cell contains a single impulse-transmitting arm called an axon, whose multiple branches connect at their tips with the intricate tracery of receiving surfaces, known as dendrites, on other neurons.

The junction between the axon of one neuronal cell and the dendritic branch of another is known as the synapse: a point where transmission of the electrical nerve impulse is facilitated by chemicals such as acetylcholine, a neurotransmitter, and enzymes like acetylcholine esterase, which modify the chemical transmitters.

If a composite could be made of all the theoretical models for the brain's role in learning and memory it might go like this:

A new experience is perceived; it initiates some kind of reverberating electrical signal among a set, or web, of neurons. As the short-term memory signals leap across thousands of synapses they begin to activate chemical processes at the synaptic junctions. Within a few seconds or minutes, chemical changes begin – perhaps the synthesis of RNA (ribonucleic acid), perhaps the construction of complex molecules in the form of peptides or proteins. The chemical material forms a longer-lasting but still temporary memory trace, and perhaps in hours or days it induces profound anatomical changes in the cortex of the brain. Later, these physical-chemical changes become the soldered wiring of long-term memory. This entire memory process may take place step by step from the first act of perception, or all its stages may be activated independently, and proceed at varying speeds.

Today's blaze of activity in molecular memory research was largely ignited by a slender, soft-spoken Swedish scientist names Holger Hydēn at the Univèrsity of Goteborg. On the basis of his extraordinarily delicate microscopic dissection of single brain cells, Hydēn believes that RNA, which carries information for the synthesis of proteins, is a chemical key to memory.

Recently, he trained scores of right-pawed rats to perform unfamiliar left-pawed tasks, like reaching for food down a narrow-necked jar with a left paw. He taught scores of others to balance like circus performers on steeply tilted wires. When he sacrificed these trained animals and dissected their brain cells, Hydēn found large increases in the RNA content of both neurons and glial cells. He also found surprising changes in the chemical Composition of the RNA itself, which might indicate that the new memories were being laid down as newly synthesized and altered protein molecules.

At the University of Michigan, Dr. Bernard W. Agranoff uses schools of educated goldfish and the powerful antibiotic puromycin to test the relationship of RNA and memory. He trains his goldfish to swim back and forth across a barrier in their tank whenever a light flashes, a task they can remember for a month. Then he injects minute doses of puromycin through the skulls of the fish before, during and after training. Puromycin blocks the RNA-induced synthesis of protein and it also blocks a goldfish's memory. An injection of the antibiotic seconds after training will destroy memory at once. An injection before training does not prevent the fish from learning its task, but it does make it forget the lesson within a couple of days. An injection an hour after training has no effect at all.

Agranoff's work shows that short-term recall is extremely fragile, but that once it is consolidated into long-term memory it can withstand powerful chemical assaults. It does in goldfish, and recently other scientists have verified the fact in mice.

Many chemicals can enhance memory rather than wipe it out, and these drugs also help zero in on memory's molecular aspects. At Irvine, near Los Angeles, Dr. James L. McGaugh and his colleagues of the University of California have found that drugs such a strychnine, picrotoxin and Metrazol, which stimulate the central nervous system, help maze-trained mice to consolidate short-term recall into long-term memories.

Oddly enough, McGaugh finds, these drugs work most effectively in strains of mice bred for their stupidity at running mazes. They are far less effective in genetically brighter mice.

Krech, who keeps an alert and imaginative eye on the research in this field, draws a lesson from McGaugh's work. "Here," he says, "we have a chemical memory pill" which not only improves memory and learning, but which can serve to make all mice equal when God – or genetics – hath created them unequal! It's most pregnant with social implications, and promises and forebodings for the future."

While some workers inject their animals with chemicals to strengthen memory and others study brain function by wiping memory out, a different experimental strategy has been explored for the past 15 years by a Berkeley team of many talents. It was originally led by Krech and Dr. Mark R. Rosenzweig, also a psychologist; Dr. Edward L. Bennett, biochemist, and Dr. Marian C. Diamond, neuroanatomist. Krech has recently focused his research on the memory-transfer controversy, but his colleagues are continuing the team's earlier work.

Experiments by this Berkeley group show that the memory-building experiences of an abundant, intellectually challenging life can enrich an animal's brain structure and chemistry – and that isolation and environmental impoverishment can impair it. Some of their experiments even indicate that rats in an "enriched" environment may learn to solve many kinds of problems more effectively than deprived ones, and that animals reared in isolation perform more poorly.

To seek out these physical, chemical and behavioral changes, the Berkeley researchers run a 24-hour-a-day rat nursery school that might well delight the Montessori movement. In the animal rooms of Tolman Hall on the University of California campus typical groups of Berkeley rats undergoing enrichment will romp in colonies of a dozen cousins from the time of weaning on.

Their roomy cages are filled with more toys, games and gym equipment than F. A. O. Schwartz: wheels to run in, boxes to clamber over and snuggle into, chains to swing on, ladders to climb, platforms to leap. For a half-hour every day, white-coated experimenters take the rats from their cages, run them through mazes and reward them with sugar pellets.

By contrast, in rooms nearby, other rats from the same genetic strain live in what the experimenters term the Impoverished Condition. They sit alone in tiers of isolated cages with blank metal side walls, in a quiet, dimly lighted room. They are fed well, but attendants come and go silently to replenish food and water. Only once a week, for weighing, are the rats handled; otherwise they have no companionship, no stimulation, no lively memories.

These rats may live their widely varied lives for weeks or months. For some, nursery school may last all day and night; for others, only a couple of hours a day. But ultimately, as it must to all rats, comes the end of the experiment and the moment of sacrifice: a swift, clean death by decapitation; surgical excision and meticulous weighing of brain tissue; chemical and anatomical analysis of cortex, neurons and glial cells; assay of enzyme levels at synaptic junctions; microscopic studies of the cells and structures.

From their experiments over the years, the Berkeley researchers feel certain of their evidence for anatomical and chemical change. Their behavioral findings are still ambiguous in many ways, and highly tentative, they believe.

Structurally, the cerebral cortex of a rat that leads an enriched life is likely to weigh about 6 per cent more than its isolated, impoverished littermate. This cortex, where some sophisticated learning and memory functions are probably localized is also about 6 per cent thicker. It contains more protein and has 14 per cent more glial cells, which – according to researchers like Sweden's Hydén – play a major role in the consolidation of permanent memory. At the synapses, the enriched rat brain shows greater quantities of enzymes like acetylcholine esterase that facilitate the transmission of impulses from neuron to neuron. Finally, the enriched rat's neurons are larger, and may be more intricately and richly endowed with dendrite branches.

In its most recent experiments, the Berkeley group has discovered that some of these significant changes in brain physiology may show up after only 30 days of enrichment, and many occur with only two hours a day of stimulation, rather than 24. Impoverishment may be even more important than enrichment for some of the anatomical differences, the group believes.

For several years now, Dr. Rosenzweig and his colleagues have been testing the hypothesis that problem-solving abilities should improve along with the structural and chemical changes inside the brains of environmentally enriched rats. But as Dr. Rosenzweig says: "The work has turned out to be difficult and inconclusive thus far."

In one typical test, invented years ago by Krech, rats must learn to scoot down a series of four double alleys, choosing the correct path each time, to reach a food reward. Sometimes the correct paths are all illuminated by a light bulb; when a rat learns that pattern, the experimenters reverse the lights, and make all the correct alleys dark. On this kind of learning and problem-solving apparatus rats whose lives are enriched for many weeks are often significantly better performers than their deprived relatives. But a rat's I.Q. is not that simple. Under other conditions, when enrichment and deprivation start in adulthood rather than infancy, the impoverished rats mysteriously score better at the light-dark reversal test than their enriched brothers.

The years of experiments at Berkeley, with no fewer than 10,000 rats, certainly demonstrate the remarkable plasticity of the brain – how it can grow richer anatomically and chemically in an enriched, active environment, and how its growth may be stunted in an impoverished world without stimulus or challenge. Whether learning and memory can also be improved by enriching a rat's environment is still a puzzle.

It was during these Berkeley experiments that Krech and other brain researchers were first thrown into an uproar by the astonishing reports of a young, free-swinging psychologist named James V. McConnell at the University of Michigan. McConnell worked with planarians, the tiny spade-headed flatworms that abound in ponds and quiet streams. However lowly their intelligence, McConnell was able to train them to solve simple mazes. Then he minced up a batch of his trained worms, fed them to naive, uneducated planarians – and announced that the uneducated cannibals had apparently eaten an education! Their heady diet enabled them to "remember" the mazes which their newly digested cousins had acquired before being eaten. This was the first of the memory-transfer experiments. Since then there have been many more.

At Baylor University, for example, Dr. Georges Ungar claims he has moved up the evolutionary ladder from worms to mammals. He trains rats to become completely accustomed to a frightening, clanging noise; then he extracts from their brain tissue a broth containing proteinlike polypetide molecules, and injects the broth in the abdomens of Swiss mice. The mice would normally quiver and crouch in fear at Ungar's clanging noises, but given a shot of "trained" brain broth, they don't turn a hair, Ungar has reported.

Ungar sees polypeptide molecules as the possible chemical of memory. To Dr. Allan L. Jacobson, center of a recent scientific storm over interanimal memory transfer, the molecule is brain RNA.

At the University of California at Los Angeles, Jacobson and his colleagues have reported they can transfer specific skills from rat to rat by injections of "trained" RNA alone. In one experiment, for example, a group of rats is trained to press a lever for a food pellet whenever a light blinks. Another group learns to press the lever only when a click sounds. RNA injections from the brains of light-trained rats, according to Jacobson, will cause untrained rats to respond to the light signal but not to clicks. RNA injections from click- trained rat brains transfer the click response to untrained animals, but they will not respond to lights, Jacobson reported.

All this is highly debatable, and not long ago 22 scientists from nine different research institutes signed a joint report stating they had tried to duplicate the Jacobson results in 18 sets of virtually identical experiments – and that all had failed. But even there neither prosecution nor defense rests. A leading signer of the mass dissent was Dr. William L. Byrne of Duke University. Yet two months later, Byrne and David Samuel of Israel's Weizmann Institute reported they had accomplished a successful transfer experiment of their own, using brains of rats trained to press levers with their paws. "Memory transfer is a real phenomenon," Byrne now maintains.

But at Irvine, in California, McGaugh is far more cautious. "Despite recent reports of successful transfer effects," he says, "any conclusions and interpretations must be presented with extreme caution. The recipe for a successful transfer experiment has not yet been prepared. The memory molecule has not yet been found."

Now the story of "psychoneurobiochemistry" shifts back to Berkeley, where Krech, who has long held a posture of "stiff-necked disbelief" in the memory-transfer concept, has begun to change his mind. Today he flings caution aside and says: "Well, maybe – I don't believe it, but, well, maybe. . . "

Krech is an eager intellectual speculator; he bubbles with ideas, conjectures, strong opinions. In his office, in debate or discussion, he juts his clipped, graying beard forward assertively; throws his head back; slants his eyes through dark-rimmed glasses beneath a high, broad forehead; tilts and sways in a wooden swivel chair at an ancient roll-top desk; scribbles sentences, words and diagrams on piles of pink scratch paper to punctuate his arguments, and litters the desk top with discarded paper scraps as he talks.

The implications of interanimal memory transfer are extraordinarily important, Krech feels, and that is why he has begun to discuss them publicly, and to devise key experiments to test new aspects of the theory. "If it should come to pass that they are right," he says of the True Believers in transfer, "then we will indeed have made a huge jump forward. For we would have a most effective behavioral assay method, with our available biochemical techniques, to zero in on this marvellous brain goulash which can transfer information from one brain to another. We will be able to isolate and identify in detail all the proteins, enzymes, RNA's or other macromolecules which constitute the so-called chemical memory pellets. After that..."

To Krech, his recent efforts at memory transfer are far from definitive research, but rather a "quick, furtive hop" into an intriguingly significant new field. His own approach, as the experiments demonstrate, is not to try transferring entire memories, but rather to explore a theory that intense experiences which produce strong memories lay down both structural and chemical changes in the brain, and that both kinds of change are required for memory. The chemical changes may decay with time, but the structural changes should be permanent. And so, if Krech's furtive hop is successful, the experiments may establish that once the permanent structural changes of memory are fixed – in a rat at any rate – a fresh supply of undecayed "memory molecules" from another rat may act as a booster shot to revive the signals along the first rat's old, inactive memory circuits.

At this point Krech believes he has a "smidgeon of evidence" to support the theory. But Edward Bennett, the biochemist who has worked with him for years and is now doing the chemistry in the new transfer experiments, is more cautious about the meaning of their work. "Until other laboratories can routinely reproduce transfer effects in identical experiments," he insists, "reported positive results will be of little scientific value." Krech agrees that transfer has yet to be demonstrated, but he and Bennett are pushing on with their experiments.

In the laboratory, scientific caution may give Krech pause; in the arena, never. He is known, at Berkeley and at campus seminars across the country, as a contentious, independent advocate of varied controversial causes, and as a brilliant teacher. One of his former graduate students, now an outstanding brain researcher himself, says devotedly: "Krech is a kind of Talmudic scholar in psychology. He's an authoritarian father figure. He's tough, critical, impatient. Some people dislike him, many fight with him. But I think that if God had a sense of humor, He'd be a Krech."

Krech's colleagues sometimes criticize him for his candor and his willingness to extrapolate, to draw human, social and even political inferences from his work with lowly rats. More conservative members of the Faculty Club dispute his outspoken views on war, on the campus draft resistance, on Academic Senate meetings (he opposes the first, publicly supports the second, sometimes upsets the third). None of this bothers Krech a bit.

"Frankly," he says, "I'm not always considered too kosher among my colleagues because I'm concerned about the social problems that flow from our knowledge, and I say so publicly. I have fears about science; I feel that scientists can't be allowed to run the world, and if what I say weakens the awe so many people hold toward scientists, that's all right with me. There's too much awe toward science already, and the scientists have fallen victim to their own propaganda about the scientific method.

"Scientific reports almost never describe all the blind alleys, all the hunches that didn't work out, all the false starts and experimental errors. The tables and graphs are always nice and tidy, just as if nature itself handed out the data in tabular form. But nature doesn't give up its data easily; scientists have to seize it, and while they're grabbing they select and leave a lot behind. Biological science is largely hit-or-miss, not mathematics."

In 1968 a subcommittee of the United States Senate held hearings in Washington on a proposal by Senator Walter F. Mondale (D., Minn.) to create a new national commission on the social implications of advances in the health sciences. Among the advances cited were feats like heart transplants, the imminence of "genetic engineering" and the foreboding possibility of chemical controls over learning. While other distinguished scientists told the Senators that science needed only to be left alone with plenty of financial support, Krech testified far differently.

"Brain research," he declared, "is immeasurable more significant for the future of man than anything else now going on in science.

"It is my judgment that within 5 to 10 years there will be available a regimen combining psychological and chemical measures which will permit us to exercise a significant degree of control over the development of man's intellectual capacities."

This can mean, Krech has said many times, a future of enormous promise: chemical therapy for many of the mentally retarded and senile; chemical release for those who suffer from crippling memories of the past; teamwork among chemists, psychologists and educators for what Krech terms the "first business of society – the development of the mind of the child, the shaping of its strengths and weaknesses, the potentiation of its inherent possibilities."

But these models of brave new worlds, Krech feels, hold danger as well as promise. When a man can raise or lower I.Q. and memory and learning capacity at will, when he can selectively strengthen clusters of special abilities like verbal skills, or mathematical reasoning, or artistic talent, disturbing questions will intrude.

"Who is to decide what happens to whom?" Krech asks. "The parent? The huckster? The pediatrician? The school board? And what will be the effects upon our society of either increasing the I.Q. level of all men, or of increasing the distance between the brighter and the duller? What new demands will these possibilities place upon our already inadequate educational facilities? What political changes will they lead to? Remember, the changes induced in the brain may be irreversible ones."

To Krech these questions demand thoughtful reflection and decisions by all citizens, not by scientists or politicians alone. Krech said it to the Senators, and he repeats the message whenever he can:

"The brain reasearcher, whether he be a psychologist or electrophysiologist or biochemist, has neither the wisdom nor experience nor knowledge to tell society: 'Don't worry your unscientific heads about this; I will save society – from *me!*"

And he adds: "The issues I raise are much too persuasive and too profound to permit the physician to scribble the necessary social policy on his prescription pad. Here we, *all* of society, must provide our experts with political and moral and ethical guidance."

Brain reasearch, Krech feels, is a vital part of man's quest for basic knowledge about himself, and no one can halt that ancient drive, no matter where it leads. Its practical yield can be at least as promising as the present revolution in biochemical genetics, which is already pointing toward new therapies for hereditary disease and cancer. But, uncontrolled, the results of brain research could raise problems more harrowingly difficult than the ones that began with Hiroshima. They can be discerned right now in what is happening to the goldfish, rats and mice in today's laboratories. They will soon, within a very few years, face all citizens with decisions of almost unbelievable complexity. And citizens together, guided but not dominated by the experts, must decide for themselves how those problems will be solved.

9.

Hallucinogens of Plant Origin

Richard Evans Schultes

Over the millennia, man has selected "races" or "varieties" of this cultigen *Cannabis sativa* with desired characteristics: some for stronger fibers; some for higher oil content; others for greater narcotic potency. AUTHOR

An outstanding mark of this century will certainly be the growth in use, abuse, and misuse, in sophisticated cultures, of hallucinogenic substances of vegetal and synthetic origin (*1*).

Primitive cultures, where sickness and death are usually ascribed to a supernatural cause, have long accorded psychoactive plants a high, even sacred rank in their magic, medical, and religious practices, because their ethnopharmacology often values the psychic effects of "medicine" more than the physiological. Ethnobotanical studies have recently advanced our understanding of known hallucinogenic plants and have uncovered new ones, some of which have yielded compounds (Fig. 1) of extraordinary chemical and pharmacological interest, even of promise in modern medicine (*1-5*).

The pace of research into hallucinogens in dying or disappearing primitive cultures; the success of studies into the plants and their constituents; and the confusion generated by casual or frivolous interests in some segments of our society — all seem to justify an ethnobotanical summary based upon the premise that, even though only an interdisciplinary consideration can adequately cope with this fast-growing field, the starting point for understanding naturally occurring psychoactive substances must be an appreciation of the identification and aboriginal significance of the plants involved. My own ethnobotanical research which, since 1936, has taken me into remote areas of the New World to study native narcotics, has convinced me that there exists an appreciable number of hallucinogenic plants still unknown to science, and that we can no longer afford to ignore reports of aboriginal uses merely because they fall beyond the limits of our credence. Primitive cultures are fast disappearing and, with them, native knowledge of plant properties that could help us along paths of academic and practical achievement.

In view of the number of plant species, variously estimated at from 400,000 to 800,000, those that have been used as hallucinogens are few; probably no more than 60 species of cryptogams and phanerogams. Only 20 may be considered important. Even more interesting is the unexplained concentration of the majority of the hallucinogens in the New World. In both hemispheres, there are plants with hallucinogenic properties which apparently have never been employed as narcotics, but even if these were added to those that man has bent to his purpose, the

Reprinted from *Science*, vol. 163, pp. 245-254, January 17, 1969.

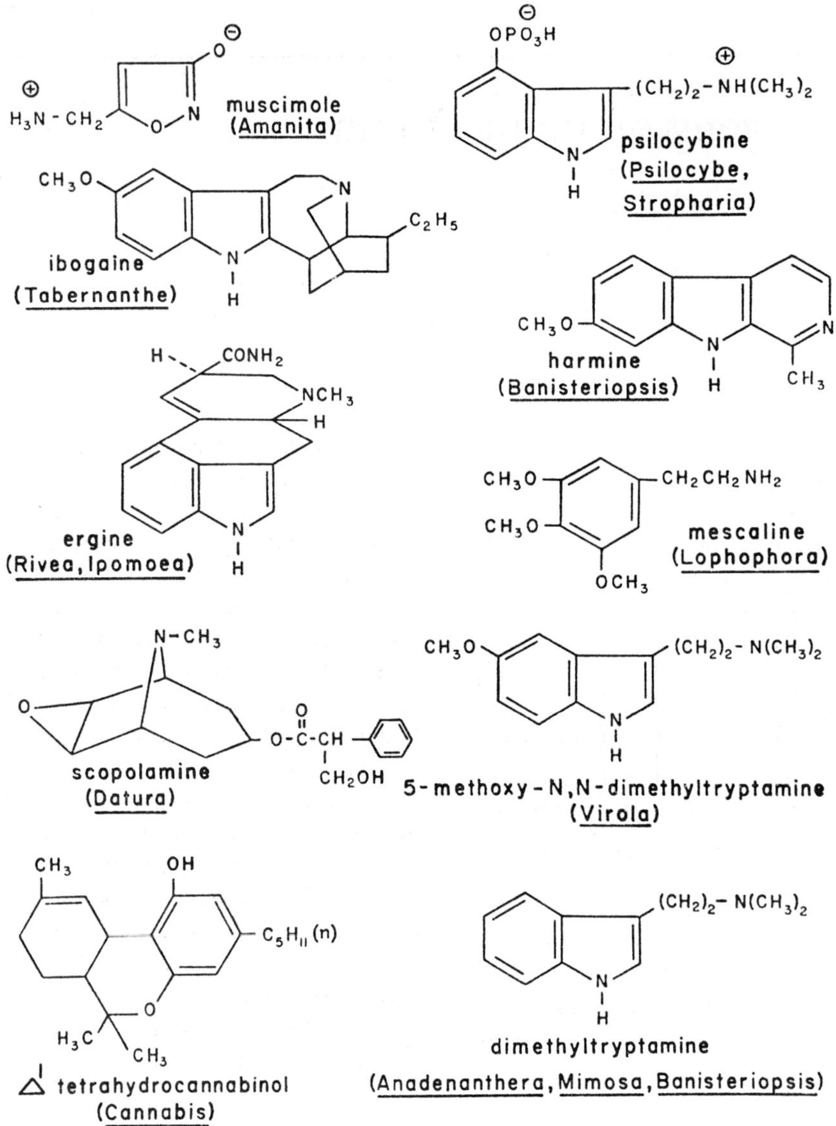

Figure 1 Main hallucinating constituents of psychotomimetic plants. (Reprinted from *Annual Review of Plant Physiology*, vol. 21, p. 574, 1970. Courtesy of the Botanical Museum, Harvard University, and publishers of *Annual Review of Plant Physiology*.

number is very small. Since most hallucinogens owe their activity to alkaloids and at least 5000 higher plants are alkaloidal, the small number of hallucinogens is even more astounding. Glycosides, resinoids, essential oils, and other organic constituents may also be responsible, so the limited number of hallucinogens must be considered challengingly provocative (*6*).

Hallucinogens occur nearly throughout the plant kingdom. Although most are spermatophytes, some of the biologically, chemically, and sociologically most fascinating are cryptogams.

Agaricaceae

The presence of toxic constituents in so many basidiomycetes led to the early discovery of hallucinogenic properties in the mushrooms. It has even been postulated that mushrooms were anciently and widely valued in primitive religions; that the very concept of deity arose from their effects; and that their present disjunct ritualistic use is relict.

The hallucinogenic use of the fly agaric (*Amanita muscaria*) by primitive tribesmen in Siberia came to the attention of Europeans in the 18th century. This fungus – widespread in north-temperate parts of both hemispheres – has long been recognized as toxic; its name refers to the European custom of employing it to poison flies. In recent times, its use as an inebriant has been known in only two centers: extreme western Siberia, among Finno-Ugrian peoples, the Ostyak and Vogul; and extreme northeastern Siberia, among the Chukchi, Koryak, and Kamchadal. Tradition established the use of fly agaric by witch doctors of the Lapps of Inari in Europe and of the Yakagir of northernmost Siberia. Formerly, the narcotic employment of *Amanita muscaria* was apparently more widespread, and it has even been suggested that the ancient giant berserkers of Norway induced their occasional fits of savage madness by ingesting this mushroom (*7*).

In Siberia, several mushrooms, often an expensive article of trade, suffice to cause an intoxication. They may be taken as extracts in cold or warm water or milk, either alone or with the juice of *Vaccinium uliginosum* or *Epilobium angustifolium*. Sometimes, a dried mushroom may simply be held, moistened, in the mouth. Among the Koryak, the women chew the mushrooms and roll them into elongated pellets which the men swallow. Many of these peoples have discovered that the intoxicating principles are excreted unaltered in the urine, almost as hallucinogenic as the original plant material. This discovery has given rise to the custom, sometimes ritually executed, of the inebriate's drinking his own or another's urine when he feels the intoxication waning, thus repeatedly effecting a continuation of the narcosis.

Effects of *Amanita muscaria* vary appreciably with individuals and at different times. An hour after the ingestion of the mushrooms, twitching and trembling of the limbs is noticeable with the onset of a period of good humor and light euphoria, characterized by macroscopia, visions of the supernatural and illusions of grandeur. Religious overtones – such as the urge to confess sins – frequently occur. Occasionally, the partaker becomes violent, dashing madly about until, exhausted, he drops into a deep sleep.

Since 1869, when muscarine was isolated, most workers have assumed that the toxicity and hallucinogenic properties of *Amanita muscaria* could be attributed to this alkaloid. Studies have shown, however, that muscarine is a minor constituent which could not be alone responsible. The same is true of the trace amounts of bufotenine reported in the carpophores. Recent pharmacological tests show that the central nervous system activity is due primarily to muscimole, an unsaturated cyclic hydroxamic acid, and to amino acids, ibotenic acid, and the less active muscazone. Since ibotenic acid is a precursor for muscazone, the variation in intoxication potential of the fly agaric may be due to fluctuations in the ratio between these two constituents. There is evidence that still other as yet uncharacterized principles may take part in the toxicity of this species (*8*).

About 3500 years ago, Aryan peoples swept from the north into India, bringing with them the cult of a plant called *soma*. They deified the plant as a holy inebriant, drinking its juice in religious rites. More than 1000 hymns to soma have come down intact in the Rig Veda. For over 2000 years, during which period the Aryans abandoned the original plant and forgot it, the identity of soma has been a deep mystery. For a century Western civilization has been puzzled by the identity of soma. More than 100 species have been suggested as its source, but none has gained acceptance. *Ephedra*, a gymnosperm, and *Periploca* and *Sarcostemma* (Asclepiadaceae) have been principal contenders – all fleshy, leafless or nearly leafless desert vines. Some botanists have felt that soma could have been *Cannabis*, others that it was wholly mythical and never was derived from a plant. The most recent identification of soma as *Amanita muscaria* appears to be the first that satisfies all of the many descriptive details and evidence gleaned from the Vedic hymns, and none seems to contradict it. If correct, it represents a meaningful contribution to ethnobotany in view of the extraordinary religious and social role of soma as emphasized in one of the earlier texts in the Indo-European world (*9*).

Another center of hallucinogenic use of mushrooms lies in Mexico. Archeological "mushroom stones" indicate that a sophisticated mushroom cult existed in Guatemala 3500 years ago. Early Spanish chroniclers wrote in detailed opposition to the diabolic mushrooms of the Aztec, *teonanacatl* ("food of the gods"), eaten ceremonially for divination, prophecy, and worship; but since four centuries failed to produce evidence of such use of mushrooms, the suggestion that the chroniclers had confused the dried mushrooms with the dried crowns of the hallucinogenic peyote cactus was accepted. Only during the past two decades have ethnobotanical studies elucidated the extent of modern use in southern Mexico of at least 20 species of mushrooms in four genera among nine tribes. These mushrooms belong to *Conocybe, Panaeolus, Psilocybe,* and *Stropharia,* with *Psilocybe mexicana* perhaps the most important (*10*). Many, if not all, contain psilocybine, an extraordinarily biodynamic tryptamine with a phosphorylated side chain, and an unstable derivative psilocine (*11*).

Psilocybe yungensis has been suggested as the identification of a "tree fungus" reported by early missionaries as the source of an intoxicating beverage of the Yurimagua of Amazonian Peru. No evidence, however, points to the present use of an hallucinogenic mushroom in that area (*5*).

Among spermatophytes, hallucinogens are found throughout the dicotyledons, but the monocotyledons are notably poor in psychotomimetic properties.

Zingiberaceae

There have been vague reports of the hallucinogenic use of *Kaempferia galanga* in New Guinea, but, as yet, ethnobotanical and phytochemical corroboration are wholly lacking (*12*).

Moraceae

Perhaps one of the oldest known hallucinogens is *Cannabis sativa,* a member of a monotypic moraceous genus, sometimes, together with the hops plant, set aside in a distinct family Cannabinaceae (*13*). One of the most ancient of cultivated plants, this dioecious, weedy annual that may grow to a height of 15 feet (5.5 meters), is native probably to Central Asia. It is the source of hemp fiber, a seed oil, and a narcotic. Over the millennia, man has selected "races" or "varieties" of this cultigen with desired characteristics: some for stronger fiber; some for higher oil content; others for greater narcotic potency. Selection for narcotic activity has been especially notable in regions like India, where the inebriating properties had religious significance. Futhermore, it is recognized that the strength of the intoxicating principles in any given "race" of *Cannabis* will decrease as the plant is grown in more northern latitudes.

Hemp was reported in a Chinese document 3500 years ago. The Assyrians used the plant in 9th century B.C. as an incense. The Sanskrit Zend-Avesta first mentioned its intoxicating resin in 600 B.C. Herodotus wrote that the Scythians burned its seeds to produce a narcotic smoke. In Thebes, it was made into a drink with opium-like properties. Galen reported general use of hemp in cakes which, eaten to excess, were narcotic. In 13th century Asia Minor, the hashishins were political murderers who, excited to their nefarious work by ingesting hashish, a form of cannabis, would carry out murder for pay; from this Arabic term comes our word assassin.

Cannabis is probably the most widely disseminated hallucinogenic plant, now known in virtually all inhabited parts of the world, escaping easily from cultivation and growing spontaneously. Consequently, its narcotic use in sophisticated societies, especially in urban areas, has recently increased, presenting major legal, moral, social, and health dilemmas to European and American authorities. Its employment in primitive societies may shed much light on problems resulting from its use and abuse in more advanced communities.

Recent study has established that the euphoric and psychotomimetic activity of *Cannabis sativa* is due primarily to Δ^1-tetrahydrocannabinol (*11*) found concentrated in a resin abundant in the unripened fruits and adjacent leaves and normally absent from or sparse in staminate plants. Although much is known about the effects in man of crude *Cannabis* preparations, little is yet understood of the biological activity of pure tetrahydrocannabinol. Despite its long history as a therapeutic agent, especially as a sedative in the treatment of hysteria, it has disappeared from the pharmacopoeias of most nations, since it is now considered to be medically valueless. The intoxication from abusive use of *Cannabis* preparations exhibits high variability but is characterized usually by an early onset of a dreamy state with confusion of thoughts, euphoric exaltation, and extreme happiness often alternating with abnormal depression. Visual and auditory hallucinations are common. Time perception is almost always altered. There may be nausea, dizziness, delusions of persecution and suffering, and rarely suicidal tendencies. Although not addictive, *Cannabis* may be definitely habituating. Dulling of the intellect and psychotic disturbances are often associated with excessive and continued use, but occasional use does not appear to be exceptionally deleterious.

Methods of taking *Cannabis* vary widely. In the New World, *marihuana* or, in Brazil, *maconha* – the dried, crushed flowering tops and leaves – are smoked, usually mixed with tobacco, in cigarettes. In parts of primitive Africa, *Cannabis* enjoys an important role in religion and magic. *Hashish,* the resin from pistillate flowers, is eaten by millions, especially in Moslem areas of North Africa and the Near East. It is in India, however, that *Cannabis* assumes extraordinary religious significance and where, as a result, man has selected "races" characterized by high concentrations of tetrahydrocannabinol. The ancient Atharva-Veda called it a "liberator of sin" and "heavenly guide," and it is still used in temples as a sacred plant. Three *Cannabis* preparations are commonly employed narcotically in India. *Bhang,* the weakest, is the dried plant gathered green, powdered, and made into a drink with water or milk or, with sugar, into candies called *majun*; opium and *Datura* may often be added. *Ganja,* usually smoked with tobacco, but sometimes eaten or drunk as an infusion, consists of dried pistillate tops with exuded resin carefully gathered from cultivated or wild "races" especially rich in tetrahydrocannabinol. *Charas,* pure resin removed from leaves and stems, also from especially cultivated, strongly narcotic "races," is normally smoked. *Cannabis* is the drug of the poor in India, where, in addition to its religious use, it is valued in folk medicine and as an aphrodisiac; and hedonistically as an euphoric narcotic, especially in activities requiring endurance or physical effort.

Although the marked increase in smoking marihuana in the United States poses a variety of problems, much of the drug illicitly used at the present time in the United States is weak, since it consists not of the pure resin, but of crushed leaves, twigs, and tops of plants notably low in tetrahydrocannabinol. These plants grow spontaneously, spread mainly from hemp formerly cultivated in plantations for fiber, at one time a major agricultural industry in North America. Marihuana smuggled into the country from Mexico or other tropical areas represents usually a stronger and potentially more troublesome narcotic.

The Moraceae likewise provides one of the most poorly understood hallucinogens *Olmedioperebea sclerophylla,* a jungle tree, the fruits of which reputedly were the source of an intoxicating snuff employed formerly by Indians of the Pariana region of central Amazonia. No chemical study of this plant has been published, and direct observation of the preparation and use of the snuff has been impossible (see *2, 3*).

Aizoaceae

More than 225 years ago, the Hottentots were reported chewing the vision-inducing narcotic *khanna,* a name now applied in South Africa to species of *Mesembryanthemum,* several alkaloidal species of which can produce torpor when ingested. No direct evidence, however, connects the Hottentot khanna with *Mesembryanthemum,* and other plants – *Cannabis* and *Sclerocarya caffra* (Anacardiaceae) – have been suggested (*14*).

Myristicaceae

An important hallucinogenic snuff of sundry tribes of northwesternmost Brazil and adjacent Columbia and Venezuela is prepared from the reddish bark resin of several jungle trees of *Virola* – *V. calophylla, V. calophylloidea,* and *V. theiodora.* The use of this snuff, called *yakee, paricá, epená,* or *nyakwana,* centers among the Waiká of Brazil and Venezuela, where it is made by drying and pulverizing the resin. Taken in excessive doses through a bamboo or bone snuffing tube and acting almost immediately, it sometimes has additional ingredients – the bark-ashes of *Elizabetha princeps* (Leguminosae) or *Theobroma subincanum* (Sterculiaceae) and, occasionally, powdered leaves of *Justicia pectoralis* var. *stenophylla* (Acanthaceae). Recent study has shown that *Virola* snuff contains high concentrations of tryptamines, especially 5-methoxy-*N,N*-dimethyltryptamine; and preliminary evidence suggests that tryptamines may be present also in *Justicia.* In Colombia, the snuff is taken only by witch doctors; among the Waiká, all adult males may employ it hedonistically at any time, and it is used excessively in endocannibalistic ceremonies. Effects include initial excitability, numbness of the limbs, twitching of the facial muscles, nausea, hallucinations, and, finally, a deep, disturbed sleep. Macroscopia is frequent and enters into Waiká beliefs about the spirits resident in the drug (*15*).

It is interesting that primitive American cultures have discovered these properties in *Virola,* since the related Asiatic *Myristica fragrans* – the common nutmeg – is hallucinogenic and is thought to have been employed narcotically in southeastern Asia. It is occasionally so employed in sophisticated circles in Europe and in the United States. The toxicology of nutmeg is still not completely elucidated but the effects seem to be attributable to several phenylisopropylamines among the many components of nutmeg oil (*16*).

Leguminosae

Several other South American snuffs are prepared from leguminous trees, wild or cultivated, of the genus *Anadenanthera (Piptadenia)* (*17*). The best known is *yopo* of the Orinoco area of Colombia and Venezuela, made from the toasted beans of *A. peregrina*, usually with an alkaline admixture of ashes of calcined shells and causing, when blown into the nostrils, an intoxication marked by a temporary fury and trance accompanied by visual hallucinations and eventual stupor (*18*).

This snuff, known as *cohoba*, was once employed widely in the Caribbean. The earliest report dates from Columbus' second voyage in 1496. Many early writers confused cohoba with tobacco snuff (*19*). Five indoles have been isolated from seeds of this tree, chief of which are bufotenine and *N,N*-dimethyltryptamine (*20*). Anthropologists have often attributed most intoxicating South American snuffs to *A. peregrina*, even those employed in regions where the tree is unknown, but its use is circumscribed mainly to the Orinoco and peripheral areas (*21*). There is indirect, though weak, evidence that *A. colubrina* might have been the source of the narcotic snuff *vilca* or *huilca* of southern Peru and Bolivia and *cébil* of northern Argentina. This species, closely related to *A. peregrina*, possesses the same psychoactive constituents (*22*).

Yurema, an hallucinogen of the Karirí, Pankararú, and other Indians of eastern Brazil, prepared from *Mimosa hostilis*, forms the center of a cult using an infusion of the root to bring on glorious visions of the spirit world (*23*). The active principle has been identified as *N,N*-dimethyltryptamine, an indole occurring in the closely allied genus *Anadenanthera*.

Sophora secundiflora, a shrub of dry areas of the American Southwest and adjacent Mexico, bears dark red beans known as "mescal beans" or "red beans." They contain a toxic pyridine – cytisine – causing nausea, convulsions, hallucinations, and occasional death from respiratory failure. A report by the Spanish explorer Cabeza de Vaca mentioned them as articles of trade among Texas Indians in 1539. The Stephen Long Expedition in 1820 reported the Arapaho and Iowa using large red beans as a medicine and narcotic. They have been found in archeological sites, all dated before A.D. 1000, sometimes with evidence of possible ritualistic use of the beans.

Indian groups of Mexico and Texas formerly ingested these seeds in a ceremonial Red Bean Dance, and Plains Indians employed them as an oracular or divinatory medium, for visions in initiation rites, and as a ceremonial emetic and stimulant. The bean is still often worn as part of the ornamental dress of the leader of the peyote ritual, indicating possibly its earlier use as a narcotic, a role which it lost with the arrival of the much safer hallucinogenic cactus (*2-4,24*)

The seeds of some species of *Erythrina* resemble the mescal bean and may have been valued as hallucinogens in Mexico, where they are called *colorines* and are often sold in markets mixed with *Sophora secundiflora*. Some species contain toxic indole or isoquinoline derivatives.

Several species of *Rhynchosia* may also have been employed as narcotics in ancient Mexico (*4*). In Oaxaca, the red and black beans of *R. phaseoloides* and *R. pyramidalis* are known as *piule*, a name locally signifying also the seeds of hallucinogenic morning glories of the region, and Oaxacan natives today recognize *Rhynchosia* seeds as toxic. An as yet uncharacterized alkaloid has been isolated from the genus. *Rhynchosia* seeds have been identified on Aztec paintings, together with mushrooms, suggesting that they may have been valued as hallucinogens.

A recent report of the hallucinogenic use of *Cytisus (Genista) canariensis* by Yaquí medicine men in northern Mexico has been experimentally substantiated (*25*). This species, member of an alkaloid-rich genus, is native to the Canary Islands, not Mexico, and contains cytisine, the psychoactive constituent of mescal beans.

Zygophyllaceae

An herb native to dry areas from the Mediterranean east to northern India, Mongolia, and Manchuria – *Peganum harmala* – possesses definite hallucinogenic properties in the alkaloid content of its seeds: harmine, harmaline, harmalol, and harman, bases known from at least eight families. This and other species of *Peganum* are highly valued in folk medicine. Although there have been vague reports of its hallucinogenic employment, its actual narcotic use to induce visions is open to doubt (*26*).

Malpighiaceae

The natives of South America have discovered in the related family Malpighiaceae a number of bizarre hallucinogens that likewise owe their activity to harmala alkaloids – several species of *Banisteriopsis (27)*. From *Banisteriopsis caapi, B. inebrians*, and *B. rusbyana* and the related *Tetrapteris methystica*, Indian tribes in the western Amazon of Brazil, Bolivia, Colombia, Ecuador, and Peru, the Orinoco headwaters in Venezuela, and the Pacific coast of Colombia prepare a psychotomimetic decoction or infusion variously known as *ayahuasca, caapi, natema, pinde,* or *yajé*. Usually, only one species is employed; but, in the westernmost Amazon, the preparation may involve the bark of *B. caapi* or *B. inebrians* and the leaves of *B. rusbyana*. Recent studies show that leaves of *B. rusbyana* are added to reinforce the hallucinogenic effects of the drink prepared from the bark of other species of the genus. The identification of yajé as *Prestonia amazonica,* although widely accepted, is wholly discredited (*28*). Other plants, some highly toxic, are sometimes added to *Banisteriopsis* drinks – *Alternanthera, Psychotria, Nicotiana, Datura,* and *Malouetia. Tetrapteris methystica,* with no admixture, is the source of a potent hallucinogenic drink in the Rio Tikié in Brazil; although no chemical studies on this species have been made, there is every reason to presume that its effects are attributable also to harmala alkaloids.

The chemistry of the malpighiaceous hallucinogens has, in general, been more critically investigated than the botany, yet failure of chemists to insist upon rigorous botanical determination of material analyzed has created chaos (*29*). An alkaloid in *B. caapi* was shown in 1928 to be identical with harmine isolated a century ago from *Peganum harmala.* Subsequent work has proven that the active bark constituents are harmine, harmaline, and tetrahydroharmine – all with a β-carboline skeleton, while leaves of *B. rusbyana* contain in addition dimethyltryptamines. Harmine has been found in *Cabi paraensis*, an eastern Amazon plant allied to *Banisteriopsis,* but, although valued in popular medicine, it is apparently not used as an hallucinogen.

Ayahuasca- or caapi-intoxication varies widely but normally begins with giddiness, nervousness, nausea, profuse perspiration leading to lassitude, listlessness, and detachment. The onset of color vision starts generally with a bluish or purplish aureole and, while sometimes these are the only colors experienced, red, green, and orange imagery is often rich, and lightning-like flashes of light are common; synesthesias may or may not be experienced. Muscular coordination is rarely altered; in fact, many Indian caapi ceremonies involve dancing. Occasionally, a deep sleep interrupted by dreams culminates the intoxication. The native takes caapi for divination, prophecy, diagnosis and treatment of diseases, communion with ancestors, preparation for war, male adolescent rites, and many other purposes. The drug represents, obviously, an ancient culture trait, and some Amazonian peoples treat it as a sacred, almost deified, element.

Caapi and ayahuasca were discovered more than a century ago by the British explorer Richard Spruce. Notwithstanding a century of botanical study, this complex of narcotics offers still a most fertile field for interdisciplinary investigation which must be done soon because of the pace of acculturation in many areas of South America.

While *Banisteriopsis* is normally taken as a beverage, there is evidence that in the northwesternmost Amazon it may be used as a snuff as well: harmine, harmaline, and tetrahydroharmine have been reported from snuff powders prepared from a vine said to be the source also of a drink in the Rio Negro basin; botanically identifiable material, unfortunately, is lacking.

Cactaceae

The famous *peyote* – "prototye" of New World plant hallucinogens – is *Lophophora williamsii*: a grey-green, spineless cactus with a small dome-shaped head bearing tufts of whitish hair and a long carrot-like root (*30*). The small chlorophyllous crown is sliced off and dried to form hard, brownish disks, known as mescal buttons. Keeping indefinitely without loss of intoxicating properties and transportable over long distances, mescal buttons are usually softened in the mouth and swallowed directly without chewing. Native to the deserts of central and northern Mexico, peyote claims centuries of use as a narcotic and was basic to pre-Columbian religious practices of the Aztec and other Mexican Indians. Much is known about the ancient religious use of peyote because of the many Spanish writings, all unsympathetic, attributing diabolic properties to the cactus. One religious questionnaire compared the ingestion of peyote with cannibalism. Primitive peyote religious dances still survive among the Cora, Huichol, and Tarahumare of northern Mexico. About 90 years ago, Plains tribes from the United States, especially Kiowa and Comanche, discovered the peyote cult on raids in Mexico, brought the sacred plant back and elaborated a wholly new cult – a combination of pagan and Christian elements – which spread rapidly throughout the Plains area, then to the Southwest and north as far as Saskatchewan. By 1922, the cult numbered 13,300 adherents and, for protection against fierce and unjust persecution from missionary and political circles, was legally incorporated into the Native American Church, now claiming 250,000 followers, which teaches reverence, high moral values, and abstinence from alcohol. Tribes near the source of peyote make ceremonial collection trips to get it; distant tribes secure their supply through the post by purchase.

Peyote was described as a new species of *Anhalonium* in 1888, and the first pharmacological studies were published the same year. In 1892, the anthropologist Lumholtz discovered it among the Huichol and Tarahumare and sent dried mescal buttons to Harvard University, upon which a definitive botanical identification was made. It represented a species first described in 1845 as *Echinocactus williamsii,* although later study showed that it deserved recognition as a distinct monotypic genus *Lophophora.*

Peyote contains at least 15 β-phenethylamine and isoquinoline alkaloids (*31*). All are active, but the most important one is mescaline. Peyote intoxication is highly complex and variable. Its most spectacular phase, for which mescaline is responsible, comprises the kaleidoscopic play of visual hallucinations in indescribably rich colors, yet auditory and tactile hallucinations and a variety of synesthesias are among the other effects. Consequently, the intoxications induced by mescaline and by peyote itself are very different, but they have unfortunately been confused in the literature. Most psychological experimentation has been based on mescaline, whereas native religious rites have experiences resulting from the effects of all of the alkaloids together. While the visions are important in native cults, peyote is likewise revered as a sacred medicine, and curing rites are common in the ceremony.

Other cactaceous plants have been employed as hallucinogens, and evidence suggests that still others do possess psychoactive constituents. The Tarahumare use, in place of *Lophophora* for special religious and medicinal purposes, several species of *Mammilaria* and *Echinocactus*. Various species of *Ariocarpus, Astrophytum, Aztekium, Dolichothele, Obregonia, Pelecyphora,* and *Solisia* are popularly classed as peyote in Mexico, possibly because they resemble the true peyote but perhaps equally well suggesting a use similar to that of *Lophophora*. Alkaloids similar to or identical with those of *Lophophora* have been found in some of these Mexican genera (*6*). A large columnar cactus, *Trichocereus pachanoi,* the "San Pedro" of northern Peru, forms the basis of an hallucinogenic drink taken by witch doctors for diagnosis, divination, and to make oneself owner of another's identity. Several species of *Trichocereus* contain mescaline and other alkaloids (*31*).

Apocynaceae

Why the family richest in alkaloids – Apocynaceae – should be so sparingly represented among the hallucinogens remains an enigma. The natives of the wet tropical forests of West Africa, especially of Gabon, chew the yellowish root of *Tabernanthe iboga* to offset hunger and fatigue (*32*). In large doses, iboga causes excitement, mental confusion, and a drunken madness characterized by prophetic utterances. Its many uses in folk medicine in primitive African societies may be attributed to its 12 alkaloids, the principal one of which – the indole, ibogaine – has cocaine-like effects. Although ethnobotanical data reporting its use to induce visions are few and vague, there is some evidence that its ceremonial role, central to the Bwiti cult, seems to rest in part on its hallucinogenic properties.

The literature abounds with reports that *Prestonia (Haemadictyon) amazonica* is employed, under the name yajé, in the Amazon as an ingredient of an hallucinogenic brew—reports stemming from misinterpretation of botanical field data, careless identification, and even guesswork. Although this interesting genus of 30 tropical American vines is often considered toxic by the natives, little is known of its chemistry, only one species having been reported to contain an alkaloid of still undetermined identity. An analysis of leaves erroneously determined as *P. amazonica* (but belonging probably to *Banisteriopsis rusbyana*) indicated *N,N*-dimenthyltryptamine, an indole as yet unrecorded from the Apocynaceae.

An unidentified lacticiferous jungle liana, the root bark of which is employed, without the admixture of any other plant material, by remote Tanimuka Indians in Amazonian Colombia in preparing a strongly hallucinogenic drink used in male adolescent rites, may represent an apocynaceous species (*5*).

Convolvulaceae

The early Spanish chroniclers of Mexico reported on numerous occasions the religious use of the lentil-like seeds of the Aztec *ololiuqui,* a sacred, hallucinogenic vine with cordate leaves. Several illustrations of the plant – the best in a voluminous study of the medicinal plants, animals, and stones of "New Spain" by Hernández, personal physician to the King of Spain who worked in Mexico from 1570 to 1575 – leaves no doubt that ololiuqui represented a morning glory. Most of the chroniclers were ceclesiastical authorities who railed against this "diabolic seed," and Christian persecution drove the native cults into hiding.

Corroboration of the identity of ololiuqui waited for 400 years, since no morning glory was found employed in pagan religious rites. The apparent absence of hallucinogenic use of a convolvulaceous plant, together with the fact that no intoxicating constituent was known to exist in the family, led ethnobotanists to assume that ololiuqui must have been one of the several narcotic species of *Datura* – despite the insistence of reliable Mexican botanists that the plant was a morning glory. This theory gained quick acceptance and persists even today, partly because tubular flowers of the Convolvulaceae and those of *Datura* are rather similar and partly because the syndromes of ololiuqui- and *Datura*-intoxication coincided well. Only in the late 1930's was actual voucher botanical material of a morning glory employed as an hallucinogen collected in Mazatec country in Oaxaca, and the accuracy of the ancient reports seemed to be vindicated by modern fieldwork (*33*). Later, another psychotomimetic morning glory – *badoh negro* of the Zapotec of Oaxaca – was found to represent *Ipomoea violacea* (*34*).

Among the natives of Aztec Mexico, ololiuqui was used for divination perhaps even more than peyote and teonanacatl. Hernández wrote that: " . . . when the priests wanted to commune with their gods . . . " they ate ololiuqui seeds, and " . . . a thousand visions and satanic hallucinations appeared to them. . . ." Believed to possess a deity of its own, this plant was an ingredient also of magical ointments and enjoyed an exalted place in Aztec medicine. Modern Indians grind the seeds on a stone, soak them in water or alcoholic drinks, and filter them; ingest the filtrate, since the hard impervious testa may otherwise allow the seeds to pass intact through the digestive tract.

Chemical elucidation of the activity of these convolvulaceous hallucinogens awaited another quarter of a century. A few seeds of *Rivea corymbosa* were early subjected to preliminary pharmacological study and shown to be capable of inducing a "half narcosis" in frogs, but no active crystalline principle could be isolated, despite the ability of psychiatrists to produce hallucinations in man with the seeds. Finally, in 1960 and subsequently, chemical analysis proved that the psychotomimetic constituents of the seeds of these morning glories are alkaloids identical with or related to those known from ergot (*Claviceps purpurea*), a fungus of long medical history in Europe – chanoclavine, elymoclavine, ergine, ergometrine, isoergine, and lysergol. The principal components of these indoles are *d*-lysergic acid, amide acid, and *d*-lysergic acid methylcarbinolamide, ergoline derivatives closely allied to lysergic acid diethylamide, the famous hallucinogen LSD *(11)*.

Recent further phytochemical work has indicated relatively large amounts of ergoline alkaloids in additional species of *Ipomoea* and in other genera such as *Argyreia* and *Stictocardia (35)*. The unexpected discovery of lysergic acid alkaloids known only from lower fungi in one of the evolutionarily most advanced families of the Metachlamydeae is chemotaxonomically of great interest.

Labiatae

The discovery of narcotics in the Mint family in both hemispheres is interesting. Crushed leaves of *Salvia divinorum* yield a drink employed by the Mazatec of Oaxaca, Mexico, in their divinatory rites, especially when sacred mushrooms or morning glory seeds are unavailable *(36)*. Known locally as *hierba de la Virgen* or *hierba de la Pastora,* this mint may represent the hallucinogenic *pipiltzintzintli* of the ancient Aztec, in which case it bespeaks a long history of narcotic use. The Mazatec likewise are said to employ as hallucinogens the leaves of *Coleus pumila* and *C. blumei*, both Old World introductions from southeastern Asia. Chemists have yet to find a toxic principle in *Salvia divinorum*, although investigators have experimentally substantiated the psychoactive effects of the plant. Chemical studies of the two species of *Coleus*, at least on the basis of the material growing in southern Mexico, have not been carried out, although other Old World species valued in folk medicine have been investigated. An interesting intoxicant of Turkestan tribesmen, *Lagochilus inebrians (37)*, toasted leaves of which are taken with honey or sugar, contains a polyhydric alcohol, responsible apparently for its narcotic effects.

Solanaccae

One of the richest alkaloidal families, the Solanaceae, contains a number of plants that have been put to use as hallucinogenic agents.

Perhaps the most important genus is *Datura*, certain species of which have been recognized since classical times as poisons and narcotics, eventually entering into religious and magical rites or valued in witchcraft and sorcerers' potions *(2-4,14,-38)*. Parts of the plants, especially of *D. fastuosa*, were smoked for pleasure, often with *Cannabis*, alcohol, or tobacco, for their hallucinatory properties in Asia and Africa. *Datura* was even more widely prized in the New World. Pre-Columbian Mexicans valued *D. meteloides* and *D. innoxia*, among other species, as medicinal plants and for use in divination as hallucinogens. This use of *toloache (Datura)* has persisted in northern Mexico even today in the hinterlands, despite its extreme toxicity. It is still known in the American Southwest, where many tribes (Navaho, Yokut, Yuman, Paiute, and others) employ the seeds, roots, and leaves, usually as a beverage, either in adolescent or divinatory rites, for diagnosis or directly for visionary effects; the Zuñi use *Datura* extensively as a narcotic, and their priests, who commune with the feathered kingdom at night, put powdered root into their eyes, ears, and mouth or, to commune with the dead or to bring rain, chew the root. The Indians of northeastern North America made limited narcotic use of jimson weed, *D. stramonium*.

In South America, *Datura* is represented by a borescent species forming a subgenus, *Brugmansia*, native to the Andean highlands from Colombia to Chile – *D. arborea, D. aurea, D. candida, D. dolichocarpa, D. sanguinea, and D. vulcanicola;* and in the warmer lowlands, *D. suaveolens*. Handsome trees with large, showy flowers and well known in horticulture, they are chromosomally aberrant cultigens, unknown in the wild state, and have been associated with man from earliest times, probably because of their medicinal and narcotic properties. Some have been of inestimable importance to ancient Andean civilizations.

The preparation and use of *Datura* differ widely in areas of South America, but it is frequently taken in the form of pulverized seeds dropped into fermented *chicha* or as an infusion; leaves or twigs likewise may be used. The intoxication is marked by initial violence so furious that the partaker must be restrained pending the onset of a deep, disturbed sleep during which hallucinations interpreted as spirit visitations, enabling the witch doctor to diagnose disease, discover thieves, and prophesy the future of tribal affairs and aspirations, are experienced. The Jivaro value *Datura* for correcting refractory children who are given the seeds in the hope that the spirits of their forefathers may come to admonish them. The ancient Chibcha of the Bogotá area gave wives and slaves of deceased chieftains potions of *Datura* to induce stupor prior to their being buried alive with their departed master. The Siona of today frequently mix *Datura* leaves with *Banisteriopsis* in preparing yajé.

Accurate information on the species used by tribes for special purposes is not available, but since most, if not all, species of *Datura* of North America and the Old World, as well as the *Datura* tree species of South America, contain similar tropane alkaloids – hyoscyamine, scopolamine, and atropine – this does not pose the problem that it might with certain other hallucinogens.

The Kamsá of Sibundoy in the Colombian Andes are extremely narcotic conscious, employing several species and numerous named clones of these species, vegetatively perpetuated and so highly atrophied that they may possibly represent incipient varieties. Some of these clones or "races" are such monstrosities that their identification to known species has defied efforts, although the Indians have very definite native names for them. Differing in their narcotic strength, they are valued by witch doctors for different purposes. An extreme variant of an indeterminate species of tree *Datura* has been described as a distinct genus *Methysticodendron amesianum*. Known by the native name *culebra borrachera*, this beautiful tree, endemic apparently to the high Sibundoy Valley, is stated by the Indians to be far more potent than any *Datura*. Whether or not this tree represents a distinct genus or an extremely monstrous cultivar of a *Datura* is still open to debate. Its chemical constitution includes *l*-scopolamine and hyoscyamine with very minor amounts of other alkaloids (*2–4*).

Although the narcotic use of *Datura* has been recognized from early times, the genus still constitutes botanically, ethnobotanically, and phytochemically a field for productive investigation.

Evidence for the former narcotic use on the eastern slopes of the northern Andes and elsewhere of *Brunfelsia*, a genus phylogenetically close to *Duboisia*, is quite real but is not yet corroborated by field evidence and observation (*5–6*). Numerous herbarium collections of *Brunfelsia* from Bolivia, Brazil, Colombia, Ecuador, and Peru indicate a broad spectrum of therapeutic uses in folk medicine, especially to relieve rheumatic pains, and one species, *B. hopeana*, is official in the Brazilian pharmacopoeia. Several herbarium collections refer to its use as a narcotic. *Brunfelsia*, a tropical New World genus of up to 25 species, is known to be toxic, and since several alkaloids, probably tropanes of indeterminate structure, and the coumarin compound scopoletine have been isolated from this genus, it probably does possess hallucinogenic and otherwise biodynamic constituents.

A century ago, attention was drawn to the narcotic *árbol de los brujos* or *latué* – *Latua pubiflora* – a spiny solanaceous shrub endemic on the Chilean coast and employed by the Indians as a virulent poison capable of causing delirium and hallucinations. Very little of a botanical and ethnobotanical nature is known about latué, and its chemistry is not well understood (39).

Henbane (Hyoscyamus niger) has been known from early classical times in the Mediteranean area as a poison and narcotic capable of causing permanent insanity. Fantastic visual hallucinations comprise one characteristic of the intoxication. This plant, containing atropine and scopolamine, became important in the conjuration of demons, prophecy, and soothsaying in European sorcery of the Middle Ages. *Hyoscyamus muticus,* of Africa and Asia, with similar effect, is smoked as an inebriant in India. *Belladonna (Atropa belladonna)* and *mandrake (Mandragora officinalis),* both dangerous solanaceous poisons due to tropane alkaloids, may likewise have been occasionally employed in classical times or Europe of the Middle Ages for their hallucinating properties (*14*).

Compositae

The Mexican *Calea zacatechichi*, the leaves of which are taken in infusion for divination by the Chontal Indians of Oaxaca, is the most recently discovered hallucinogenic plant (*40*). Called *thle-pela-kano* or "leaf of god," it is believed to clarify the senses. Few chemical studies appear to have been made on this shrub.

In conclusion, Lewin's appraisal of man's search for hallucinogens may be appropriately pertinent (*14*): "The passionate desire which . . . leads man to flee from the monotony of everyday life . . . has made him instinctively discover strange substances. He has done so, even where nature has been most niggardly in producing them and where the products seem very far from possessing the properties which would enable him to satisfy this desire."

References and Notes

1. A. Hoffer and H. Osmund, *The Hallucinogens* (Academic Press, New York, 1967).
2. R. E. Schultes, *Psyched. Rev.* **1**, 145 (1963).
3. — — — *Harvard Rev.* **1**, 18 (1963).
4. — — — *Planta Med.* **13**, 125 (1965).
5. — — — *Lloydia* **29**, 293 (1966).
6. — — — in *Ethnopharmacologic Search for Psychoactive Drugs,* D. Efron, Ed. (U.S. Public Health Service Publication No. 1645, Washington, D. C., 1967), p. 33.
7. V. P. Wasson and R. G. Wasson, *Mushrooms, Russia and History* (Pantheon, New York, 1957); R. G. Wasson, in *Ethnopharmacologic Search for Psychoactive Drugs,* D. Efron, Ed. (USPHS Publication No. 1645, Washington, D. C., 1967), p. 405.
8. C. H. Eugster, in ibid., p. 416; P. G. Waser, in ibid., p. 419.
9. R. G. Wasson, *Soma: Divine Mushroom of Immortality* (Harcourt, Brace & World, New York, in press).
10. R. Heim and R. G. Wasson *Les Champignons Hallucinogènes du Mexique* (Editions, Museum d'Histoire Naturelle, Paris, 1958).
11. A. Hofmann, in *Chemical Constitution and Pharmacodynamic Action,* A. Burger, Ed. (Dekker, New York, 1968), p. 169.
12. J. Barrau, *J. Agr. Trop. Bot. Appl.* **9**, 245 (1962).
13. R. J. Bouquet, *Bull. Narcotics* **2,** No 4, 14 (1950).
14. L. Lewin, *Phantastica – Narcotic and Stimulating Drugs* (Kegan, Paul Trench, Trubner, London, 1931).
15. R. E. Schultes and T. Holmstedt, *Rhodora* **70,** 133 (1968).
16. A. T. Shulgin, T. Sargent, C. Naranjo, in *Ethnopharmacologic Search for Psychoactive Drugs,* D. Efron, Ed. (USPHS Publication No. 1645, Washington, D.C., 1967), p. 202; A. Weil,*Econ. Bot.* **19**, 194 (1965).
17. R. E. Schultes, in *Ethnopharmacologic Search for Psychoactive Drugs,* D. Efron, Ed. (USPHS Publication No. 1645, Washington, D.C., 1967), p. 291.
18. S. H. Wassen, in ibid., p. 233.
19. W. E. Safford, *J. Wash. Acad. Sci.* **6**, 547 (1916).

20. B. Holmstedt and J. Lindgren, in *Ethnopharmacologic Search for Psychoactive Drugs,* D. Efron, Ed. (USPHS Publication No. 1645, Washington, D.C., 1967), p. 339.

21. J. M. Cooper, *Bull. Bur. Amer. Ethnol.* **5**, No. 143, 525 (1949).

22. S. von R. Altschul, in *Ethnopharmacologic Search for Psychoactive Drugs,* D. Efron, Ed. (USPHS Publication No. 1645, Washington, D.C., 1967), p. 307.

23. O. GonCalves de Lima, in *Arqu. Inst. Pesqui. Agron.* **4**, 45 (1946).

24. W. LaBarre, *The Peyote Cult* (Shoe String Press, Hamden, Conn., 1960).

25. J. Fadiman, *Econ. Bot.* **19**, 383 (1965).

26. D. Porter, "The taxonomic and economic uses of*Peganum* (Zygophyllaceae)," term paper, Botanical Museum, Harvard University (1962).

27. R. E. Schultes, *Harvard Univ. Bot. Mus. Leafl.* **18**, 1 (1957).

28. — — — and R. F. Raffauf, *Harvard Univ. Bot. Mus. Leafl.* **19**, 109 (1960).

29. V. Deulofeu, in *Ethnopharmacologic Search for Psychoactive Drugs,* D. Efron, Ed. (USPHS Publication No. 1645, Washington, D.C., 1967), p. 393.

30. A. Rouhier, *La Plante Qui Fait les Yeax Émerveillés – le Peyotl* (Gaston Doin, Paris, 1927): R. E. Schultes, *Harvard Univ. Bot. Mus. Leafl.* **4**, 129 (1937); ibid. **5**, 61 (1937): *Amer. Anthropol.* **40**, 698 (1938); *Cactus Succulent J.* **12**, 177 (1940); W. LaBarre, *Curr. Anthropol.* **1**, 45 (1960).

31. C. Friedberg, *J. Agr. Trop. Bot. Appl.* **6**, 439 (1959).

32. V. E. Tyler, *Lloydia* **29**, 275 (1966); H. G. Pope, *Econ. Bot.,* in press.

33. R. E. Schultes, *A Contribution to our Knowledge of Rivea corymbosa, the Narcotic Ololiuqui of the Aztecs* (Botanical Museum, Harvard Univ., Cambridge, Mass., 1941).

34. T. MacDougall, *Bol. Centro Inv. Antropol. Mexicanas* **6**, 6 (1960).

35. A. der Marderosian, *Amer. J. Pharm* **139,** 19 (1967); *Lloydia* **30**, 23 (1967).

36. R. G. Wasson, *Harvard Univ. Bot. Mus. Leafl.* **20**, 77 (1962).

37. A. Bunge, *Memoires des Savants Etrangers St. Petersburg* **7**, 438 (1847).

38. W. E. Safford, Annual Report 1920 (Smithsonian Institution, Washington, D. C., 1922), p. 537.

39. A. Murillo, *Plantes Médicinales du Chili* (A. Roger y F. Chernoviz, Imprimerie de Lagny, Paris, 1889), p. 152.

40. B. MacDougall, *Gard. J.* **18**, 105 (1968).

10.

Cannabis

Andrew T. Weil

As the use of *Cannabis* becomes more and more extensive in western industrialized societies, our primitive knowledge of the drug becomes more embarrassing and more unfortunate because it renders us unable to make informed, rational decisions about the problem. AUTHOR

It is astonishing that so little is known about Cannabis in the face of all the controversy that has raged about this drug for so many years. In the form of marihuana, Cannabis has been used in the United States for more than 50 years, but reputable experimental studies of it done in the US can be counted on the fingers of one hand. The world scientific literature on Cannabis is vast, but almost all of it is based on rumor, anecdote, and second hand observation. The simplest kinds of data about the pharmacological effects of the drug have not been collected.

In December 1968 Norman Zinberg, Judith Nelsen, and I reported in *Science* the results of a human study of marihuana conducted at the Boston University School of Medicine. This study was the first human work with the drug since 1946 in the US and, to our knowledge, the first "double blind" human study ever done on Cannabis – neither the experimenter nor the subject knew whether the subject had taken a real drug or an inactive placebo until the testing was concluded. Our results have been widely publicized and seem to have stirred up further argument over the harmlessness or harmfulness of marihuana – a question that has become so lively in the US that nearly everyone feels entitled to express an opinion on it. The resulting confusion has made it even harder to sort fact from rumour and prejudice.

The hemp drugs are prepared from *Cannabis sativa,* which is also the source of hemp fibre and of a drying oil. Cannabis is called Indian hemp to distinguish it from other kinds of hemp (like Manila hemp), but it is probably native to China. It now grows wild over an enormous range of climatic zones and, in fact, is one of the hardiest roadside weeds in the US. The form of the plant varies greatly: some are low and sparse; others, growing in more favourable locations, are three metres high with robust stalks and spreading branches. When the plants are in flower, one can learn quickly to distinguish male plants from females.

The intoxicating properties of hemp reside in a sticky resin exuded by the tops of the plants, particularly the females. Male plants contain some resin but are cultivated mainly for fibre, not for drugs. The resin itself, when prepared for smoking or eating, is known as hashish; it is several times more potent than marihuana or other preparations of the whole Cannabis plant.

When a pharmacologist says that drug A is more potent than drug B, he means that it takes more of drug B to produce a given effect, but that both drugs can pro-

---

Reprinted from the September 1969 issue of *Science Journal* by courtesy of the author and *Science Journal*, London, England.

duce the same effect. For example, dextro-amphetamine is three or four times as potent a stimulant as dl-amphetamine, but if someone takes 15 mg of the dl-form, he will be as stimulated as if he had taken 5 mg of dextro-amphetamine. On the other hand, when a pharmacologist says that drug A is more powerful than drug B, he means that the effects of A are qualitatively greater than those of B regardless of dose. For example, morphine is a more powerful analgesic than aspirin.

I have said that hashish is more potent, than marihuana, meaning that if someone consumes enough crude Cannabis, he will experience effects as marked as those of hashish; there are no power differences between the two. Unfortunately, many persons confuse potency with power in talking about psychoactive drugs. Because LSD is a very potent hallucinogen producing effects in doses as low as 0.1 mg, an assumption is often made that it is also the "strongest" or "worst" drug of this class. But mescaline – a considerably less potent drug with an effective dose in the range of half a gramme – has as much hallucinogenic power as LSD. And if one talks not about hallucinations but rather about ability to induce changes in normal consciousness, the hemp drugs may also be as powerful as these hallucinogens. The point to remember is not that hemp is "milder" than LSD but rather that because hemp drugs (including hashish) are so much less potent than LSD, users rarely if ever consume enough to become as intoxicated as they can become on so-called stronger drugs. This distinction is important because synthetic drugs derived from hemp are now available in pure form and may soon turn up in illicit traffic. Then users may be able to consume enough to experience the extremes of power of Cannabis.

The potency of whole Cannabis depends mainly on resin content. Some preparations of the plant, like the *ganja* of India, are more intoxicating than marihuana because they are taken from plants selected for high resin content and include only the resin-rich flowering tops. Others, like Indian *bhang*, Moroccan *kif*, and South African *dagga*, are nearly equivalent to marihuana.

An important and unanswered question is how and why samples of Cannabis differ so widely in pharmacological activity. Cannabis users know that different batches vary considerably in intoxicating effect. Most of them also claim that different kinds of Cannabis cause different symptoms. If such claims actually have a pharmacological (rather than psychological) basis, they suggest that different forms of Cannabis might differ in quality of resin, not simply in content.

Botanists feel that only one species of hemp exists, but they have not done nearly enough work on the phytochemistry of this species. American users seek out "Acapulco Gold" as the strongest variety of Mexican marihuana, but no one knows whether Acapulco Gold is really obtained from a special variety of Cannabis, whether it is simply any marihuana with a high proportion of flowering tops as opposed to lower leaves (and thus with a high resin content), or whether it is just any marihuana a dealer can get a high price for. An important research priority, therefore, is better botanical and phytochemical characterization of hemp and its intoxicating preparations.

Cannabis resin has been subjected to extensive chemical analysis, but until very recently the structures of its pharmacologically active constituents were unknown. A number of inactive compounds were isolated over the years; including cannabinol, cannabidiol, cannabigerol, and cannabidiolic acid (reported to have sedative and anti-bacterial actions). Pharmacologists have long thought that the inebriating properties of hemp were confined to a particular distillation fraction (the so-called "red oil") containing a series of isomers of tetrahydrocannabinol or THC, but the exact structures of these compounds were not known. In the past few years two isomers of THC have been isolated from hashish and their structures have been elucidated by modern spectrographic methods. They are named 1-Δ^1 *trans*-THC and 1-Δ^6-*trans*-THC. (Because of a confusion in nomenclature, the ^{1}compound is sometimes labelled Δ^9.) The Δ^1 isomer is the major constituent. It appears to be quite active and is currently being advertised as "the psychotomimetically active constituent of hashish", although the evidence for this claim is far from substantial.

In 1965, R. Mechoulam and Y. Gaoni, working at the Weizmann Institute of Science in Israel, reported the first total synthesis of Δ^1-THC; the compound is now available and will soon be distributed by the National Institute of Mental Health as a research drug. I will refer to the Δ^1 compound hereafter simply as THC. The tetrahydrocannabinol compounds are unique in the growing index of psychoactive drugs. They do not fall into any of the large families of these drugs like the tryptamines or the substituted phenylethylamines and, in fact, unlike them contain no nitrogen. The most urgent priorities in chemical research on Cannabis are the identification of other isomers in the plant and the determination of their role, if any, in Cannabis intoxication.

The animal pharmacology of Cannabis can be reviewed quickly. Effects of hemp in animals are largely confined to the central nervous system. The commonest observable signs are vomiting and diarrhoea, fine tremors, and muscular incoordination. Hashish has also been reported to cause fluctuations in blood sugar, lowered body temperature, increased heart rate, depressed respiration, increased urine output and conjunctival vascular congestion. It has been difficult to establish a lethal dose of crude Cannabis; huge doses have been given to dogs without causing death. I.C. Chopra and R.N. Chopra report a lethal dose for cats of 10 g/kg of body weight which is equivalent to 0.7 kg for a 70 kilo man, although such an extrapolation from cats to man must be regarded very cautiously.

Hashish extract has no effect on the gross behaviour of rats and mice. It does not affect simple learning in rats, but does decrease aggressiveness induced by isolation in rats. In dogs, Cannabis extracts induce "dreamy" somnolent states that have reminded some researchers of the later phases of human intoxication.

And last, but not least, hemp does not cause addiction in animals. Here I must interrupt to say a word about the serious problems created by the use of vague words like addiction. As we get to know more about the non-therapeutic uses of psychoactive drugs, it becomes clear that drug habituation is a complex psychological and physiological interaction, not easily described in simple terms. One aspect of this complexity is that each drug or class of drugs is associated with its own particular habit. Thus the habit of a heroin user is quite different from that of an alcoholic, which in turn is different from that of a cocainist. And all of these are unlike the dependence that can develop on Cannabis. A new terminology, recently introduced, is that of drug dependence of a specified type: for example, dependence of the morphine/heroin type or dependence of the Cannabis type. It is a much better way of talking about habitual uses of drugs.

Two of the most important physiological characteristics of common kinds of drug dependence are tolerance and withdrawal. Tolerance is the need for larger doses of a drug to produce a given effect as the drug is taken more and more frequently. It is a prominent feature of dependence on a number of drugs including the opiates, alcohol, barbiturates and, especially, the amphetamines, and it can develop over a short period of time. It probably reflects an adaptation of cellular biochemistry to the continued presence of these drugs in the system and may be a clinical manifestation of the induction of degradative enzymes in the liver. Withdrawal is any syndrome that consistently appears when the intake of certain drugs that have been used regularly is cut off; by definition, it is relieved when administration of the drug is reinstituted. Withdrawal, like tolerance, characterizes the dependence on such drugs as opiates, alcohol, barbiturates and amphetamines.

A more precise statement about Cannabis is that tolerance and withdrawal do not occur in animals receiving hemp drugs. This finding does not mean that dependence on these drugs cannot occur; it does mean that dependence of the morphine/heroin or alcohol/barbiturate type does not occur with Cannabis.

There is little else to learn from the animal studies. The animal pharmacology of hemp has been beset from the start by formidable problems, chief among which has been the inability of biochemists to identify with certainty an active principle. If the THC proves to be the sole active constituent, animal pharmacologists will be able to administer exact doses of a known compound. Until now they have had to work with mysterious red oils and other unknown components of the resin – all of them highly insoluble in water, difficult to administer and chemically heterogeneous. At the moment, however, there is no firm evidence that THC reproduces the action of whole Cannabis. The Δ isomer is also pharmacologically active, and other minor constituents also share in the total syndrome of Cannabis intoxication. In addition, THC itself is depressingly difficult to work with, for it is unstable and insoluble in water, and its metabolic fate is completely obscure.

Further, it has been difficult to find a suitable assay for Cannabis intoxication in animals. Unlike most psychoactive drugs, Cannabis has a minimum of physiological actions. There is none of the stimulation of sympathetic nervous system seen with LSD or mescaline, no predictable changes in reflexes and few alterations of any measurable body functions except heart rate. Nor is there any way of detecting any Cannabis component in the body fluids. And the most interesting effects of the drugs in humans – such as altered perceptions of time, space and auditory stimuli – are not easily investigated in animals.

Finally, the effects of Cannabis in animals vary wildly, not only from species to species but also from individual to individual within the same species. Rabbits react differently from dogs and, worse, dog A reacts differently from dog B. As a consequence of all these pharmacological misfortunes, most good animal pharmacologists have given up the study of Cannabis in frustration, and the scientific literature, predictably, is nearly worthless. At the moment, the only profitable area of study in animals is clarification of the metabolic fate of THC and other Cannabis components in the body. Experiments on this problem will probably not be successful until supplies of radioactively labelled THC become available.

The human pharmacology of Cannabis is even less certain. One peculiarity of the intoxication caused by hemp in man is the relative absence of physiological signs. Only two effects of Cannabis on the body have been established beyond question: conjunctival vascular congestion and tachycardia. My colleagues and I observed both of these signs in all of our research subjects, and they have been noted in most previous experiments. The drug probably has a few other effects yet to be demonstrated in the laboratory; it appears to decrease markedly the flow of tears and saliva, for example. This antisecretory action – presumably a result of blockade of the para-sympathetic nervous system – is important because it places Cannabis (in terms of its peripheral effects) for the first time with a known class of drugs, the prototype of which is atropine. Since atropine causes tachycardia by inhibiting transmission in the vagus nerve, it is tempting to speculate that the increase in heart rate that follows Cannabis smoking may also be an atropine-like effect.

Research on the psychopharmacology of Cannabis has been painfully inadequate, but we can see already that there is an enormous discrepancy between the subjective and objective effects of hemp on the mind. Persons who smoke the drug say they experience major changes in consciousness and mental functioning, but they do not have anything to show for it. Testing done to date reveals that subjects who take Cannabis for the first time have minimum subjective effects but significant impairment of performance on a variety of psychological tests; the impairment is greater as doses of drug are increased and tests are made more complicated. If one studies users of the drug, however, a different picture emerges. These subjects become high but do not show performance decrements. In fact, it appears that once a person becomes accustomed to the effects of Cannabis, he can compensate fully for the drug's influence on performance of tasks of ordinary complexity.

This finding is consistent with the testimony of users, who claim that a marihuana high is much more manageable than an alcohol or hallucinogen high. It also sets Cannabis apart from other psychoactive agents. Inexperienced users of intoxicants always show greater disruptive effects and always learn to adapt to the drugs as use increases in frequency. But most drugs impose absolute limits on the degree of adaptation possible. An alcoholic will perform much better than a non-drinker when the two are equally intoxicated on alcohol, but he can never fully compensate for the limits set by alcohol on the speed of nervous functioning. With Cannabis, compensation is 100 per cent, implying that hemp does not interfere with the general efficiency of the nervous system.

Also unlike most psychoactive drugs, Cannabis appears to exert no clinically significant actions on lower brain centres. Rather, its effects are confined to the cortex and such higher mental functions as memory, thought and perception. This locus of action may explain why it is possible for first users to ignore the mental effects of the drug completely. In drinking alcohol for the first time, a person may not have a pleasurable experience but, if he drinks enough, he will be unable to escape the fact that he is intoxicated; he will feel his nervous system behaving abnormally when he tries to walk or talk. But with Cannabis a first time smoker can consume enormous amounts of drug and have the physical signs of intoxication without any mental effects at all. A related observation is that drinkers have a hard time acting sober whereas Cannabis users seem to be able to suppress the effects of their drug without much difficulty.

Many people do not realize that no one has yet shown any specific, objective way in which a person high on marihuana is different psychologically from one who is not. Whenever a difference has shown up on a psychological test, it has been part of a non-specific, across the board performance decrement. This is seen in most subjects, expecially with non-users or with unusually high doses, on complicated tests. Such decrements are seen with many drugs that influence alertness or other general mental functions; they are not helpful in explaining the nature of the changes in consciousness experienced by users on usual doses. A simpler way of stating the problem is that the only way of knowing that a person is high on Cannabis is if he tells you he is – hardly a satisfactory or useful criterion for psychopharmacologists.

Therefore, a priority goal in Cannabis research is the discovery of some specific objective change in mental function associated with the acute action of the drug. We have some clues as to where we might find such a change; the most promising ones point to a disturbance in a particular kind of memory. It seems valid to distinguish three kinds of memory in man, which may be separated anatomically in the brain and which are affected differently by certain diseases and certain drugs. The first kind of memory has been called immediate and seems to cover events of the past few seconds. It is as if all information coming into the brain is held in some storage location for a very short time before a decision is made to retain it or discard it. If it is retained it passes to a second location. (called recent memory), where it may remain for days or, perhaps, weeks. Eventually, if it is to be kept longer, it moves to a third (long term) storage location, which is the permanent memory file. Each of these locations has connections to consciousness, so that memories may be retrieved from all of them.

In senile dementia, the classic psychological change is loss of recent memory, with sparing of immediate and long term memory. A senile patient can remember a string of numbers read to him long enough to recite them back and can go into autobiographical detail about his childhood. He cannot remember the day of the week, his telephone number or the events of the previous day. By contrast, in certain forms of post-traumatic amnesia, immediate and recent memory are spared, but information filed prior to the trauma cannot be retrived from the long term memory storage. Considerable indirect evidence suggests that Cannabis interferes acutely with the retrieval of information from the immediate memory storage, so that persons high on the drug have difficulty remembering what happened in the past few seconds.

A principal manifestation of this interference is a characteristic but subtle disturbance in speech. Even experienced Cannabis users commonly report that they have difficulty talking to others when high. They describe this difficulty as a fear of not making sense and of forgetting what they are saying. Interestingly enough, very few non-users can actually detect any problem in a user's speech, but after listening to many high subjects in our laboratory, my colleagues and I are convinced it exists. The exact nature of the difficulty appears to us to be that high persons must do more work than usual to remember from moment to moment the logical thread of what they are saying. As a result, they tend to forget what they are going to say next and also tend to go off on irrelevant tangents. We are presently undertaking experiments on recorded samples of subjects speech in order to document some of these effects.

At the same time we are seeking evidence for related manifestations in other spheres of mental activity, because the effect on speech appears to us to be part of a more general effect. And, indeed, such evidence has been easy to find in the testimony of users. Again and again we have collected accounts of subjective distress users feel under the influence of the drug when they have to concentrate on any activity requiring quick access to information perceived in the immediate past. We have also been impressed that this subjective distress is rarely accompanied by objective manifestations of difficulty. In other words, it looks as if a significant disturbance of immediate memory retrieval has few noticeable consequences in terms of behaviour, although it may cause great anxiety in the subject's mind.

Psychologists have devised a number of very different tests of short term memory, some of which may measure the specific function that Cannabis appears to affect. If we can demonstrate the effect, it will be the first specific, objective mental change documented in users of the drug on usual doses and will be very helpful in understanding the condition called high, in psychological terms.

Another area in which we might look for such a change is perception because marihuana users consistently report that their perceptions alter significantly after they consume the drug. Here again is an apparent paradox since all testing done to date has failed to show any objective changes in sensory function during acute Cannabis intoxication. Resolution of the paradox probably lies in paying closer attention to what users say. There is really no indication from them that their sense organs are working differently from usual. Rather, the change might be in what they do with incoming sensory information. For example, most users claim that listening to music is much more interesting and pleasurable when they are high. They do not claim that they hear tones of lower volume or that they can better discriminate between pitches of tones. Yet all of the testing of auditory function under Cannabis has been aimed at the ear: at auditory thresholds, pitch discrimination and the like.

It would make more sense to look for effects not on the ear but rather on that part of the brain that processes auditory information. Cannabis seems to affect the secondary perception of sensory information not the primary reception of it. Unfortunately, it is considerably harder to study secondary perception because the neural organization underlying it is less accessible to direct experimentation and much less well understood. A working hypothesis is that incoming sensory information (such as auditory signals representing music) normally follows conditioned pathways through the secondary perception network in order to get to consciousness. Under Cannabis, which might interfere with this normal processing, information may take novel routes to consciousness and thus be perceived in novel ways. Such a model would explain why users often say that under Cannabis they see things for the first time"as they really are", or why they dwell on aspects of complex visual or auditory stimuli they would ordinarily ignore.

These ideas are merely speculations, but I think they suggest kinds of experiments that are more likely to provide useful information than experiments done in the past. Before we can get to them however, we must solve a number of methodological problems that remain obstacles to the study of Cannabis. Foremost among these is our primitive understanding of the chemistry of the drug. We must know how and why samples of Cannabis differ in pharmacological activity, what components are active and what part each active component plays in the intoxication. Only then will we have any confidence that the findings of one researcher are comparable to those of another. At the moment there is no assurance that one batch of Cannabis is like another batch even if both have the same concentration of Δ^1THC.

We must quickly settle the problem of whether THC reproduces the action of the whole plant because pharmacologists are now under great pressure to conduct studies with THC rather than the whole plant. If THC is not the same thing as Cannabis, then the results of these studies will have no clear significance and will not help us understand what Cannabis does to people who use it.

The reason pharmacologists prefer to work with THC is that it is a pure, homogeneous compound so that known doses can be administered. Certainly, experiments with THC can be made more precise scientifically than experiments with the crude plant. But if scientific accuracy is obtained at the expense of relevance to the real world, it is not desirable. Pharmacologists also prefer to study Cannabis not by having subjects smoke it (which is an "inaccurate" route of administration) but by have them swallow exact doses of it in one form or another. If we were confident that the actions of the drug were constant regardless of route of administration, we might go along with this trend but, in fact, Cannabis appears to be a different drug when ingested rather than smoked. When smoked in large doses Cannabis rarely produces anything more striking than marked sedation. When eaten in large doses, it can produce a syndrome closely resembling the delirium of high fever. True hallucinations occur very rarely after the drug is smoked; they are more frequent after ingestion, as are other symptoms (like disorientation) more usually encountered with the hallucinogenic drugs. Cannabis smokers almost never report hangovers (they may feel "sluggish" the next day), but marked hangovers (malaise, headache, gastrointestinal disturbances) are common in persons who eat cannabis. The effects of a single inhaled dose of hemp appear to be dissipated after three hours; effects of a single oral dose may last 12 hours or longer. All of this suggests that components of Cannabis resin that are destroyed by the heat of smoking may get into the body when the drug is ingested. At least, it suggests that researchers should be wary of giving the drug orally and concluding that their results apply to persons who smoke it.

Another serious methodological problem is our present lack of placebo for smoking good enough to fool chronic users of Cannabis. As is the case with most psychoactive drugs, the effects of Cannabis appear to depend as much on non-pharmacological factors as on pharmacological ones. For example, a particular person's reaction to Cannabis on a particular occasion depends as much on "set and setting" as on the drug. Set is the psychologist's term for an individual's expectations of what a drug will do to him; it includes much of what we commonly call personality. Setting is the total environment (physical and social) in which a drug is taken. Occasionally, the combined effects of set and setting can overshadow completely the pharmacological actions of a drug. In studying Cannabis, therefore, it is necessary to recognize setting as a critical variable and to control for set both of subjects and experimenters. The double blind method of drug testing is the best device pharmacologists have come up with to safeguard against unpredictable variations in these non-pharmacological factors; it requires a satisfactory placebo.

When research subjects are volunteers who are naive to marihuana, an effective placebo is no problem. In our own experiments, for example, portions of male hemp stalks grown for fibre production and containing no THC were quite adequate. But experienced Cannabis smokers are accustomed to judging the quality of the drug while they are consuming it; they may accept an inert substance as Cannabis but will quickly decide that it is impotent Cannabis if they do not feel themselves getting high. Precisely because the administration of Cannabis by smoking takes time (often 20 minutes or longer), it is extremely difficult to find a workable placebo that chronic users will believe is Cannabis until they have finished smoking it. Several workers are independently searching for a suitable material, and as soon as we find one, we can proceed with good double blind experiments on users.

In the meantime it would be extremely useful to have standardized supplies of Cannabis available so that all researchers could work with uniform material of known potency. The United Nations Laboratory in Geneva maintains a reference sample of Cannabis, but national laws make the procedure by which individual scientists can obtain this drug forbiddingly complicated. In the US the National Institute of Mental Health is growing Cannabis for research, but supplies are not yet available and even when they are, the laws of many states will prevent scientists from acquiring them. If we are to obtain the information we need on Cannabis, this kind of obstructive legislation must quickly be revised to permit the free conduct of scientific investigations.

A more serious practical concern is the extent to which Cannabis has become a political issue in many countries including the US and the UK. In these countries the "Cannabis Question" has given rise to such heated controversy that everyone feels qualified to express an authoritative opinion on the drug regardless of his knowledge. In such a climate, scientists often succumb to pressure from one side or the other to abandon their objective neutrality and come up with data to bolster factional arguments. Whenever this happens, it makes the acquisition of new information about Cannabis even harder.

As the use of Cannabis becomes more and more extensive in western industrialized societies, our primitive knowledge of the drug becomes more embarrassing and more unfortunate because it renders us unable to make informed, rational decisions about the problem. Educational and legal programmes designed to regulate the use of Cannabis cannot be successful unless they take into account the facts of pharmacology; I very much hope that governments will realize the necessity of encouraging the free gathering of the kinds of information I have outlined in this article.

FURTHER READING

"Clinical and Psychological Effects of Marihuana in Man," by A.T. Weil, N.E. Zinberg and J.M. Nelsen (in *Science* **162;** 1234-1242)

Hashish: Its Chemistry and Pharmacology, edited by G.E.W. Wolstenholme and J. Knight (Little, Brown, Boston, 1965)

"The Use of Cannabis Drugs in India," by I.C. Chopra and R.N. Chopra (in *Bulletin on Narcotics,* **9,** 4-29, 1957)

Cannabis: 'The Wooton Report, (HMSO, London, 1968)

11.

The "Speed" That Kills or Worse

Jonathan Black

Unlike barbiturates which are "downs," amphetamines are "ups." They are energizers that kill the desire to eat and sleep, speed up the heart and deliver a neural charge to the cortex. Precisely how they work is uncertain. **AUTHOR**

In his senior year at college, Norman found himself with an unfinished thesis and a deadline only two days away. No-Doz pried his eyes open, but his brain seemed to turn to rubber after hours of typing. So his slightly overweight girl friend came up with some little yellow diet pills, 10 milligrams apiece. With the first pill, Norman's mind clicked into gear and his fingers pattered over the keyboard as intricate insights streamed out of his head. After 10 hours he took a break and cleaned out all the drawers of his desk, arranged the pens and pencils in precise parallels, and stacked all his books so that the bottom corners were exactly even. Then he slid the pile so it coincided perfectly with the right angle at the corner of his desk. He stared at the pile for 20 minutes. Then he popped another pill, whistled through 10 more hours of typing and polished up the conclusion of his thesis with some more rather arcane insights. Norman drank half a quart of orange juice, emptied the icebox, and cleaned out all the shelves. Then he retyped some earlier pages which a dirty eraser had smudged, called up his girl and chattered gaily for 40 minutes. Then Norman passed out for 10 hours with dreams of an A in his head. He got a B-plus.

Norman was speeding, but well under the limit. "Speed" is the street name for amphetamines and amphetamine-related drugs, drugs which act as stimulants to the central nervous system. Norman's flirtation with speed, and his drug-induced reactions, are not at all unusal. Amphetamines are commonly used by students and housewives and G.I.'s and truck drivers and tax lawyers. Increasingly, however speed is also being abused by students and housewives and G.I.'s and truck drivers and tax lawyers – not to mention members of the drug culture.

Unlike barbiturates, which are "downs," amphetamines are "ups." They are energizers that kill the desire to eat and sleep, speed up the heart, and deliver a neural charge to the cortex. Precisely how they work is uncertain. One theory holds that they act directly on nerve-circuit receptors; a more widely held view is that amphetamines act indirectly by releasing chemicals at the synapse – the juncture to two nerve ends that carries the impulse from one nerve to the other.

The most basic amphetamine compound – 1-phenyl-2-aminopropane – is marketed as Benzedrine. The *d* form[1] of that

[1] A term used by organic chemists to distinguish between two molecules of identical atomic components but different structure.

compound, dextroamphetamine sulphate, is known as Dexedrine; if a methyl group is added onto the final nitrogen atom, the molecule becomes *d*-desoxyn, or Methedrine. The two *d*-amphetamine compounds are more potent that Benzedrine, but there is no chemical evidence to support the widespread street myth that Methedrine is any more powerful than Dexedrine.

Cocaine, an alkaloid extracted from dried coca leaves, produces almost identical symptoms, though of shorter duration, when taken intravenously. The opiate derivatives, like heroin and morphine, differ in that they produce drowsiness and lethargy. Marijuana can bring on an alert euphoria that approximates the effects of small doses of amphetamines.

Psychedelic or hallucinogenic drugs, like LSD and mescaline, are much more selective than amphetamines in their stimulation of the central nervous system. However, amphetamines may also produce hallucinations at certain stages of intoxication or withdrawal. And other drug confections, brewed up by underground "cooks" and labeled with such initials as D.O.M. and S.T.P., often mimic speed and bridge the gap between amphetamines and hallucinogens.

Amphetamines are a sprawling family marketed by several large pharmaceutical companies under about 75 brand names for over-the-counter sales. With a prescription, amphetamines are completely legal and very cheap – about 5 cents per 5-milligram tablet. Besides selling as Methedrine, Desoxyn, Benzedrine and Dexedrine, they may be combined with barbiturates under such trade names as Desbutal and Dexamyl or, abroad, Preludin and Ritalin. On the street and in the black market, speed is "splash," "crank," "rhythm," "meth" or "crystal" (the last term referring to the pure powdered form of methamphetamine hydrochloride).

Thousands and probably millions use amphetamines without becoming wild-eyed "speed freaks." They drop pills to finish papers, wax floors, lose weight, write songs, sing songs, have conferences, sculpt, wake up and think more clearly. When amphetamines were first synthesized in 1927, they were intended solely as a decongestant in the treatment of bronchial asthma. They are now prescribed for a minimum of other uses: to treat obesity, mild depressions and rare occurrences of narcolepsy, and to treat hyperkinetic brain-damaged children. As antidepressants, amphetamines are beneficial only for a matter of days. Then a tolerance builds, and larger doses only bring on the risk of dependence and deeper depression. As diet pills, amphetamines can produce a maximum weight loss of 5 to 10 pounds over several weeks, most studies report, although some doctors question even that figure, crediting it to a placebo effect. Any attempt to lose more weight by using speed can open the Pandora's box of amphetamine abuse.

The term "abuse" is more accurate than "addiction"; in fact, amphetamines appear not to be physically addictive. Because the body quickly builds up a tolerance to the drug, however, doses must be escalated to maintain the high. Eventually, a pill-popper may be drawn to snorting (sniffing amphetamine powder) or "shooting" (injecting speed into the veins). Speed freaks may experience the physical withdrawal cravings common to narcotics addicts, but for the most part it is the psychological vise of the drug that turns users into abusers and abusers into "addicts." Speed stretches the emotional sine curve; euphoric peaks are followed by troughs that are unbearable wells of despair and depression. The desire to duplicate the first and the need to avoid the second – the "crash" – leads to the amphetamine habit.

The speed epidemic blossomed about three years ago in San Francisco's Haight-

Ashbury district and quickly popped up in the nation's other drug haven, New York's East Village. Quiet flower children became ravaged scarecrows. The cannibalism of speed was easy to spot: emaciated bodies cocked in twisted postures; caved-in jaws, grinding and grinding; pockmarked skin, torn and scratched and white, and a constant talking, talking, talking.

Amphetamine abuse has leveled off in the Haight and the East Village. Nationwide, however, speed abuse is increasing, and the amphetamine industry booms. Ten billion pills are produced annually, about 50 tablets for every man, woman and child in the United States. That is about three times the production of barbiturates. Amphetamine drugs account for about 8 per cent of all prescriptions filled in the United States, according to one estimate.

At least half of that production is diverted into illicit sales, via fake prescriptions, fake doctors and manufacturers who just want to unload their pills. A recent investigation by the House Select Committee on Crime revealed that a pharmaceutical firm had shipped 1.5 million amphetamines to a Mexican "wholesaler" whose address turned out to be the 11th hole of the Tijuana Country Club. "If it's that easy to fool a pharmaceutical firm," commented Representative Jerome R. Waldie, Democrat of California, "I fear even to hazard a guess as to how many millions of pills find their way into the black market."

The quantity of illegally manufactured speed rivals the amount that is produced legally and then diverted illicitly. On the "home factory" level, the Food and Drug Administration reports, "speed is the most popular drug of clandestine chemists." Of 36 underground drug labs raided since last July, the F.D.A. says the majority were involved in the manufacture of meth. In a year-end survey of 10 national districts, the Bureau of Drug Abuse Control of the F.D.A. reported illegal speed "readily available" in five major areas: New York, Chicago, Minneapolis-St. Paul, Dallas and Seattle. Only three areas reported any scarcity, and there were no statistics from California, the speed capital of the world. Dr. David Smith of the Haight-Ashbury Free Medical Clinic, a street-speed expert, says: "There isn't a place in the United States where speed doesn't exist."

Manufacturers range from the small street dealer with a "bathroom" lab, selling ounces to support a habit, to the large distributer with a weekly profit measured in the thousands of dollars. Predictably, the quality of underground speed varies. Police raided an East Village lab in New York recently and discovered a $500,000 operation churning out meth. "He was making Ajax," complained one former customer. "One shot and you turned into a raisin."

Death from an overdose of speed is rare. The record for injection of speed within a single 24-hour period goes to one young man who survived 15,000 milligrams or about 1,000 times any medically sanctioned dosage. It is possible to "overamp," of course, but there is only scattered evidence to support the caution "Speed kills" directly. Speed freaks do commit suicide (a celebrated Lower East Sider recently "fell" off the roof of a six-floor tenement), contract hepatitis and other deadly liver diseases, or occasionally OD (overdose) on heroin or barbiturates while attempting to "come down" from too much speed.

Speed maims and destroys, however. Large doses create extreme suspiciousness and hyperactivity, which often combine to produce a hostile, belligerent temper. "There are no gentle folk on speed," says one ex-user. "They all turn vicious." Dr. John Kramer of the California Rehabilitation Center in Corona reports that "most high-dose amphetamine users describe involvement in episodes in which murder or mayhem was avoided by the slimmest of margins." And a study in the Haight described the difficulties many users have in obtaining speed, and the necessity for "rip-offs" (thefts): "This kind of activity related to the marketplace, coupled with the paranoia induced by long runs on speed, accounts for the excessive amount of violence which prevails in a speed-using community."

In small doses amphetamines do enable greater concentration. But the line between "enough" and "too much" is fragile, and even a single pill can befuddle. "I recall taking one 5-milligram tablet of dextroamphetamine," says Dr. Sidney Cohen of the National Institute of Mental Health, "and I was high and jittery for eight hours." Often, speed produces a grotesque hyperconcentration. Dr. John Griffith, a leading amphetamine researcher, reports: "A graduate student hired as a temporary employee was discovered to be a chronic amphetamine user. This was first suspected when, after being asked to comment on a minor research project, he turned out a 453-page book which was largely unintelligible."

Even the increase in mechanical dexterity under amphetamines is more than offset by the bizarre perceptions that guide that dexterity. One speed freak, attempting to fix a radio, dismantled 12 other perfectly good radios and could'nt put any of them back together. Subjective "brilliance" is often warped by obsessive compulsions. For every apocryphal college student who memorized an astronomy book on speed while coasting to a Phi Beta Kappa, there is the sad story of his friend who wrote a stunning final exam, but all on the first line of his blue book. A speed freak in the quiescent phase may string beads interminably, pour out endless writings, empty a purse for hours, or sit curled in a chair and fill every corner of a piece of paper with tiny, intricate, repeating curlicues.

Speed becomes most dangerous when ingested by snorting or shooting. The size of intravenous doses is awesome: single injections of as much as 500 and 1,000 milligrams per syringe are often repeated every few hours. The advantage in shooting speed, besides the enormous amount of the drug that can be slammed through a needle, is the enveloping "rush" that comes as the drug wallops into the central nervous system. In his "Underground Journal," Herbert Huncke, guru to the early Beats, writes:

"I felt it almost instantly, and the rush to my head like a short circuit. My body began to pulsate or grow tiny antennae all quivering in anticipation – and my aliveness took on new substance and was alerted to the waves of energetic forces sweeping around me. My personal force developed an awareness and it was as though I would receive communication at telepathic level. The surroundings were already capable of absorbing force, drive, spirit, and the flow of magic sensed, never seen. I was thoroughly elated."

Speed is a tempting siren. Unlike first experiences with many hard drugs, those with speed are almost always pleasurable. Many users step up the dosage simply to duplicate that first gorgeous rush. Then, as the body builds up tolerance, the dosage must be increased still more. On high doses, sleep becomes impossible, and the somewhat irregular life style of the casual amphetamine abuser is swallowed up into the splintered world of the speed freak in a debilitating series of "runs" and "crashes."

The crash can trigger a state of intense irritability or despair. To avoid it, the speeder may take some "downs" in an effort to sleep – or take more speed. This solution can lead to typical runs of from several days on up to one or two weeks. During that period it is impossible to sleep or eat–food literally cannot be swallowed – and speed's cannibalism of the body begins: "I knew I was dying," one speeder recalls. "My tongue was shredded, my lips were cracked, the whole inside of my mouth was eaten out. My skin was itching all over, my fingernails were flaking off, and my hair and teeth seemed to be falling out."

"Speed is a hundred times more dangerous than heroin," says one amphetamine researcher at Rockefeller University. "A lifetime intravenous drip of pure heroin would produce no medical damage."But speed knocks out the body's defenses by pushing it beyond the limits of endurance with hyperactivity while depriving it of rest and nourishment. Common speed damage includes colds, infection, muscle tremors, cardiac problems, nausea, cramps, respiratory difficulties and hepatitis—the last not only a frequent result of using unsterile needles but probably stemming in part from the toxicity of Methedrine itself. Malnutrition symptoms appear without exception, and there is a high incidence of minor accidents: cuts, scratches, sprains and bruises.

Besides the physical destruction, there is the psychic damage. Although amphetamines provide a brief glow of potency and confidence, the other end of the seesaw is nothingness:

"I make lots of plans, but I don't ever carry any of them out," says one speed freak. "I get into this megalomaniac bag about five hours into speed, and I'll build these mountainous castles in my minds, all the far-out things I'm going to do, and all the money I'm going to make. I'll be driving a Rolls-Royce and have two speed labs going at once, a heroin refining plant, my own private two-engine plane. I'll be running the Mafia. And when I start to come down I realize that none of it is going to exist and it's like someone pulled the bottom out of your brain. I feel empty and suicidal in about four or five hours."

Heavy speeding invariably produces two psychic nightmares: hallucinations and paranoia. The hallucinations go beyond gentle spiritual probes and sumptuous psychedelic distortion; they are bizarre and mystical and violent, mimicking the traits of a schizoid psychosis:

"I saw energy patterns in the air," a speeder said, "and eventually began to see people no one else saw. They'd walk into shadows of buildings and disappear. Two out of every three people I saw didn't exist. I'd see them come dancing down the street, out from cars. I refused to believe I was hallucinating—I never hallucinated on acid or D.M.T. or S.T.P., and I was dropping two and three tabs [amphetamine tablets] a day – so I started reading up on antimatter physics. I developed a theory of antimen, A-men, and I was in contact with them, I was their leader. I began building machines out of television sets so I could contact them on other planets and in other galaxies, but men with motorcycles would drive around my house and up the walls and order me to stop."

One common hallucination is that one's body is infested with sores and vermin. "When I was speeding," says one resident at Phoenix House, a therapeutic drug community on Manhattan's Upper West Side, "I'd look at my skin and it was all moving and crawling. There were millions of bugs and pimples and scars. I'd pick at them, tear at my skin, and later I'd look at my skin and I'd destroyed myself." More compulsive speed freaks commonly attack themselves with tweezers, and have been known to cut away pieces of flesh.

Paranoia seems to occur almost without exception. This is not the everyday paranoia about tapped wires and plainclothes policemen and the unhidden stash, but a full-blown terror of The Plot, of pursuit by criminals, gangs, the Mafia, assassins, the devil, friends and relatives. It is the terror of buildings falling, rooms and tunnels collapsing, the terror that everyone knows exactly what you are thinking all the time, that you are being watched and followed and discussed, that you are the center of attention of complete strangers. "Two guys would be walking toward me on the sidewalk," says one speed freak, "and one would say to the other, 'You better not do that,' but he was talking to me. And two more strangers would come by, and one would say, 'I don't think that's a very good idea,' and he was looking at me." Eventually the paranoia causes such eccentricity of behavior that the speed freak is likely to be carted off to Bellevue, as in the case of one terrified man who was recently found by police cowering in a truck under the West Side Highway, or another who was attempting to stop traffic on the Brooklyn-Queens Expressway under the delusion that he was being pursued by the Mafia and had to get help.

One of the most intriguing qualities of speed in large doses is its ability to trigger symptoms mirroring those of paranoid schizophrenia. It was in 1958 that Dr. P.H. Connel published a classic monograph, "Amphetamine Psychosis," reporting his study of 42 patients at five British hospitals. Connell's conclusion: "The clinical picture is primarily a paranoid psychosis with ideas of reference, delusions of persecution, auditory and visual hallucinations in a setting of clear consciousness. . . . The mental picture may be indistinguishable from acute or chronic paranoid schizophrenia."

Since then, scholarly debate has been quietly raging in medical journals over two questions: Is amphetamine psychosis actually created by the drug, or does it only precipitate an undetected schizophrenia? Second, is the amphetamine psychosis a "true" psychosis, or are there differences, such as a lack of auditory hallucinations, or a lack of genuine thought disorders? Bellevue speeders between 1966 and 1968 tended to confirm Connell's findings. Hallucinations and thought disorders were correlated with urine tests, which can determine the amount of amphetamines in the blood; it appeared that hallucinations peaked in the most advanced states of toxicity.

Recently, experimenters have attempted to induce the psychosis in controlled hospital settings. At Bellevue, for example, four experienced amphetamine abusers, all judged nonschizophrenic, were given closely monitored doses of methamphetamine, and their reactions and behavior recorded. One subject was given 595 milligrams over a 46-hour period. Here are some descriptions of his reactions, as logged by the experimenters:

"Slight headache. Talking about manliness and Thomas Wolfe. 'He had being a man in the palm of his hand.' That book was his guts,' etc.

"Tearfully discusses mother's promiscuity, abandonment by parents and grandfather, wife's infidelity, and homosexual attacks by relatives.

"Hostile to nursing staff, and feels that their attitude is disparaging. Unaware of his own provocativeness.

"Agitated philosophic diatribe with riddles that make little sense.

"Returns from patients' movie after a half-hour, frightened. Feels he can judge at a glance which patients are dangerous and 'sensed the presence of danger' at the movie.

"Speaking of revelations but is unable to explain what has been revealed.

"Writing and talking excitedly, constantly.

"Preaching loud and constantly. Content as above.

"Preaching at maximum volume. Unable to cooperate with interviews or psychological testing. Amphetamine cut."

Not all speed freaks manifest the acute dislocations of amphetamine psychosis. A few have worked out weekend "binges," and hold down 9-to-5 jobs in between. Others simply resist the temptation of massive doses, or of upping the ante by snorting or shooting – happily protected by their middle-class squeamishness about needles. That inhibition, however, is beginning to disappear with the enveloping drug promiscuity, and it is also possible to become quite adequately freaked out by dropping a handful of mother's little diet pills.

In addition, speed encourages the use of other drugs, not just out of a desire for further experimentation, but specifically to come down from over-amping. Barbiturates are widely used, but frequently orally taken doses may not act strongly enough, and the speeder turns to shooting heroin or barbiturates. When injected, barbiturates are peculiarly dangerous because they can collapse veins and, if veins are missed, can produce dangerous abscesses.

When the ups and downs, the physical deterioration and the paranoia become too overwhelming, many speed freaks try to stop. For some, it can be surprisingly easy. For others, it is impossible; even the worsening trip is not enough to offset the psychological dependence induced by the drug.

Those who can stop usually recover remarkably, although it may take months. There are some, however, who feel that something in their heads has been permanently affected, and they usually say it is the memory. There has been experimental evidence that animals have suffered permanent brain damage from heavy amphetamine doses, and studies following a "speed epidemic" in Japan several years ago point to possibility of similar brain-cell changes in man. Any number of speed freaks feel a vague emotional certainty that they are branded for life, such as one girl at Phoenix House who described her unresolved drug odyssey:

"I've done everything, shot everything, but I flipped out on speed. I would do three days of speed, and then shoot heroin to come down, and then do more speed to get up. But even when you're clean, you still carry around the habit. My nerves are so shot, sometimes in encounter [group therapy] all I can do is shake."

It becomes more and more difficult to characterize the typical amphetamine abuser, though certain occupational groups stand out. One study found widespread amphetamine abuse highest among medical personnel, with housewives and "nocturnal workers" ranking second and third. The study also found amphetamine abuse high in various underworld callings – among petty thieves, convicts, and prostitutes. "Successful" criminals, however, were reported unlikely to be involved with speed. At Bellevue, the mean age in a survey of 60 amphetamine abusers was 25. Most had police records, and about one third had heavy drinking histories. "They're sure as hell not flower children," says Bellevue's speed expert, Dr. Burton Angrust. "They're tough and psychotic. Delinquents. What we used to call hoods."

A report on the Interstate Commerce Commission by Ralph Nader and his Raiders finds an alarming dependence on habit-forming drugs (mostly Bennies) among truckers: "Almost 80 per cent of the drivers in our survey described the use of such drugs as 'widespread.' They disclose that 'Hey, you got a pill?' is now a most common question among drivers at truck stops and terminals. Occasionally, drivers have prescriptions; more often, they get the pills from the black market, from truck stops, and from other drivers. Some use their wives' reducing pills. . . . After getting hopped up on pills, some drivers often need liquor to relax in order to sleep."

Although it is difficult to document, one subgroup, homosexuals, appears to be overrepresented among speedsters. The Bellevue survey noted an abnormally high ratio of homosexuals among amphetamine-abuse patients. It seems more likely that speed appeals to the gay world, rather than that the drug has any power of sexual conversion. Any harassed "outcast" group might turn to speed, with its promise of confidence and power. The psychotic drug paranoia of speed may replicate the paranoia inflicted on homosexuals by straight society; perhaps it is easier to live with the paranoia induced by drugs than that induced by sex. And in the freaked–out world of heavy speed use, being gay is at worst a sprout of eccentricity in a riotous garden of psychosis.

Speed is still not a ghetto drug, however, and probably will never be. Heroin makes the junkie oblivious to rats and roaches and garbage; speed only makes him acutely aware of his surroundings. Financial considerations, too, prevent speed from becoming popular in the ghetto. Like the heroin addict, the speed freak often works up a costly habit, and the poor must hustle to pay for drugs. Although speed and junk are in the same street price range, the junkie is relatively cool between hits. But the speed freak is generally so completely disorientated and trapped on his private, paranoiac 24-hour treadmill that hustling is out of the question.

While some speed freaks get high alone, many are part of groups. One reason is that the life style of the heavy speed user does not exactly coincide with that of the nonspeed world. Speeders have their own language, their own rituals, their own disjointed schedule. Also, speed for many is a social drug; the changes it induces often demand an audience. Finally, the dedicated speed freak inevitably reaches a state of such acute psychic dislocation that he becomes incoherent and intolerable to anyone but another incoherent, intolerable speed freak.

New York City probably has had more than its share of off-beat amphetamine sets in the "cultural underground." One, the Theater of the Ridiculous, paid in–group tribute to the speed legacy with a recent production "from the Crystal Amphetamine Room of the Heroin Hilton." There is also the Warhol set, with its coterie of superstars, its tapes, movies, performances, and most recently, the book "a."

And for jet-set exotica, there was the group of celebrated actors, writers, musicians and public figures who poured in and out of the Frankenstein lab of "Dr. Feelgood," one of New York's fabled Park Avenue "vitamin doctors." There speed and any number of novel concoctions were injected by the glassful. Ultimately, the bizarre activities reached such a level that even Dr. Feelgood's most obsessive patients were forced to drop out. "My last shot," recalls a survivor of one of those final sessions, "was a blood red thing about a foot long. I went blind for two days, and when my eyesight finally came back, I threw away all my speed and hungup my works on the living room lampshade."

Two years ago, Dr. Donald Louria, author of the "The Drug Scene," predicted it was only a matter of years before speed became America's No. 1 drug problem. He now says he has had to qualify that ominous prophecy because of the unforeseen phenomenon of multidrug abuse. The most surprising development over the last few years has been the "give-me-Librium-or-give-me-meth" syndrome within the drug culture, the supermarket spree from drug to drug.

Drugs are being brewed together in unholy combinations and cut with virtually anything that comes to hand; more and more new drugs, especially psychedelics, are being churned out in labs. And certain drugs – once untouchable – are becoming fashionable. "The social stigma of the junkie is gone," said one staff member at Phoenix House. "A year ago, heroin was a dumb thing. Now junk's cool."

Nevertheless, despite the multidrug spree, the Louria prediction about speed's growing popularity was not far wrong. "I emphasize that there has been a major and substantial increase in amphetamine use and abuse in the drug subculture, on college campuses and on Wall Street," testified the Haight-Ashbury Clinic's David Smith before the Senate Special Subcommittee on Alcoholism and Narcotics. "And if you had to pick out a drug with which you should be most concerned, it should be amphetamine, because this is a drug that is completely out of control in our society."

Speed is every bit as lethal as heroin, although its toll may not show up as directly as in the weekend body counts of heroin overdoses. But, because amphetamines have a twilight medical legitimacy, most drug laws are lenient about speed. In New York City, for instance, the possession of any amount of speed – even with acknowledged intent to sell – is punishable only as a misdemeanor with a maximum of one year in jail. With heroin, possession of one-eighth of an ounce is a felony, and possession of more than one pound can mean a life sentence. For the year 1969, 77.8 per cent of all drug arrests made by the New York City Narcotics Bureau were for heroin, and only 0.2 per cent for amphetamines and hallucinogens.

The United States is not the only country to endure the amphetamine carnage. Japan's troubles have been mentioned; England has had a seesawing heroin/speed problem and in Canada the use of amphetamines is growing. Sweden's troubles with speed have reached crisis proportions in recent years. In 1965, Sweden attempted to halt the growing illegal use of speed by making Preludin available to everyone, free and without a prescription. Instead of stemming the tide of amphetamine abuse, legalization doubled the number of addicts in two years. Cases of hepatitis increased more than twofold. In 1967, the experiment was ended, but Sweden's speed problem remained, and there is no end in sight.

Right now the United States seems to be on a heroin binge. But drug use is faddish and cyclical, and speed is probably more popular now than ever. "In the winter," explained one East Villager, "I switch to heroin so I won't catch pneumonia. In the spring I go back to speed."

V. Homeostasis

The evidences of rhythms and their effects on internal processes in man and other organisms has led recent investigators to reconsider the classical concepts of homeostasis. Imposed upon Claude Bernard's classical concept of the constancy of the internal environment is that of a rhythmicity controlled by biological clocks. Major homeostatic systems such as temperature, and hormonal and neural control, clearly exhibit such behavior. The essay by John D. Palmer, professor of biology at New York University, clearly describes how many biological clocks affect human activity. To understand man's fundamental nature we must discover how this internal machinery works.

12.

The Many Clocks of Man

John D. Palmer

. . .studies of time-dependent processes in man are only in the initial, descriptive stages; we are discovering and describing more and more new rhythms. Unfortunately, as yet we have few concrete notions as to how the internal machinery . . . actually functions. AUTHOR

In most cases the intervals of time we recognize are arbitrary, man-made designations. Periods such as seconds, hours, or weeks, plus the more esoteric intervals – weekends, the "cocktail hour," vacation time – are obvious cerebral artifacts of our present civilization. Mixed in with all the arbitrary units of time is the period of a day. A day is the interval between successive sunrises and is generated by the rotation of the earth on its axis in relation to the sun. At present, days are 24 hours long, but just a short 370 million years ago (a fleeting instant in the 4.6 billion years that the earth has existed) the earth gyrated more rapidly, and days were only 22 hours long.

At first thought one might decide that the 24-hour day is just another arbitrary interval of time chosen by man because of its ease of recognition and its convenient duration. We sleep at night; work and eat in the daytime. If no other clues were available to know what to do next, we would need only to look for the sun. Still, we know that men isolated from all view of the sun and from man-made timepieces – either deep in a cave or in soundproof laboratory quarters – still continue to settle down for the "night," to eat, and to regulate their activities around a 24-hour schedule. And while this evidence is often said to reflect our adjustment to a heavenly time interval, it may also be interpreted to mean that this 24-hour rhythm is an intrinsic characteristic of man's inner workings.

This period is also found to be a constant feature of plants and animals. As shown by literally thousands of observations, organisms maintained under laboratory conditions in which all time clues are eliminated still continue to measure out 24-hour periods with surprising accuracy, leading some researchers to conclude that a period of approximately 24 hours is a fundamental attribute of protoplasm. It may be that a period of 24 hours is a kind of absolute in the biotic kingdom. Scientists are now aware that within the bodies of all organisms is a "living clock" signaling off these periods.

One concrete example of these rhythms is the sleep-wakefulness cycle in man. About ten years ago, a young speleologist became obsessed with the idea that it was scientifically important for him to live in an ice-filled cave, sans clocks, for a protracted period of time. For sixty-three days he lived 375 feet below ground where the temperature held constant at 32° F., the relative humidity remained unchanged at 100 percent, and the darkness was complete save for a small, battery-powered light. Each time he awoke, ate, or prepared to retire, he called over a field telephone to a surface camp, where the times of the calls were recorded. The inexorable cold and dampness reduced his body temperature to less than 97° F., and he was constantly threatened by avalanches and cave-ins – still he held out for the sake of science and whiled away his time writing a best seller on his subterranean adventures. Throughout his underground stay he tried mentally to keep track of the passage of time on the surface. When the men in the surface camp informed him on September 14 that his experiment was over, he estimated the date at August 20. His judgment of the passage of time had been exceedingly sluggish. Mentally, he had lost 25 days! However, his living clock (as evaluated by the times of his retiring/awakening phone calls) had ignored his mental confusion and guided his body functions all the while, measuring off periods of activity and sleep that totaled just longer than a day: 24 hour and 31 minutes on the average.

A slight deviation from an exact period of 24 hours is the rule rather than the exception when plants and animals are maintained in strictly unvarying conditions. However, a slight inaccuracy should not dethrone the 24-hour period as an absolute, for when one contemplates the possible intervals from less than nanoseconds to the life-span of the universe, a "near-miss" to 24 hours should be leniently accepted. After all, even the atomic clock of the Bureau of Standards, the paragon of accuracy, mysteriously slows down every sunrise.

As seen in Figure 1, because the period of the cave dweller's rhythm was slightly longer than 24 hours, his sleep-wakefulness cycle fell out of phase with the actual day-night cycle. Only once again during the experiment did it come into phase with the day-night cycle outside, bringing about a rather interesting result. "Daily" during his underground sojourn in this quasi limbo, he entered limited scientific observations and numerous complaints in a log. Save for one entry, the diary is a hodgepodge of chronicled discomforts, misadventures, perpetual intestinal uprisings, cave-ins, and real and imagined terrors. In this particular entry, however, the diary tells us that "for the last few days I have felt very optimistic, I suffer less from the cold; I am better adapted to conditions." During this optimistic period (days 36-39 on the graph) his sleep-wakefulness rhythm was again in phase with normal day-night cycles in the French Alps outside the cave.

Since this pioneering venture, more sophisticated, more comfortable, and considerably less dangerous observations of man in isolation have been carried out, especially at the Max Planck Institute in Germany. In these experiments, light and temperature (the major time-signaling cues of our environment) can be held rigorously constant, while subjects remain in isolation for many weeks.

Despite the relative comfort of the modern experimental setup (including kitchens and baths), it is still difficult to obtain a suitable number of volunteers. Few people are willing to subject themselves to the rigors of prolonged isolation and the indignities of continuous medical measurements. Luckily, money can entice some into cooperating, and in the end investigators turn to that always popular experimental subject – the graduate student. His captive willingness is enhanced by his penury, and in exchange for three meals a day, the quiet of constant conditions in which to study, and a temporary escape from the pressures of graduate school, he is more than willing to provide periodic blood and urine samples and to sit impaled on rectal temperature probes for days on end. This traditional exploitation has an additional advantage – the student, as a burgeoning scientist, can relate his

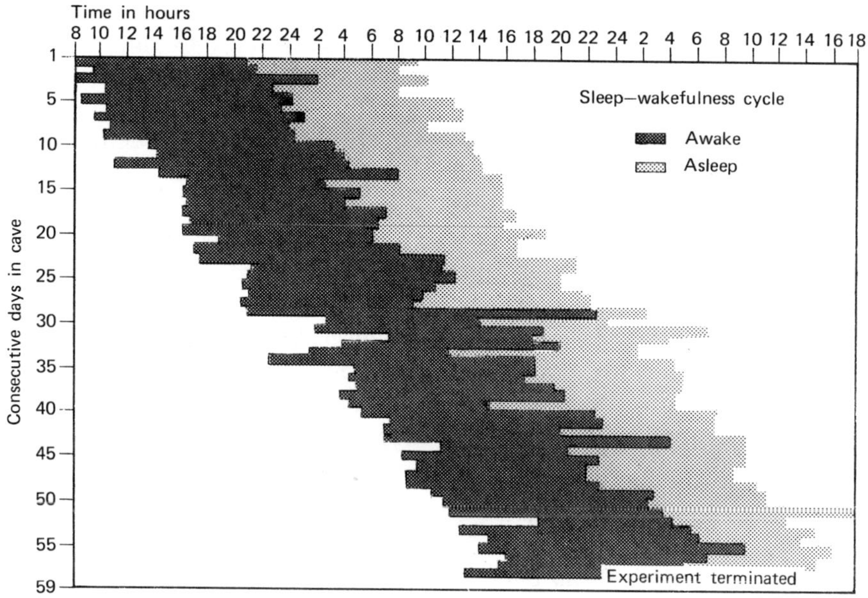

Figure 1

experience to the investigator in a meaningful way after the experiment is over.

In the last few years more than fifty subjects have lived for various lengths of time in these bunkers, and in all cases their internal clocks continued to govern their sleep-wakefulness pattern and other body rhythms in close accordance with the 24-hour time period.

The sleep-wakefulness rhythm does not appear to be present at birth. In a study in which parents of newborns were asked to jot down the times that their babies awakened or fell asleep, it was found that not until the third week of life were signs of a rhythm apparent. During the next few weeks the neonates' nocturnal sleep time increased to an average of ten hours while the day sleep decreased to slightly longer than three and a half hours, beginning to approximate an adult pattern. It is, of course, several years before all day sleep is abandoned. In most of these studies the children were raised in the usual pattern: the parents, for convenience' sake, actively labored to develop sleeping patterns in their newborns that would be similar to their own. However, interesting data came from one set of indulgent parents who allowed their child to determine its own sleeping pattern. In this case, a pattern of sleep geared to a 24-hour day did not develop until the child's eighteenth week of life.

Apparently the major stimulus for the development of a sleep-wakefulness pattern is the parents' concern for getting the child's schedule to conform to their own. However, maturation of the child also plays an important role, as is shown in studies of premature babies. Rhythmicity in these tiny infants develops much more slowly than it does in full-term neonates, despite hospital pressures, to get them to conform to a daily pattern.

The living clock may also function in a related aspect of sleep: the ability of some people to awaken at a predetermined time each morning without the aid of an alarm clock. All men are not equally endowed with this capacity, and those who have it show different degrees of accuracy. Various idiosyncratic rituals have been developed by some people to impress the waking time upon their minds: one expert stamps out the desired hour on the floor – much like the old counting horse of vaudeville – before retiring; and another, a politician, slowly raps out the waking hour with his forehead against the bedpost while patriotically whistling "The Star-Spangled Banner."

Temperature Rhythms

The body temperature of man is also something of an absolute. Under the control of an elaborate thermostat located in the brain, body temperature is regulated at about 98.6° F. Like the interval of 24 hours, this too is a "relative" absolute in that body temperature is not perpetually 98.6° F., but varies by a few degrees around this average. In 1842 it was discovered that in the small hours of the morning one's temperature is low, rising to a maximum during midday or afternoon. An afternoon temperature slightly over 99° does not necessarily mean that one is slightly feverish, while the same temperature in the morning could indicate febricity.

Since body temperature depends on the balance between heat production and loss, one immediately thinks of the effects of muscular activity, food intake, and sleep as the causal factors of the rhythm. While all these factors exert profound influences on body temperature, they are not the causes per se of the rhythm. Studies in which subjects are confined to bed for several days, eating identical meals at regularly spaced intervals or fasting, show that these treatments do not abolish the rhythm or decrease its amplitude. One man, completely paralyzed with poliomyelitis for sixteen months, still displayed a normal temperature rhythm.

Because body temperature is so easily measured, it is one of the most commonly studied human rhythms. Like other body rhythms is has been shown to persist whether in the constant conditions of deep caves or in the laboratory. For example, before the start of their lonely confinement in the experimental bunkers described earlier, rectal temperature probes were implanted in each volunteer so that deep body temperature could be continuously recorded during the times of wakefulness and sleep without disturbing the subject. Even in these static conditions all subjects displayed distinct temperature rhythms. A particularly interesting case is shown in Figure 2. This subject displayed a sleep-wakefulness rhythm with a period of 33 hours, but a temperature rhythm with a period of a more normal 24.8 hours. Therefore, the individual "lost" over 5 days during his stay in constant conditions, while his thermostat lost only half a day. A possible conclusion to such a finding is that the body must have multiple clocks, each controlling a specific function.

Several studies show that man's mental abilities and physical dexterity vary rhyth-

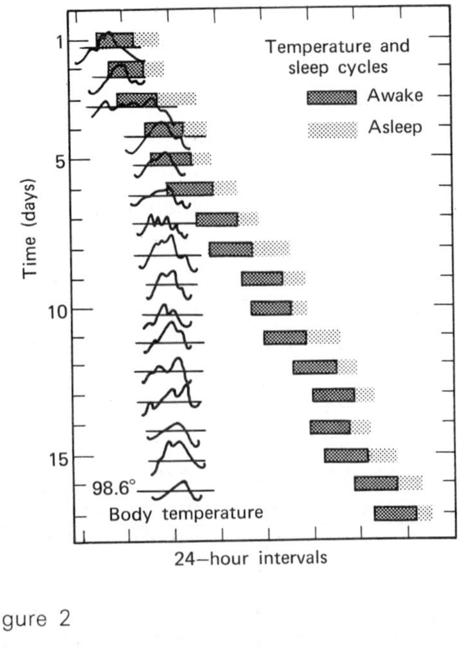

Figure 2

mically over a 24-hour span and that the forms of these rhythms are similar to the form of his personal daily temperature curve. For example, the speed of performing such simple tasks as dealing four hands of cards, sorting a deck of cards by denomination, or multiplying eight digit numbers by each other is studied at various times over the day. These, along with reaction times and steadiness of the hand, are commonly found to be rhythmic. When the subject's temperature curve is simultaneously determined, its form and the forms of the above parameters are usually similar.

Experiments such as these, together with a plethora of similar findings, suggest a causal relationship between body temperature changes and efficiency. The relationship is further strengthened by the findings that "morning" people – those who rise early and work, learn, or perform best in the morning – have temperature curves that reach their daily maximum before noon; while "evening" people have temperature curves that peak in the late afternoon or early evening. The latter case, in which peak performance is delayed until late in the day, discredits the old adage that "sleep recharges the body like a battery" and that body energy gradually runs down during the day.

Rhythmic Time Perception

A few years ago it was found that one's subjective time perception varies rhythmically. For example, if a subject is asked to estimate the passage of a 60-second interval at different times of the day, he tends to overestimate – indicate periods longer than 60 seconds – when his temperature is lowest and underestimate when it is at its peak, suggesting that the rate at which an endogenous physiological "time-perception" mechanism runs is dependent on body temperature. This is a logical conclusion, for it is well known that all metabolic processes run faster at higher temperatures. Therefore if this clock, which would be expected to be a metabolic entity, was caused to run faster at higher temperatures, then the subjective evaluation of time would be shorter, and vice versa. This deduction was proved experimentally many years ago by an eminent physiologist who capitalized on his wife's bout with influenza. Throughout her illness she indulged him by estimating 60-second intervals while he recorded her attempts with a stopwatch. The higher her fever, the quicker she supposed time to be passing. These findings have now been confirmed many times by artificially augmenting human body temperature with exposure to diathermy or drugs.

The Body's Chemical Rhythms

Many, if not all, of the myriad chemical reactions that take place within the human body are probably rhythmic. Because the end products and excesses of some of these reactions are excreted from the body in the urine, the progress of these inner body reactions can be followed by urinalysis. Just about every component easily analyzable in the urine – potassium, chlorine, sodium, phosphate, and hormones, plus pH and water volume – are found to vary rhythmically, with peaks occurring in the daytime and minima at night. These rhythms make their first appearance in the newborn between the fourth and fifteenth week of life, and are not caused by diet, activity, sleep–wakefulness, or other cycles.

An interesting and unusual series of observations on excretory rhythms has been carried out north of the Arctic Circle. On the Spitsbergen Islands, north of Norway, the sun never sets during the summer months and the days are usually overcast, making it difficult to guess the time of day from the sky. Nineteen subjects – again, mostly graduate students – assisted in an experiment in this desolate region. Before arriving in Spitsbergen many of their rhythms, especially their excretory ones, had been studied at great length. Once in Spitsbergen, complete camping equipment was issued to the students, including sham watches, which obstensibly recorded standard 24-hour days, but which actually (and unknown to the subjects) measured out 21- or 27-hour "days."

Two camps, completely isolated from one another in uninhabited territory, were set up and the subjects, unaware of the timepiece subterfuge, were instructed to carry out their daily activities within the framework of time signaled by their watches. Their routine was interrupted every few hours to take oral temperatures and urine samples (in which potassium, sodium, chlorine, and water content were measured).

It should be pointed out that living within these abnormal time schedules meant that every eight 21-hour days were equal to seven real days; and every eight 27-hour days were equal to nine real days. Halfway through either set of eight experimental days the subjects were exactly 180° out of phase with real time – they were up and active at what would have been nighttime back home. The results of the six weeks' study were surprising.

The temperature rhythm of all but one subject adjusted quickly to the 21- or 27-hour days. Less quickly, the rhythms in sodium, chlorine, and water volume also locked into the artificial days. The potassium rhythm, however, seemed less susceptible to "deceit," and in most cases, maintained its 24-hour periodicity. While some of the body's rhythms can be made to operate according to time intervals of slightly more or less than 24 hours, others (the potassium rhythm for example) cannot. The results suggest that there is probably more than one living clock to control separate body functions, and furthermore, that the dependability of each is variable.

It is biologically necessary for a rhythmic process to be adaptable to change, for man is not sedentary in his habits, but moves restlessly over the face of the earth. Today, jet travel shaves hours off, or piles hours on, the length of a single day as travelers speed eastward or westward, and these geographic relocations place considerable stress upon one's living clock. For example, a person leaving New York at 6:30 P.M. will arrive nonstop in Rome at 8:30 A.M. (local time) just as this ancient city is awakening. However, the traveler's living clock is signaling 2:30 A.M. (New York time) and is informing him that it is time to retire. The clock must readjust to the new local time, and until it does, a person will not feel up to par. It is interesting to note that lack of immediate adjustment to new time zones was first observed in 1860 in an orangutan that was being shipped from Java to Germany. The crewmen noticed that the ape tended to maintain its Java sleep pattern in spite of the ship's westward movement through consecutive time zones. Unfortunately, the observations were cut short at the Cape of Good Hope when the animal died after drinking a bottle of rum.

The diplomatic services, large corporations, and athletic coaching staffs want their representatives to be in peak form when entering into business and competitive events in other countries, and a number of studies have recently been undertaken to this end. The procedure is simple but expensive: enthusiastic volunteers are flown from the United States to all parts of the world, and the rates at which their rhythms adjust to these new localities are measured. In addition to rhythmic processes, other assessments – reaction time, subjective fatigue, and decision time – are made on each subject before and after translocation.

After westerly translocations, for example, from Italy to the United States (crossing six time zones) or from the United States to Japan (ten time zones), subjects had to force their living habits to conform with the new local time. It was found that in general the various rhythms took five to six days to completely rephase to the new local times. Reaction and decision times were impeded, and fatigue was significantly increased the first day in the new location, but settled down to near normal by the second day. Older men showed higher fatigue levels than younger subjects.

On easterly flights – from Japan to the United States and from the United States to Italy – the rhythms rephased to local time much more rapidly: most were completely adjusted to the new local times by the second day. As with westward flight, reaction and decision times were impaired and the level of fatigue increased on the first day. North–south flights in which travelers remained in the same time zone had no effect on the biological rhythms, but again, fatigue was experienced after both outgoing and return flights.

The relative ease with which rhythms adjust after eastward flights as opposed to westward flights can be seen in the following example. A businessman traveling to Rome, and unaware of biological rhythms, would schedule his flight so that he could leave New York at a convenient time and arrive in Rome on the morning of the day of his appointment. He would therefore leave New York at 8:00 P.M. so that after an eight-hour flight he would arrive in Rome at 10:00 A.M. (local time), which corresponds to 4:00 A.M. of his own "body time." Biologically he should still be asleep, but he can force himself to carry out his business (with reduced efficiency, perhaps) and still engage in after dusk social activity there. By bedtime he will be quite tired; his own clock will be registering about 7:00 P.M. (it is already adjusting), and he will easily sleep through the remainder of the Roman night. On the other hand, suppose our businessman is with an Italian firm and must come to America. Leaving Rome at 9:00 A.M. would bring him into New York at 11:00 A.M. (local time) giving him time to conclude his first day of business and join in some social activities, but he will have to force himself to stay awake much of this time. He then tries to retire with New Yorkers just after midnight, but his biological clock is now signaling about 7:00 A.M. Roman time, which is his time to awaken. Even though he has been up for 24 straight hours (assuming he did not sleep in the plane) he now finds it difficult to fall asleep, for while it is easy to avoid sleep by a conscious effort, it is impossible to force oneself to fall asleep through mental persuasion. Thus, adaption can be expected to require a longer time after a long westward flight than after an eastward flight.

Thus far I have described only a few of the rhythms known to exist in man – there are many more. The list includes pulse rate, blood output by the heart, circulating red and white blood cells, the amount of protein in the blood at any one time, circulating blood volume, blood pressure, the capacity of the lungs, cell division, a variety of psychiatric illnesses, the adaptation of the ear to new sounds, hormone secretions, the retention of memorized material, and about thirty or forty other rhythms. Surprisingly, even one-time or infrequent events in the human lifespan also appear to be subject to rhythmicity. For example, the rate of childbirth is greater between 2:00 and 7:00 A.M. and lowest between 2:00 and 8:00 P.M. (No wonder obstetricians often appear haggard.) Death rate is also highest between 2:00 and 7:00 A.M. Women tend to begin menstruation in the wee hours.

Another natural period on earth is the synodic-lunar month, the time between successive new moons. It is an interval of 29.5 days. There are many organismic rhythms – most of them reproductive cycles in lower marine animals – with periods of 29.5 days. For example, certain insects hatch out of their pupal cases into adults at the times of full moon. Summer egg production in the Mediterranean sea urchin is greatest at the time of full moon as was known by Aristotle, who recommended that gonad *aficionados* collect them at this time to obtain maximum enjoyment of the delicate ovaries.

That the moon was believed to have some influence on man is indicated by the reference of insanity as "lunacy." As it turns out there have been few scientific studies designed to examine the possibility of the effect of the moon on human life. Even the hint of interest in such an investigation would generate condemnation from fellow scientists, for one of the great triumphs of science in the past has been to abandon astrology and the notion that the movements of heavenly bodies in some way influence the lives of man. Some brave souls, however, in the interest of scientific inquiry, are willing to suffer ostracism by their colleagues and look for moon-related influences on human endeavor.

The human menstrual cycle, by its very name, implies a relationship to the month and moon. However, even elementary textbooks promulgate that the menstrual cycle averages 28 days (rather than 29.5 – the number of days in a synodic-lunar month), and so ingrained is this belief that the eight and a half million women using the "pill," regulate their menstrual cycles to 28 days. Close reexamination of the data collected by earlier workers (the same data that produced the 28-day interval for the menstrual cycle) has now shown that the true average period of the human menstrual cycle is 29.5 days – the exact length of the synodic-lunar month. It was also found that the average gestation period – the time elapsed between the day of conception and delivery – was exactly nine lunar months (266 days).

Armed with this information it is now possible to count backwards 266 days from a birth date to learn the day of conception. By examining a large number of birth dates, then, it should be possible to learn if there is a synodic-lunar monthly rhythm in the time of conception. After statistically examining the birth dates of over a quarter of a million children born in New York municipal hospitals, it was found that the birthrate, and therefore the conception rate, is highest during the three days around full moon, which gives the "moon-spoon-June" ditty some scientific validity. An increase in conception rate at full moon suggests an increase in mating activity – a phenomenon seen in many lower animals.

Ecologists were quick to postulate that the added light reflected onto earth from the full moon must be the cause for the increased mating activity, although there is no proof of this. Nevertheless the newspapers made headlines of this "fact" nine months after "Black Tuesday" (November 9, 1965), the day the east coast of the United States was incapacitated by the pandemic power failure. It was amusingly reported that just nine months after the blackout, birthrates in the New York hospitals had increased. "Therefore," the newspapers proffered, "the ecologists are wrong, it is complete darkness that stimulates mating activity." The papers may have been wrong, for, as can be seen in photos taken that night, Black Tuesday was a night of full moon.

Scientists and others have always wondered if there might not be some sort of rhythm in the human female's sexual desires. For years it has been tacitly assumed by the medical profession that the female desire for sexual union is greatest around the time of ovulation; however the little literature there is on the subject is a mixture of folklore and fact. Last year a study was done in an attempt to resolve this question. The greatest difficulty in carrying out the study was the selection of suitable subjects. All volunteers who were taking birth control pills had to be eliminated as were those who had undergone hysterectomies, those having intercourse regularly and thus often sated, those who were admittedly frigid, and those who "always felt like sex." The investigator finally assembled thirty women, who all had regular menstrual cycles and no regular sex life. In addition they were in psychotherapy and could be queried about their sexual feelings during each session with the psychiatrist. In this way a periodic desire for intercourse over 75 menstrual cycles was documented. Figure 3 shows that the women's libido was highest during the latter half of menses and during the period prior to ovulation. Other studies have shown that women also reach orgasmic climax more often around the middle of their menstrual month.

One of the most intriguing aspects of time-dependent processes is the rhythmic response of recipients to various toxins and medications. For example, if identical doses of bacterial toxins are injected into mice at various times of day, induced lethality is found to be rythmic: 30 percent of the mice are killed when injected at 8:00 P.M. while less than 20 percent die from identical injections administered at midnight. Similar studies with ethyl alcohol caused death to 60 percent of a sample mouse population when the alcohol was

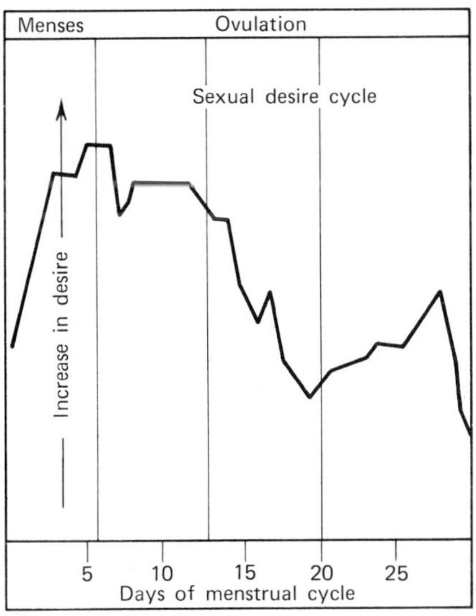

Figure 3

administered at 8:00 P.M., while 20 percent died after 8:00 A.M. injections. Many drugs used on human beings come only in alcohol solution. Sodium pentobarbital, a commonly used anesthesia, was experimentally administered to rats and mice at different times of the day; injections given at night produced intervals of unconsciousness 66 percent longer in duration than identical doses given during the daytime. Damage caused by whole body X-irradiation has also been shown to be rhythmic, nighttime exposure being the most dangerous.

It has been shown that the allergic reaction in man varies with the time of day. Injection of common house dust just under the skin causes an inflammatory reddening and a welt, the severity of which is greatest after 11 P.M. injections, and least after 11:00 A.M. Many patients with asthma have more frequent attacks at night. Another study concerned itself with the length of time that aspirin continues to circulate in the blood. It was found that it remains in circulation a shorter length of time if taken at night, indicating that the analgesic benefits are prolonged when the medication is taken during the daytime.

Some of the most useful and relevant studies on the human living clock should come from the field of medicine. The discovery that the effect of some medications varies with the time of day could alter some of the basic tenets of therapeutic medicine and the test-screening procedures of new drugs. Unfortunately, the medical profession is far behind in the study of human rhythms. There are many questions that need answering in this area. For instance, should the heartbeat rhythm be taken into account before a heart transplant is attempted? Should not surgeons be sure that a donor's kidney is in phase with the excretory rhythms of the potential recipient? Is it wise to subject "evening people," whose temperature cycles and body processes do not reach a maximum until afternoon, to early morning surgery?

Are their bodies in a state best able to survive at this hour? Should manned space shots be initiated early in the morning when most men's reaction times and work proficiency have not yet peaked for the day? Would not more world records be broken if athletes were aware of their body rhythms in performance efficiency? There is a great deal left to learn.

Because of the living clock's relentless activity, we are not the same people from one hour to the next; but at the same time each day, we are much like we were the day before and much like we will be tomorrow. Thus far, our studies of time-dependent processes in man are only in the initial, descriptive stages; we are discovering and describing more and more new rhythms. Unfortunately, as yet we have few concrete notions as to how the internal machinery of the living horologe actually functions. This mechanism must be deciphered because of its medical implications, because rhythmic behavior is a fundamental property of all life, and because of the insight it may provide into the personal and social functioning of our species.

VI. Behavior

Since students of animal behavior are beginning to apply their findings toward a better understanding of human actions, it is appropriate to examine the ideas of the world-renowned ethologist, Niko Tinbergen. This paper is the text of his inaugural address at University of Oxford, 27 February 1968. Four questions essential to our understanding of the biology of aggression are posed and considered. Above all, he says an attack must be launched upon an enemy from within, our unknown selves and our failure to admit it.

The second paper is by Lionel Tiger, anthropologist at Rutgers and author of the recent book *Men In Groups*. It considers the possible origins of sexual discrimination and is certainly relevant to the present vigorous controversy over the female's role in a male dominated society. The author builds an argument that male dominance is a product of natural selection rather than a cultural tradition. But women's lib groups take notice. If you want to change a system you have got to understand it.

13.

On War and Peace in Animals and Man

N. Tinbergen

It is the effects of our behavior that begin to endanger the very survival of our species and, worse, of all life on earth. **AUTHOR**

In 1935 Alexis Carrel published a best seller, *Man – The Unknown (1)*. Today, more than 30 years later, we biologists have once more the duty to remind our fellowmen that in many respects we are still, to ourselves, unknown. It is true that we now understand a great deal of the way our bodies function. With this understanding came control: medicine.

The ignorance of ourselves which needs to be stressed today is ignorance about our behavior – lack of understanding of the causes and effects of the function of our brains. A scientific understanding of our behavior, leading to its control, may well be the most urgent task that faces mankind today. It is the effects of our behavior that begin to endanger the very survival of our species and, worse, of all life on earth. By our technological achievements we have attained a mastery of our environment that is without precedent in the history of life. But these achievements are rapidly getting out of hand. The consequences of our "rape of the earth" are now assuming critical proportions. With shortsighted recklessness we deplete the limited natural resources, including even the oxygen and nitrogen of our atmosphere (*2*). And Rachel Carson's warning (*3*) is now being followed by those of scientists, who give us an even gloomier picture of the general pollution of air, soil, and water. This pollution is seriously threatening our health and our food supply. Refusal to curb our reproductive behavior has led to the population explosion. And, as if all this were not enough, we are waging war on each other – men are fighting and killing men on a massive scale. It is because the effects of these behavior patterns, and of attitudes that determine our behavior, have now acquired such truly lethal potentialities that I have chosen man's ignorance about his own behavior as the subject of this paper.

I am an ethologist, a zoologist studying animal behavior. What gives a student of animal behavior the temerity to speak about problems of human behavior? Of course the history of medicine provides the answer. We all know that medical research uses animals on a large scale. This makes sense because animals, particularly vertebrates, are, in spite of all differences, so similar to us; they are our blood relations, however distant.

But this use of zoological research for a better understanding of ourselves is, to most people, acceptable only when we have to do with those bodily functions that we look upon as parts of our physiological machinery – the functions, for instance, of our kidneys, our liver, our hormone-producing glands. The majority of people bridle as soon as it is even suggested that studies of animal behavior

Reprinted from *Science*, Vol. 160, pp. 1411-1418, June 28, 1968.

could be useful for an understanding, let alone for the control, of our own behavior. They do not want to have their own behavior subjected to scientific scrutiny; they certainly resent being compared with animals, and these rejecting attitudes are both deep-rooted and of complex origin.

But now we are witnessing a turn in this tide of human thought. On the one hand the resistances are weakening, and on the other, a positive awareness is growing of the potentialities of a biology of behavior. This has become quite clear from the great interest aroused by several recent books that are trying, by comparative studies of animals and man, to trace what we could call "the animal roots of human behavior." As examples I select Konrad Lorenz's book *On Aggression (4)* and *The Naked Ape* by Desmond Morris (*5*). Both books were best sellers from the start. We ethologists are naturally delighted by this sign of rapid growth of interest in our science (even though the growing pains are at times a little hard to endure). But at the same time we are apprehensive, or at least I am.

We are delighted because, from the enormous sales of these and other such books, it is evident that the mental block against self-scrutiny is weakening – that there are masses of people who, so to speak, want to be shaken up.

But I am apprehensive because these books, each admirable in its own way, are being misread. Very few readers give the authors the benefit of the doubt. Far too many either accept uncritically all that the authors say, or (equally uncritically) reject it all. I believe that this is because both Lorenz and Morris emphasize our knowledge rather than our ignorance (and, in addition, present as knowledge a set of statements which are after all no more than likely guesses). In themselves brilliant, these books could stiffen, at a new level, the attitude of certainty, while what we need is a sense of doubt and wonder, and an urge to investigate, to inquire.

Potential Usefulness of Ethological Studies

Now, in a way, I am going to be just as assertative as Lorenz and Morris, but what I am going to stress is how much we do not know. I shall argue that we shall have to make a major research effort. I am of course fully aware of the fact that much research is already being devoted to problems of human, and even of animal, behavior. I know, for instance, that anthropologists, psychologists, psychiatrists, and others are approaching these problems from many angles. But I shall try to shown that the research effort has so far made insufficient use of the potential of ethology. Anthropologists, for instance, are beginning to look at animals, but they restrict their work almost entirely to our nearest relatives, the apes and monkeys. Psychologists do study a larger variety of animals, but even they select mainly higher species. They also ignore certain major problems that we biologists think have to be studied. Psychiatrists, at least many of them, show a disturbing tendency to apply the *results* rather than the *methods* of ethology to man.

None of these sciences, not even their combined efforts, are as yet parts of one coherent science of behavior. Since behavior is a life process, its study ought to be part of the mainstream of biological research. That is why we zoologists ought to "join the fray." As an ethologist, I am going to try to sketch how my science could assist its sister sciences in their attempts, already well on their way, to make a united, broad-fronted, truly biological attack on the problems of behavior.

I feet that I can cooperate best by discussing what it is in ethology that could be of use to the other behavioral sciences. What we ethologists do not want, what we consider definitely wrong, is uncritical application of our results to man. Instead, I myself at least feel that it is our method of approach, our rationale, that we can offer (*6*), and also a little simple common sense, and discipline.

The potential usefulness of ethology lies in the fact that, unlike other sciences of behavior, it applies the method or "approach" of biology to the phenomenon behavior. It has developed a set of concepts and terms that allow us to ask:

1. In what ways does this phenomenon (behavior) influence the survival, the success of the animal?
2. What makes behavior happen at any given moment? How does its "machinery" work?
3. How does the behavior machinery develop as the individual grows up?
4. How have the behavior systems of each species evolved until they became what they are now?

The first question, that of survival value, has to do with the effects of behavior; the other three are, each on a different time scale, concerned with its causes.

These four questions are, as many of my fellow biologists will recognize, the major questions that biology has been pursuing for a long time. What ethology is doing could be simply described by saying that, just as biology investigates the functioning of the organs responsible for digestion, respiration, circulation, and so forth, so ethology begins now to do the same with respect to behavior; it investigates the functioning of organs responsible for movement.

I have to make clear that in my opinion it is the comprehensive, integrated attack on all four problems that characterizes ethology. I shall try to show that to ignore the questions of survival value and evolution – as, for instance, most psychologists do – is not only shortsighted but makes it impossible to arrive at an understanding of behavioral problems. Here ethology can make, in fact is already making, positive contributions.

Having stated my case for animal ethology as an essential part of the science of behavior, I will now have to sketch how this could be done. For this I shall have to consider one concrete example, and I select aggression, the most directly lethal of our behaviors. And, for reasons that will become clear, I shall also make a short excursion into problems of education.

Let me first try to define what I mean by agression. We all understand the term in a vague, general way, but it is, after all, no more than a catchword. In terms of actual behavior, aggression involves approaching an opponent, and, when within reach, pushing him away, inflicting damage of some kind, or at least forcing stimuli upon him that subdue him. In this description the effect is already implicit; such behavior tends to remove the opponent, or at least to make him change his behavior in such a way that he no longer interferes with the attacker. The methods of attack differ from one species to another, and so do the weapons that are used, the structures that contribute to the effect.

Since I am concentrating on men fighting men, I shall confine myself to intraspecific fighting, and ignore, for instance, fighting between predators and prey. Intraspecific fighting is very common among animals. Many of them fight in two different contexts, which we can call "offensive" and "defensive." Defensive fighting is often shown as a last resort by an animal that, instead of attacking, has been fleeing from an attacker. If it is cornered, it may suddenly turn round upon its enemy and "fight with the courage of despair."

Of the four questions I mentioned before, I shall consider that of the survival value first. Here comparison faces us right at the start with a striking paradox. On the one hand, man is akin to many species of animals in that he fights his own species. But on the other hand he is, among the thousands of species that fight, the only one in which fighting is disruptive.

In animals, intraspecific fighting is usually of distinctive advantage. In addition, all species manage as a rule to settle their disputes without killing one another; in fact, even bloodshed is rare. Man is the only species that is a mass murderer, the only misfit in his own society.

Why should this be so? For an answer, we shall have to turn to the question of causation: What makes animals and man fight their own species? And why is our species "the odd man out"?

Causation of Aggression

For a fruitful discussion of this question of causation I shall first have to discuss what exactly we mean when we ask it.

I have already indicated that when thinking of causation we have to distinguish between three subquestions, and that these three differ from one another in the stretch of time that is considered. We ask, first: Given an adult animal that fights now and then, what makes each outburst of fighting happen? The time scale in which we consider these recurrent events is usually one of seconds, or minutes. To use an analogy, this subquestion compares with asking what makes a car start or stop each time we use it.

But in asking this same general question of causation ("What makes an animal fight?") we may also be referring to a longer period of time; we may mean "How has the animal, as it grew up, developed this behavior?" This compares roughly with asking how a car has been constructed in the factory. The distinction between these two subquestions remains useful even though we know that many animals continue their development (much slowed down) even after they have attained adulthood. For instance, they may still continue to learn.

Finally, in biology , as in technology, we can extend this time scale even more, and ask: How have the animal species which we observe today – and which we know have evolved from ancestors that were different – how have they acquired their particular behavior systems during this evolution? Unfortunately, while we know the evolution of cars because they evolved so quickly and have been so fully recorded, the behavior of extinct animals cannot be observed, and has to be reconstructed by indirect methods.

I shall try to justify the claim I made earlier, and show how all these four questions – that of behavior's survival value and the three subquestions of causation – have to enter into the argument if we are to understand the biology of aggression.

Let us first consider the short-term causation: the mechanism of fighting. What makes us fight at any one moment? Lorenz argues in his book that, in animals and in man, there is an internal urge to attack. An individual does not simply wait to be provoked, but, if actual attack has not been possible for some time, this urge to fight builds up until the individual actively seeks the opportunity to indulge in fighting. Aggression, Lorenz claims, can be spontaneous.

But this view has not gone unchallenged. For instance, R. A Hinde has written a thorough criticism (*7*), based on recent work on aggression in animals, in which he writes that Lorenz's "arguments for the spontaneity of aggression do not bear examination" and that "the contrary view, expressed in nearly every textbook of comparative psychology. . ." is that fighting "derives principally from the situation"; and even more explicitly: "There is no need to postulate causes that are purely internal to the aggressor" (*7*, p. 303). At first glance it would seem as if Lorenz and Hinde disagree profoundly. I have read and reread both authors, and it is to me perfectly clear that loose statements and misunderstandings on both sides have made it appear that there is disagreement where in fact there is something very near to a common opinion. It seems to me that the differences between the two authors lie mainly in the different ways they look at internal and external variables. This in turn seems due to differences of a semantic nature. Lorenz uses the unfortunate term "the spontaneity of aggression." Hinde takes this to mean that external stimuli are in Lorenz's view not necessary at all to make an animal fight. But here he misrepresents Lorenz, for nowhere does Lorenz claim that the internal urge ever makes an animal fight "in vacuo"; somebody or something is attacked. This misunderstanding makes Hinde feel that he has refuted Lorenz's views by saying that "fighting derives principally from the situation." But both authors are fully aware of the fact that fighting is started by a number of variables, of which some are internal and some external. What both authors know, and what cannot be doubted, is that fighting behavior is not like the simple slot machine that produces one platform ticket every time one threepenny bit is inserted. To mention one animal example: a male stickleback does not always show the full fighting behavior in response to an approaching standard opponent; its response varies from none at all to the optimal stimulus on some occasions, to full attack on even a crude dummy at other times. This means that its internal state varies, and in this particular case we know from the work of Hoar (*8*) that the level of the male sex hormone is an important variable.

Another source of misunderstanding seems to have to do with the stretch of time that the two authors are taking into account. Lorenz undoubtedly thinks of the causes of an outburst of fighting in terms of seconds, or hours – perhaps days. Hinde seems to think of events which may have happened further back in time; an event which is at any particular moment "internal" may well in its turn have been influenced previously by external agents. In our stickleback example, the level of male sex hormone is influenced by external agents such as the length of the daily exposure to light over a period of a month or so (*9*). Or, less far back in time, its readiness to attack may have been influenced by some experience gained, say, half an hour before the fight.

I admit that I have now been spending a great deal of time on what would seem to be a perfectly simple issue: the very first step in the analysis of the short-term causation, which is to distinguish at any given moment between variables within the animal and variables in the environment. It is of course important for our further understanding to unravel the complex interactions between these two worlds, and in particular the physiology of aggressive behavior. A great deal is being discovered about this, but for my present issue there is no use discussing it as long as even the first step in the analysis has not led to a clearly expressed and generally accepted conclusion. We must remember that we are at the moment concerned with the human problem: "What makes men attack each other?" And for this problem the answer to the first stage of our question is of prime importance: Is our readiness to start an attack constant or not? If it were – if our aggressive behavior were the outcome of an apparatus with the properties of the slot machine – all we would have to do would be to control the external situation: to stop providing threepenny bits. But since our readiness to start an attack is variable, further studies of both the external and the internal variables are vital to such issues as: Can we reduce fighting by lowering the population density, or by withholding provocative stimuli? Can we do so by changing the hormone balance or other physiological variables? Can we perhaps in addition control our development in such a way as to change the dependence on internal and external factors in adult man? However, before discussing development, I must first return to the fact that I have mentioned before, namely, that man is, among the thousands of other species that fight, the only mass murderer. How do animals in their intraspecific disputes avoid bloodshed?

The Importance of "Fear"

The clue to this problem is to recognize the simple fact that aggression in animals rarely occurs in pure form; it is only one of two components of an adaptive system. This is most clearly seen in territorial behavior, although it is also true of most other types of hostile behavior. Members of territorial species divide, among themselves, the available living space and opportunities by each individual defending its home range against competitors. Now in this system of parceling our living space, avoidance plays as important a part as attack. Put very briefly, animals of territorial species, once they have settled on a territory, attack intruders, but an animal that is still searching for a suitable territory or finds itself outside its home range withdraws when it meets with an already established owner. In terms of function, once you have taken possession of a territory, it pays to drive off competitors; but when you are still looking for a territory (or meet your neighbor at your common boundary), your chances of success are improved by avoiding such established owners. The ruthless fighter who "knows no fear" does not get very far. For an understanding of what follows, this fact, that hostile clashes are controlled by what we could call the "attack-avoidance system," is essential.

When neighboring territory owners meet near their common boundary, both attack behavior and withdrawal behavior are elicited in both animals; each of the two is in a state of motivational conflict. We know a great deal about the variety of movements that appear when these two conflicting, incompatible behaviors are elicited. Many of these expressions of a motivational conflict have, in the course of evolution, acquired signal function; in colloquial language, they signal "Keep out!" We deduce this from the fact that opponents respond to them in an appropriate way: instead of proceeding to intrude, which would require the use of force, trespassers withdraw, and the neighbors are contained by each other. This is how such animals have managed to have all the advantages of their hostile behavior without the disadvantages: they divide their living space in a bloodless way by using as distance-keeping devices these conflict movements ("threat") rather than actual fighting.

Group Territories

In order to see our wars in correct biological perspective one more comparison with animals is useful. So far I have discussed animal species that defend individual or at best pair territories. But there are also animals which possess and defend territories belonging to a group, or a clan (*10*).

Now it is an essential aspect of group territorialism that the members of a group unite when in hostile confrontation with another group that approaches or crosses into their feeding territory. The uniting and the aggression are equally important. It is essential to realize that group territorialism does not exclude hostile relations on lower levels when the group is on its own. For instance, within a group there is often a peck order. And within the group there may be individual or pair territories. But frictions due to these relationships fade away during a clash between groups. This temporary elimination is done by means of so-called appeasement and reassurance signals. They indicate "I am a friend," and so diminish the risk that, in the general flare-up of anger, any animal "takes it out" on a fellow member of the same group (*11*). Clans meet clans as units, and each individual in an intergroup clash, while united with its fellow-members, is (as in interindividual clashes) torn between attack and withdrawal, and postures and shouts rather than attacks.

We must now examine the hypothesis (which I consider the most likely one) that man still carries with him the animal heritage of group territoriality. This is a question concerning man's evolutionary origin, and here we are, by the very nature of the subject, forced to speculate. Because I am going to say something about the behavior of our ancestors of, say, 100,000 years ago, I have to discuss briefly a matter of methodology. It is known to all biologists (but unfortunately unknown to most psychologists) that comparison of present-day species can give us a deep insight, with a probability closely approaching certainty, into the evolutionary history of animal species. Even where fossil evidence is lacking, this comparative method alone can do this. It has to be stressed that this comparison is a highly sophisticated method, and not merely a matter of saying that species A is different from species B (*12*). The basic problem is this. We interpret differences between really allied species as the result of adaptive divergent evolution from common stock, and we interpret similarities between nonallied species as adaptive convergencies to similar ways of life. By studying the adaptive functions of species characteristics we understand how natural selection can have produced both these divergencies and convergencies. To mention one striking example: even if we had no fossil evidence, we could, by this method alone, recognize whales for what they are – mammals that have returned to the water, and, in doing so, have developed some similarities to fish. This special type of comparison, which has been applied so successfully by students of the structure of animals has now also been used, and with equal success, in several studies of animal behavior. Two approaches have been applied. One is to see in what respects species of very different origin have convergently adapted to a similar way of life. Von Haartman (*13*) has applied this to a study of birds of many types that nest in holes – an anti-predator safety device. All such hole-nesters center their territorial fighting on a suitable nest hole. Their courtship consists of luring a female to this hole (often with the use of bright color patterns). Their young gape when a general darkening signals the arrival of the parent. All but the most recently adapted species lay uniformly colored, white or light blue eggs that can easily be seen by the parent.

An example of adaptive divergence has been studied by Cullen (*14*). Among all the gulls, the kittiwake is unique in that it nests on very narrow ledges on sheer cliffs. Over 20 peculiarities of this species have been recognized by Mrs. Cullen as vital adaptions to this particular habit.

These and several similar studies (*15*) demonstrate how comparison reveals, in each species, systems of interrelated, and very intricate adaptive features. In this work, speculation is now being followed by careful experimental checking. It would be tempting to elaborate on this, but I must return to our own unfortunate species.

Now, when we include the "Naked Ape" in our comparative studies, it becomes likely (as has been recently worked out in great detail by Morris) that man is a "social Ape who has turned carnivore" (*16*). On the one hand he is a social primate; on the other, he has developed similarities to wolves, lions and hyenas. In our present context one thing seems to stand out clearly, a conclusion that seems to me of paramount importance to all of us, and yet has not been fully accepted as such. As a social, hunting primate, man must originally have been organized on the principle of group territories.

Ethologists tend to believe that we still carry with us a number of behavioral characteristics of our animal ancestors, which cannot be eliminated by different ways of upbringing, and that our group territorialism is one of those ancestral characters. I shall discuss the problem of the modifiability of our behavior later, but it is useful to point out here that even if our behavior were much more modifiable than Lorenz maintains, our cultural evolution, which resulted in the parceling-out of our living space on lines of tribal, national, and now even "bloc" areas, would, if anything, have tended to enhance group territorialism.

Group Territorialism in Man?

I put so much emphasis on this issue of group territorialism because most writers who have tried to apply ethology to man have done this in the wrong way. They have made the mistake, to which I objected before, of uncritically extrapolating the results of animal studies to man. They try to explain man's behavior by using facts that are valid only of some of the animals we studied. And, as ethologists keep stressing, no two species behave alike. Therefore, instead of taking this easy way out, we ought to study man in his own right. And I repeat that the message of the ethologists is that the methods, rather than the results, of ethology should be used for such a study.

Now, the notion of territory was developed by zoologists (to be precise, by ornithologists,*17*), and because individual and pair territories are found in so many more species than group territories (which are particularly rare among birds), most animal studies were concerned with such individual and pair territories. Now such low-level territories do occur in man, as does another form of hostile behavior, the peck order. But the problems created by such low-level frictions are not serious; they can, within a community, be kept in check by the apparatus of law and order; peace within national boundaries can be enforced. In order to understand what makes us go to war, we have to recognize that man behaves very much like a group-territorial species. We too unite in the face of an outside danger to the group; we "forget our differences." We too have threat gestures, for instance, angry facial expressions. And all of us use reassurance and appeasement signals, such as a friendly smile. And (unlike speech) these are universally understood; they are cross-cultural; they are species-specific. And, incidentally, even within a group sharing a common language, they are often more reliable guides to a man's intentions than speech, for speech (as we know now) rarely reflects our true motives, but our facial expressions so often "give us away."

If I may digress for a moment: it is humiliating to us ethologists that many nonscientists, particularly novelists and actors, intuitively understand our sign language much better than we scientists ourselves do. Worse, there is a category of human beings who understand intuitively more about the causation of our aggressive behavior: the great demagogues. They have applied this knowledge in order to control our behavior in the most clever ways, and often for the most evil purposes. For instance, Hitler (who had modern mass communication at his disposal, which allowed him to inflame a whole nation) played on both fighting tendencies. The "defensive" fighting was whipped up by his passionate statements about "living space," "encirclement," Jewry, and Freemasonry as threatening powers which made the Germans feel "cornered." The "attack fighting" was similarly set ablaze by playing the myth of the Herrenvolk. We must make sure that mankind has learned its lesson and will never forget how disastrous the joint effects have been – if only one of the major nations were led now by a man like Hitler, life on earth would be wiped out.

I have argued my case for concentrating on studies of group territoriality rather than on other types of aggression. I must now return, in this context, to the problem of man the mass murderer. Why don't we settle even our international disputes by the relatively harmless, animal method of threat? Why have we become unhinged so that so often our attack erupts without being kept in check by fear? It is not that we have no fear, nor that we have no other inhibitions against killing. This problem has to be considered first of all in the general context of the consequences of man having embarked on a new type of evolution.

Cultural Evolution

Man has the ability, unparalleled in scale in the animal kingdom, of passing on his experiences from one generation to the next. By this accumulative and exponentially growing process, which we call cultural evolution, he has been able to change his environment progressively out of all recognition. And this includes the social environment. This new type of evolution proceeds at an incomparably faster pace than genetic evolution. Genetically we have not evolved very strikingly since Cro-Magnon man, but culturally we have changed beyond recognition, and are changing at an ever-increasing rate. It is of course true that we are highly adjustable individually, and so could hope to keep pace with these changes. But I am not alone in believing that this behavioral adjustability, like all types of modifiability, has its limits. These limits are imposed upon us by our hereditary constitution, a constitution which can only change with the far slower speed of genetic evolution. There are good grounds for the conclusion that man's limited behavioral adjustability has been outpaced by the culturally determined changes in his social environment, and that this is why man is now a misfit in his own society.

We can now, at last, return to the problem of war, of uninhibited mass killing. It seems quite clear that our cultural evolution is at the root of the trouble. It is our cultural evolution that has caused the population explosion. In a nutshell, medical science, aiming at the reduction of suffering, has, in doing so, prolonged life for many individuals as well – prolonged it to well beyond the point at which they produce offspring. Unlike the situation in any wild species, recruitment to the human population consistently surpasses losses through mortality. Agricultural and technical know-how have enabled us to grow food and to exploit other natural resources to such an extent that we can still feed (though only just) the enormous numbers of human beings on our crowded planet. The result is that we now live at a far higher density than that in which genetic evolution has molded our species. This together with long-distance communication, leads to far more frequent, in fact to continuous intergroup contacts, and so to continuous external provocation of aggression. Yet this alone would not explain our increased tendency to kill each other; it would merely lead to continuous threat behavior.

The upsetting of the balance between aggression and fear (and this is what causes war) is due to at least three other consequences of cultural evolution. It is an old cultural phenomenon that warriors are both brainwashed and bullied into all-out fighting. They are brainwashed into believing that fleeing – originally, as we have seen, an adaptive type of behavior – is despicable, "cowardly." This seems to me due to the fact that man, accepting that in moral issues death might be preferable to fleeing, has falsely applied the moral concept of "cowardice" to matters of mere practical importance – to the dividing of living space. The fact that our soldiers are also bullied into all-out fighting (by penalizing fleeing in battle) is too well known to deserve elaboration.

Another cultural excess is our ability to make and use killing tools, especially long-range weapons. These make killing easy, not only because a spear or a club inflicts, with the same effort, so much more damage than a fist, but also, and mainly, because the use of long-range weapons prevents the victim from reaching his attacker with his appeasement, reassurance, and distress signals. Very few air-crews who are willing, indeed eager, to drop their bombs "on target" would be willing to strangle, stab, or burn children (or, for that matter, adults) with their own hands; they would stop short of killing, in response to the appeasement and distress signals of their opponents.

These three factors alone would be sufficient to explain how we have become such unhinged killers. But I have to stress once more that all this, however convincing it may seem, must still be studied more thoroughly.

There is a frightening, and ironical paradox in this conclusion: that the human brain, the finest life-preserving device created by evolution, has made our species so successful in mastering the outside world that it suddenly finds itself taken off guard. One could say that our cortex and brainstem (our "reason" and our "instincts") are at loggerheads. Together they have created a new social environment in which, rather than ensuring our survival, they are about to do the opposite. The brain finds itself seriously threatened by an enemy of its own making. It is its own enemy. We simply have to understand this enemy.

The Development of Behavior

I must now leave the question of the moment-to-moment control of fighting, and, looking further back in time, turn to the development of aggressive behavior in the growing individual. Again we will start from the human problem. This, in the present context, is whether it is within our power to control development in such a way that we reduce or eliminate fighting among adults. Can or cannot education in the widest sense produce nonaggressive men?

The first stop in the consideration of this problem is again to distinguish between external and internal influences, but now we must apply this to the growth, the changing, of the behavioral machinery during the individual's development. Here again the way in which we phrase our questions and our conclusions is of the utmost importance.

In order to discuss this issue fruitfully, I have to start once more by considering it in a wider context, which is now that of the "nature-nurture" problem with respect to behavior in general. This has been discussed more fully by Lorenz in his book *Evolution and Modification of Behaviour* (*18*); for a discussion of the environmentalist point of view I refer to the various works of Schneirla (see *19*).

Lorenz tends to classify behavior types into innate and acquired or learned behavior. Schneirla rejects this dichotomy into two classes of behavior. He stresses that the developmental process, of behavior as well as of other functions, should be considered, and also that this development forms a highly complicated series of interactions between the growing organism and its environment. I have gradually become convinced that the clue to this difference in approach is to be found in a difference in aims between the two authors. Lorenz claims that "we are justified in leaving, at least for the time being, to the care of the experimental embryologists all those questions which are concerned with the chains of physiological causation leading from the genome to the development of . . . neurosensory structures" (*18*, p. 43). In other words, he deliberately refrains from starting his analysis of development prior to the stage at which a fully coordinated behavior is performed for the first time. If one in this way restricts one's studies to the later stages of development, then a classification in "innate" and "learned" behavior, or behavior components, can be considered quite justified. And there was a time, some 30 years ago, when the almost grotesquely environmentalist bias of psychology made it imperative for ethologists to stress the extent to which behavior patterns could appear in perfect or near-perfect form without the aid of anything that could be properly called learning. But I now agree (however belatedly) with Schneirla that we must extend our interest to earlier stages of development and embark on a full program of experimental embryology of behavior. When we do this, we discover that interactions with the environment can indeed occur at early stages. These interactions may concern small components of the total machinery of a fully functional behavior pattern, and many of them cannot possibly be called learning. But they are interactions with the environment, and must be taken into account if we follow in the footsteps of the experimental embryologists, and extend our field of interest to the entire sequence of events which lead from the blueprints contained in the zygote to the fully functioning, behaving animal. We simply have to do this if we want an answer to the question to what extent the development of behavior can be influenced from the outside.

When we follow this procedure the rigid distinction between "innate" or unmodifiable and "acquired" or modifiable behavior patterns becomes far less sharp. This is owing to the discovery, on the one hand, that "innate" patterns may contain elements that at an early stage developed in interaction with the environment, and, on the other hand, that learning is, from step to step, limited by internally imposed restrictions.

To illustrate the first point, I take the development of the sensory cells in the retina of the eye. Knoll has shown (*20*) that the rods in the eyes of tadpoles cannot function properly unless they have first been exposed to light. This means that, although any visually guided response of a tadpole may well, in its integrated form, be "innate" in Lorenz's sense, it is so only in the sense of "non-learned," not in that of "having grown without interaction with the environment." Now it has been shown by Cullen (*21*) that male sticklebacks reared from the egg in complete isolation from other animals will, when adult, show full fighting behavior to other males and courtship behavior to females when faced with them for the first time in their lives. This is admittedly an important fact, demonstrating that the various recognized forms of learning do not enter into the programing of these integrated patterns. This is a demonstration of what Lorenz calls an "innate response." But it does not exclude the possibility that parts of the machinery so employed may, at an earlier stage, have been influenced by the environment, as in the case of the tadpoles.

Second, there are also behavior patterns which do appear in the inexperienced animal, but in an incomplete form, and which require additional development through learning. Thorpe has analyzed a clear example of this: when young male chaffinches reared alone begin to produce their song for the first time, they utter a very imperfect warble; this develops into the full song only if, at a certain sensitive stage, the young birds have heard the full song of an adult male (*22*).

By far the most interesting aspect of such intermediates between innate and acquired behavior is the fact that learning is not indiscriminate, but is guided by a certain selectiveness on the part of the animal. This fact has been dimly recognized long ago; the early ethologists have often pointed out that different, even closely related, species learn different things even when developing the same behavior patterns. This has been emphasized by Lorenz's use of the term "innate teaching mechanism." Other authors use the word "template" in the same context. The best example I know is once more taken from the development of song in certain birds. As I have mentioned, the males of some birds acquire their full song by changing their basic repertoire to resemble the song of adults, which they have to hear during a special sensitive period some months before they sing themselves. It is in this sensitive period that they acquire, without as yet producing the song, the knowledge of "what the song ought to be like." In technical terms, the birds formed a *Sollwert (23)* (literally, "should-value," an ideal) for the feedback they receive when they hear their own first attempts. Experiments have shown (*24*) that such birds, when they start to sing, do three things: they listen to what they produce; they notice the difference between this feedback and the ideal song; and they correct their next performance.

This example, while demonstrating an internal teaching mechanism, shows, at the same time, that Lorenz made his concept too narrow when he coined the term "innate teaching mechanism." The birds have developed a teaching mechanism, but while it is true that it is internal, it is not innate; the birds have acquired it by listening to their father's song.

These examples show that if behavior studies are to catch up with experimental embryology our aims, our concepts, and our terms must be continually revised.

Before returning to aggression, I should like to elaborate a little further on general aspects of behavior development, because this will enable me to show the value of animal studies in another context, that of education.

Comparative studies, of different animal species, of different behavior patterns, and of different stages of development, begin to suggest that wherever learning takes a hand in development, it is guided by such *Sollwerte* or templates for the proper feedback, the feedback that reinforces. And it becomes clear that these various *Sollwerte* are of a bewildering variety. In human education one aspect of this has been emphasized in particular, and even applied in the use of teaching machines: the requirement that the reward, in order to have maximum effect, must be immediate. Skinner has stressed this so much because in our own teaching we have imposed an unnatural delay between, say, taking in homework, and giving the pupil his reward in the form of a mark. But we can learn more from animal studies than the need for immediacy of reward. The type of reward is also of great importance, and this may vary from task to task, from stage to stage, from occasion to occasion; the awards may be of almost infinite variety.

Here I have to discuss briefly a behavior of which I have so far been unable to find the equivalent in the development of structure. This is exploratory behavior. By this we mean a kind of behavior in which the animal sets out to acquire as much information about an object or a situation as it can possibly get. The behavior is intricately adapted to this end, and it terminates when the information has been stored, when the animal has incorporated it in its learned knowledge. This exploration (subjectively we speak of "curiosity") is not confined to the acquisition of information about the external world alone; at least mammals explore their own movements a great deal, and in this way "master new skills." Again, in this exploratory behavior, *Sollwerte* of expected, "hoped-for" feedbacks play their part.

Without going into more detail, we can characterize the picture we begin to get of the development of behavior as a series, or rather a web, of events, starting with innate programing instructions contained in the zygote, which straightaway begin to interact with the environment; this interaction may be discontinuous, in that periods of predominantly internal development alternate with periods of interaction, or sensitive periods. The interaction is enhanced by active exploration; it is steered by selective *Sollwerte* of great variety; and stage by stage this process ramifies; level upon level of ever-increasing complexity is being incorporated into the programing.

Apply what we have heard for a moment to playing children (I do not, of course, distinguish sharply between "play" and "learning"). At a certain age a child begins to use, say, building blocks. It will at first manipulate them in various ways, one at a time. Each way of manipulating acts as exploratory behavior: the child learns what a block looks, feels, tastes like, and so forth, and also how to put it down so that it stands stably.

Each of these stages "peters out" when the child knows what it wanted to find out. But as the development proceeds, a new level of exploration is added: the child discovers that it can put one block on top of the other; it constructs. The new discovery leads to repetition and variation, for each child develops, at some stage, a desire and a set of *Sollwerte* for such effects of construction, and acts out to the full this new level of exploratory behavior. In addition, already at this stage the *Sollwert* or ideal does not merely contain what the blocks do, but also what, for instance, the mother does; her approval, her shared enjoyment, is also of great importance. Just as an exploring animal, the child builds a kind of inverted pyramid of experience, built of layers, each set off by a new wave of exploration and each directed by new sets of *Sollwerte,* and so its development "snowballs." All these phases may well have more or less limited sensitive periods, which determine when the fullest effect can be obtained, and when the child is ready for the next step. More important still, if the opportunity for the next stage is offered either too early or too late, development may be damaged, including the development of motivational and emotional attitudes.

Of course gifted teachers of many generations have known all these things (*25*) or some of them, but the glimpses of insight have not been fully and scientifically systematized. In human education, this would of course involve experimentation. This need not worry us too much, because in our search for better educational procedures we are in effect experimenting on our children all the time. Also, children are fortunately incredibly resilient, and most grow up into pretty viable adults in spite of our fumbling educational efforts. Yet there is, of course, a limit to what we will allow ourselves, and this, I should like to emphasize, is where animal studies may well become even more important than they are already.

Can Education End Aggression?

Returning now to the development of animal and human aggression, I hope to have made at least several things clear: that behavior development is a very complex phenomenon indeed; that we have only begun to analyze it in animals; that with respect to man we are, if anything, behind in comparison with animal studies; and that I cannot do otherwise than repeat what I said in the beginning: we must make a major research effort. In this effort animal studies can help, but we are still very far from drawing very definite conclusions with regard to our question: To what extent shall we be able to render man less aggressive through manipulation of the environment, that is, by educational measures?

In such a situation personal opinions naturally vary a great deal. I do not hesitate to give as my personal opinion that Lorenz's book *On Aggression,* in spite of its assertativeness, in spite of factual mistakes, and in spite of the many possibilities of misunderstandings that are due to the lack of a common language among students of behavior – that this work must be taken more seriously as a positive contribution to our problem than many critics have done. Lorenz is, in my opinion, right in claiming that elimination, through education, of the internal urge to fight will turn out to be very difficult, if not impossible.

Everything I have said so far seems to me to allow for only one conclusion. Apart from doing our utmost to return to a reasonable population density, apart from stopping the progressive depletion and pollution of our habitat, we must pursue the biological study of animal behavior for clarifying problems of human behavior of such magnitude as that of our aggression, and of education.

But research takes a long time, and we must remember that there are experts who forecast worldwide famine 10 to 20 years from now; and that we have enough weapons to wipe out all human life on earth. Whatever the causation of our aggression, the simple fact is that for the time being we are saddled with it. This means that there is a crying need for a crash program, for finding ways and means for keeping our intergroup aggression in check. This is of course in practice infinitely more difficult than controlling our intranational frictions; we have as yet not got a truly international police force. But there is hope for avoiding all-out war because, for the first time in history, we are afraid of killing ourselves by the lethal radiation effects even of bombs that we could drop in the enemy's territory. Our politicians know this. And as long as there is this hope, there is every reason to try to learn what we can from animal studies. Here again they can be of help. We have already seen that animal opponents meeting in a hostile clash avoid bloodshed by using the expressions of their motivational conflicts as intimidating signals. Ethologists have studied such conflict movements in some detail (*26*), and have found that they are of a variety of types. The most instructive of these is the redirected attack; instead of attacking the provoking, yet dreaded, opponent, animals often attack something else, often even an inanimate object. We ourselves bang the table with our fists. Redirection includes something like sublimation, a term attaching a value judgment to the redirection. As a species with group territories, humans, like hyenas, unite when meeting a common enemy. We do already sublimate our group aggression. The Dutch feel united in their fight against the sea. Scientists do attack their problems together. The space program – surely a mainly military effort – is an up-to-date example. I would not like to claim, as Lorenz does, that redirected attack exhausts the aggressive urge. We know from soccer matches and from animal work how aggressive behavior has two simultaneous, but opposite effects: a waning effect, and one of self-inflammation, of mass hysteria, such as recently seen in Cairo. Of these two the inflammatory effect often wins. But if aggression were used successfully as the motive force behind nonkilling and even useful activities, self-stimulation need not be a danger; in our short-term cure we are not aiming at the elimination of aggressiveness, but at "taking the sting out of it."

Of all sublimated activities, scientific research would seem to offer the best opportunities for deflecting and sublimating our aggression. And, once we recognize that it is the disrupted relation between our own behavior and our environment that forms our most deadly enemy, what could be better than uniting, at the front or behind the lines, in the scientific attack on our own behavioral problems?

I stress "behind the lines." The whole population should be made to feel that it participates in the struggle. This is why scientists will always have the duty to inform their fellowmen of what they are doing, of the relevance and the importance of their work. And this is not only a duty, it can give intense satisfaction.

I have come full circle. For both the long-term and the short-term remedies at least we scientists will have to sublimate our aggression into an all-out attack on the enemy within. For this the enemy must be recognized for what it is: our unknown selves, or, deeper down, our refusal to admit that man is, to himself, unknown.

I should like to conclude by saying a few words to my colleagues of the younger generation. Of course we all hope that by muddling along until we have acquired better understanding, self-annihilation either by the "whimper of famine" or by the "bank of war" can be avoided. For this, we must on the one hand trust, on the other help (and urge) our politicians. But it is no use denying that the chances of designing the necessary preventive measures are small, let alone the chances of carrying them out. Even birth control still offers a major problem.

It is difficult for my generation to know how seriously you take the danger of mankind destroying his own species. But those who share the apprehension of my generation might perhaps, with us, derive strength from keeping alive the thought that has helped so many of us in the past when faced with the possibility of imminent death. Scientific research is one of the finest occupations of our mind. It is, with art and religion, one of the uniquely human ways of meeting nature, in fact, the most active way. If we are to succumb, and even if this were to be ultimately due to our own stupidity, we could still, so to speak, redeem our species. We could at least go down with some dignity, by using our brain for one of its supreme tasks, by exploring to the end.

References

1. A. Carrel, *L'Homme, cet Inconnu* (Librairie Plon, Paris, 1935).
2. AAAS Annual Meeting, 1967 [see*New Scientist* **37**, 5 (1968)].
3. R. Carson, *Silent Spring* (Houghton Mifflin, Boston, 1962).
4. K. Lorenz, *On Aggression* (Methuen, London, 1966).
5. D. Morris, *The Naked Ape*, (Jonathan Cape, London, 1967).
6. N. Timbergen, *Z. Tierpsychol.* **20**, 410 (1964).
7. R.A. Hinde, *New Society* **9**, 302 (1967).
8. W.S. Hoar, *Animal Behaviour* **10**, 247 (1962).
9. B. Baggerman, in *Symp. Soc. Exp. Biol.* **20**, 427, (1965).
10. H. Kruuk, *New Scientist* **30**, 849 (1966).
11. N. Tinbergen, *Z. Tierpschol.* **16**, 651 (1959); *Zool. Mededelingen* **39**, 209 (1964).
12. — — — *Behaviour* **15**, 1-70 (1959).
13. L. von Haartman, *Evolution* **11**, 339, (1957).
14. E. Cullen, *Ibis* **99**, 275 (1957).
15. J.H. Crook, *Symp. Zool. Soc. London* **14**, 181 (1965).
16. D. Freeman, *Inst. Biol. Symp.* **13**, 109, (1964); D. Morris, Ed., *Primate Ethology* (Weidenfeld and Nicolson, London, 1967).
17. H.E. Howard, *Territory in Bird Life* (Murray, London, 1920); R.A. Hinde *et al., Ibis* **98**, 340-350 (1956).
18. K. Lorenz, *Evolution and Modification of Behaviour* (Methuen, London, 1966).
19. T.C. Schneirla, *Quart. Rev. Biol.* **41**, 283 (1966).
20. M. D. Knoll, *Z. Vergleich. Physiol.* **38**, 219 (1956).
21. E. Cullen, *Final Rept. Contr. AF 61 (052)-29,* USAFRDC, 1-23 (1961).

22. W.H. Thorpe, *Bird-Song* (Cambridge Univ. Press, New York, 1969).

23. E. von Holst and H. Mittelstaedt, *Naturwissenschaften* **37**, 464 (1950).

24. M. Konishi, *Z. Tierpsychol.* **22,** 770 (1965); F. Nottebohm, *Proc. 14th Intern. Ornithol. Congr.,* 265-280 (1967).

25. E.M. Standing, *Maria Montessori* (New American Library, New York, 1962).

26. N. Tinbergen, in *The Pathology and Treatment of Sexual Deviation,* I. Rosen, Ed. (Oxford Univ. Press, London, 1964), pp. 3-23; N.B. Jones, *Wildfowl Trust 11th Ann. Rept.,* 46-52 (1960), P. Sevenster, *Behaviour, Suppl. 9,* 1-170 (1961); F. Rowell, *Animal Behaviour* **9**, 38 (1961).

14.

The Possible Biological Origins of Sexual Discrimination

Lionel Tiger

But what has all this to do with women today? AUTHOR

The Biological Basis of Behaviour

For too long many social scientists have paid inadequate attention not only to the work and thought of behavioural biologists but also to the biological aspects of human behaviour. Sociologists, political scientists and economists have used a model of man in which biological factors were unimportant, or at least residual; no ready means of assimilating information about other living systems than the human has been available to researchers into human social action.

It thus remains the case (though this will change rapidly in the next decade) that undergraduate and post-graduate students in the social sciences almost never are required to demonstrate any serious proficiency in biology. Though they may be subjected to an immense and sophisticated burden of methodological and mathematical instruction and they will learn very well indeed how to collect and handle data, they are hardly likely to learn what data to gather in the first place – in the sense that students of other animals are guided by biological principles to determine what behaviour patterns and what events in the lives of animals reflect a species' central problems, concerns and adaptations.[1]

One serious social and scientific consequence of this is that – as usual – women have suffered particular deprivation. Because of the blurring of biological distinctions, even the categories "maleness" and "femaleness" have been inadequately treated – perhaps much less than the less significant and less provocative categories "rich" and "poor". The overt and surprising reason for this is that it has been tacitly assumed that in all spheres except the explicitly reproductive males and females are much the same. To suggest otherwise implies biases of a quasi-racial kind. Moreover, there was also an implication that female equality in practice might be eroded by the assumption of female difference in theory.

My argument in this essay is that it is necessary to see biological factors as of prime importance in discussing maleness and femaleness. This involves understanding human evolution, human neurophysiology, cross-cultural regularities – in general, the biological infrastructure of human social relationships. Knowing as we do the importance of socio-sexual differences in many mammalian reproductive systems, perhaps it will not surprise if, employing this perspective, we learn that

Reprinted by permission of the author and publisher from *Impact of Science on Society, Vol. XX,* No. 1, 1970

[1] For an American illustration of this point see: 'Written Doctoral Examinations in Sociology at Some Leading Universities", *The American Sociologist,* Vol. 4, No. 3, August 1969, p. 189-226.

we are a species boasting considerable male-female differences which extend and ramify more widely through our human societies than many of our theories of socialization, organization and action in general would have us recognize and predict.

In this brief essay I want to comment about differences in male-female social organization. The suggestion is that, not unreasonably, these social differences bear relationship to biological realities of the human species which in turn are functions of human evolution and of a particular set of primate adaptations; these adaptations include both those uniquely our own and those we share with some other primates (with whom we find we share behavioural characteristics, just as earlier we were surprised to find that we share significant physical ones). The broad outlines of the argument are more fully presented in my book *Men in Groups In Groups*[2]; here, I want to discuss the significance this argument may have for our social theories and action.

In essence, the position derives from the following set of ethnological observations and theories. The behaviour of animals evolves just as physical structure does, and both structure and function (which is a systematic way of saying behaviour) reflect animals' evolutionary adaptation to their material and social environments. This adaptation is always mediated by processes of sexual selection which Darwin first explicitly described. The social behaviour of animals is, hence, not altogether sudden. It is not, in other words, only the expression of particular local and ephemeral circumstances. Even relatively complex social propensities can be programmed genetically – just as the extraordinarily complex life cycle, for example, is programmed by genetic codes and finds its expression in a reciprocity between the particular "programmed energy" of an organism and the concatenation of habits, practices, situations and likelihoods which is a social culture.[3]

As far as humans are concerned, the problem of isolating genetically determined, behaviour patterns is magnified because we are fond of creating elaborate and sometimes secretive social systems.

Hunting, the Master Pattern

It could be said that our most elaborate biological adaptation is to create culture. But if culture is our most complex concoction, it remains unchallenged by anthropologists and biologists that hunting is the master pattern of the human species. It is the organizing activity which integrated the morphological, physiological, genetic and intellectual aspects of the individual human organisms and of the population who compose our single species. Hunting is a way of life, not simply a "subsistence technique", and it involves commitments, correlates and consequences spanning the entire biobehavioural continuum of the individual and of the entire species of which he is a member.

"That man achieved a worldwide distribution while still a hunter reflects the enormous universality of this kind of behavioral adaptation . . . he practiced hunting for 99 percent of his history . . . ."[4]

In other words, our practice of hunting was the infra-structural condition of our specialized evolution, and though it is tempting to see in this only a confirmation of theories of human bloodiness and evil,

[2] Lionel Tiger, *Men in Groups,* Random House, New York; Thomas Nelson, London, 1969; Robert Laffont, Paris, in preparation in French.

[3] For a succinct description of these principles, see: Konrad Lorenz, "The Evolution of Behavior," *Scientific American,* December, 1958.

[4] William S. Laughlin, "Hunting: An Integrating Biobehavior System and Its Evolutionary Importance," in: R.B. Lee and E.I. DeVore(eds.), *Man the Hunter,* p. 304, Chicago, Aldine, 1968.

it remains the case that hunting was a co-operative activity and that in the acts of pursuit and slaughter there was selective advantage to those individuals able to work together, attuned to each others' needs, resources and states, and willing to mould their individual behaviour to the collective pattern of the group.[5]

The critical subordinate section of this argument (for our purpose here) is that hunting was an all-male enterprise and that just as there was selection *for* co-operative hunting males, there was selection *against* both those females willing to hunt and those males agreeable to female participants in their hunts. The reasons for this proposition depend on a variety of individually disputable bits of evidence. But their over-all implications seem to point forcefully in the direction of an increased differentiation of male-female behaviour through evolution at the same time as there was probably a decreased physical differentiation.

Of course, differences in running methods, throwing skills, temperature adaptability, effects of physiological changes, etc., were clearly both the cause of and the effect of differentiated hunting experience. At the same time, the intriguing hypothesis remains that an important feature of this behavioural differentiation was the development of differing interests in and capacities for social bonding. Just as selection for reproduction operated by establishing and consummating bonds between males and females (and this was broadly programmed — for example, to follow puberty), so selection as a result of hunting depended on the readiness of the organisms concerned to form male-male bonds for these purposes while at the same time rejecting male-female ones during the hunting period. Presumably this applied too during those preliminary and celebratory-recriminatory events which were probably essential to the prosecution of strenuous tasks involving considerable technical and co-operative skills.

The fact that we have been a hunting species for probably one million years and possibly up to 26 million underlines the significance of this hypothesis of male-male linkages for the establishment of species-specific patterns of behaviour. These must remain influential today, if only as parameters in terms of which the force and effect of the patterns may be altered or vitiated by cultural factors and by the existential circumstances of particular individuals.

Biological Determination of Women's Role

But what has all this to do with women today?

At one level the role of biology in constraining the lives of women is obvious. Reproduction involves pregnancy and at least a short period of commitment to highly demanding offspring whose physical needs are not only inconsiderately recurrent but also who make psychological claims on mothers' time and energy. These claims can be avoided only by: (a) inadequate mothering, whose effects on the entire life span of the offspring we are now beginning to appreciate not only from data on humans but in essence from such studies on non-human primates as the Harlows'; (b) by a system of servants which relieves mothers of many of the burdens of rearing children (but not of having and dealing with servants); (c) a communalized system of child-rearing ranging from the kibbutz to a system of crèches and day nurseries — the economic rationale for this, however, is not only to make it easy for women to have children but to make it possible for them to work

[5] A recent discussion of this form of co-operation is by Robert Bigelow, *The Dawn Warriors: Man's Evolution Toward Peace,* New York, Atlantic-Little, Brown, 1969.

in the economy more directly than as managers of private households and mothers; and finally (d) a cultural system in which children are raised by kin of either mother or father (social or biological) but in which, in any event, women do raise children at various points in their lives, even if these are not their own offspring.

This much is obvious, and though societies respond in various ways to the problem of aiding and mollifying the stresses of child-rearing, it remains the case that this process constitutes a real (if widely approved) impediment to following the same career pattern as men. A number of communities have provided for this by offering facilities for mothers which allow them time off from work – either on a short-term or long-term basis – opportunities for retraining after an absence, the maintenance of pension and other rights, etc., and the retention of "equity" in a career line in an organization, industry, or in the community at large. Despite these provisions, however, females typically constitute a more floating segment of the labour market, work at lower rates of pay and in lower-level positions, and are more likely to be dismissed by organizations retrenching their personnel.

In general, it remains the case that females, in virtually every society, are to a large degree excluded from positions of power and substantial reward. They are clearly subdominant and even where they are educated as well as males and possess equal economic resources, they fail to achieve posts, properties or honorific awards in any degree comparable to those of males. Moreover, they find themselves largely outside the major political, economic and military decision-making processes of our time.

Of course, this is not to recommend this situation, but to identify once again the gap between the ideology of sexual equality and the reality of an only tentative and sporadic movement to this equality.

My point is that the reason for this hiatus between wish and reality is not simply the result of male churlishness, chauvinism or fear. Nor is it similar to the differences between the privileges and opportunities of different economic classes or races in stratified societies. Nor is it solely the result of a coercive process of socialization which condemns disadvantaged females to live equably and with misguided self-satisfaction. Simple-minded though it may seem, perhaps the difficulty females have faced arises in good measure because the rhetoric and the dream of equality have allowed communities to avoid coming to terms with real differences between the sexes – differences which go beyond the explicitly reproductive, and which have to do with the conduct of social affairs on even its most abstract and complex levels.

To help understand this suggestion in its proper context, it may be useful to look briefly at the social behaviour of other primate males and females, bearing in mind that we are primates ourselves – though different from all the others – and that it is possible that we share some ancient core patterns of "genetically programmed behavioural propensity" just as we clearly have in common certain evident physical structures and processes.

Primate Sex Bonds and Group Structure

An obvious feature of primate behaviour to which primatologists pay attention early in their research is male-female differentiation. In some species there is relatively little, except for the bearing and immediate rearing of children. This is particularly so among arboreal creatures for whom the problems of defense are more easily solved by fleeing up a tree than by generating defensive social organization. On the other hand, among terrestrial groups such as the south African baboon, the demands of defense on the savannah have led to the evolutionary selection of males about twice as large as females, with large jaws and sharp canines, and in general physically equipped to defend the females and young of their communities.

A correlate of these defense patterns – which so far as we know involve males only – is that the males, both the dominant ones and the sub-adult ones, form "bonds" – groups in which the individuals regard themselves as more significant to each other than to non-bond members. These are, in a real sense, "personal" relationships, as distinct from aggregation-type encounters in which there is no real element of choice.

What is fascinating about these male bonds is that they seem to be associated with political dominance. This is, in turn, directly linked with the dominant males having the greatly preponderant sexual access to oestrus females. Thus, there is the clear implication that selection of males willing and able to form bonds with other males is a constant feature of these primates' reproductive function.

This introduces quite a new element into the whole matter of social bonds and their relationship to reproduction. Not only does a male animal have to want to and be capable of consummating an encounter with a fertile female, but to reproduce himself he must also be able to engage in relatively very long-term and elaborate social relationships with several other males. Though sub-adult males do form weak bonds (rather like human boys' gangs – and, as in the human, there are no female gangs), only dominant adult males seem to be able to form these with political effectiveness. The subdominant males appear to be incapable of forming strong bonds because even four or five of them are unable to combine to overthrow the leadership of the two or three dominants from whose bonds a great deal of super-individual power is generated.

Thus a picture emerges from generalizing about the terrestrial primates; it features the importance of bonds among males for the process of selection. An additional point of equal relevance is that the stability, order and defense of the community depend on the male-bonded individuals: politics and reproduction are closely linked. *Hence the Darwinian processes of natural selection involve a combination of sexual competence with females and social competence with males.* This in turn appears to stabilize communities, provide models for the young males, and seems, indeed, to conduce to the "health" of females as well as dominant males.

(In one as yet unpublished study, it is noted that in a group of rhesus monkeys in which there was no male, the females were incapable of "governing" the group and social tension and disorganization were constant. The introduction of but one adult male into the group corrected the situation immediately, and a more normal political and social pattern quickly returned.)

What is relevant in all this to our concern here is that primate females seem biologically unprogrammed to dominate political systems, and the whole weight of the relevant primates' breeding history militates against female participation in what we can call "primate public life."

This is not only to say that female primates have no social bonds. Of course they do. First of all, they form intense bonds with their offspring and this bond is as crucial to group survival as the male-male bond seems to be.

Moreover, as we get more and better data about primate life it becomes clearer that there is something similar among some primates to "kinship systems" among humans.[6] Particular individuals born of certain females recognize certain inter-generational relationships between each other, and in terms of the group as a whole there is a tendency for the offspring

[6] Vernon Reynolds, "Kinship and the Family in Monkeys, Apes and Man," *Man*, Vol. 3, No. 2, June 1968.

of high-status females to become high-status themselves. Rudimentary but functioning "class structures" then appear to arise – not in connexion with any particular property or other resource but in terms of social relationships themselves.

That the young of dominant females should be more competent, more confident, and more capable of approaching the dominant males on whom "advancement" depends, should not surprise us altogether. But that this is also a function of rather elaborate group-kinship structures suggests that even such complex patterns as these may be broadly rooted in a biological foundation and that the contribution of females to these systems is meaningful not only for their offspring but for the entire group and – presumably for the species as a whole.

Further, that this female participation seems to tend toward the formation of stratified rather than egalitarian communities must be an item to consider among those schools of thought which are based on the broad belief that more primitive states of society – such as animals enjoy – are open and pleasant by comparison with human ones, and that the contribution of females to social procedures must be generally wholesome, egalitarian, and contrary – so runs the diagnosis – to the dismal and unhappy hierarchical structures which males create and endlessly refine.

The Genetic Foundation of Masculine Dominance

Let us return to the human case directly. In humans the bonding propensity of males – if it exists – would have been given an additional evolutionary emphasis by the function of hunting. It is important to remember that among non-human primates there is little if any differentiation between the sexes in the food-gathering activity. This is crucial, because if hunting in the human species was for males only, then a pre-existent male bonding pattern which we might have inherited along with the primates may have been strongly and unambiguously accentuated by our special human innovation of co-operative hunting. In other words, while in other primates the sexual division of labour had chiefly to do with defense and politics, in the human case this was expanded to include economics too, and herein may lie some of the resistance which human communities appear to show still to even the most sophisticated and ardent efforts to achieve sexual equality.

Every human community displays some sexual division of labour. The allocation of tasks may vary enormously. In one society a particular job will be for men and in another the same job for females. Some jobs will be done by both. But the significant regularity is that there is always some distinction between male and female work for some jobs and on some grounds. Sometimes these are linked to obvious physical factors: they involve speed, danger, muscular strength, etc. None the less, there is often no technological justification for the sexual distinction, and one is driven to the conclusion that the pattern of sexual division of labour may relate not only to real differences in skills, aptitudes and interests, but to a core pattern of the human primate: that in some circumstances, particularly those defined as dangerous, important for the community, or involving matters of high moment, males will exclude females from their groups and engage in male bonding undisturbed.

That this may be both a deliberate and an infra-social, broadly unconscious pattern–in the same sense as the male-female bond, based on sexual attraction and reproduction, is both conscious and infra-social – underlines the difficulty of doing something about this; it aggravates the difficulty of knowing precisely how to go about obtaining female equality in the labour, political, and associated spheres.

In other words, I am suggesting that a species-specific pattern of *Homo sapiens* is the creation of particular bonds between males, that these bonds are intrinsically related to political, economic, military, police and other similarly power- and dominance-centered social subsystems, that equal female colleagues – even one – could interfere with these bonding processes, that one reflection of this principle is the constant division of labour by sex, and that while conscious social management of these processes may of course alter or reverse them, the propensity to behave in this way will continue to manifest itself in each new generation until genetic change "breeds it out" – a process which even under current circumstances is very improbable in any foreseeable future.

Of course, all this is impossible to prove in the sense that an exact and reproducible cause gives rise to an exact and reproducible effect. However, biologically speaking, a species is an experiment without a control group – except in so far as it bears systematic and acceptable comparison with other species[7] – and in the human case we can use cross-cultural data to point toward items of behaviour which are common to all cultures, thus species-specific, and those which are clearly culture-specific. Hence, I have argued that the ubiquity of the male dominance and female exclusion patterns which can be identified is a serious indication of the possibility that these patterns may originate in our genetic codes, and in the interactions between genetic code and social group and particular circumstances in which individual codes work themselves out.

It should not surprise us that maleness and femaleness as biological categories have elaborate effects on even complex technologically based behaviour. Both are clearly biological features of the core of individual beings, and while there are many similarities between males and females, it is scientifically parsimonious to attend to the possibility that behavioural differences in other spheres are as significant as those in the reproductive.

At the outset of this essay I suggested that social scientists have paid inadequate attention to biological science and noted that this was particularly marked in the matter of male-female differences. It may be of interest to try to overcome this inadequacy and try to follow out some of the consequences of a revised view of control-by-genetic-process of broad social patterns and particularly of the effect of genes on the situation of women as well as on our attitudes.

How Theoretical Equality Causes Factual Inequality

My basic proposition takes the form of a paradox: that the understandable and universally acceptable notion that males and females are equal and should have all equal rights of law, economy, politics, etc., has contributed to the practical inequality of females.

Theoretical sexual equality has forced rejection of any concern about sexual differences. The practical result of this has been the continued deprivation of females and the slowing-up of a process of opening opportunities to women in present structures, of changing the structures and of adding new ones to accommodate women. At the moment, it is women who must accommodate themselves, and they are being asked to compete with men in male-oriented institutions. The net result of this is their continued deprivation and a recently increased resentment and anxiety.

A number of obvious examples come

[7] Niko Tinbergen, "On Aims and Methods of Ethology," *Zeitschrift für Tierpsychologie,* Vol. 20, No.4, 1963.

quickly to mind. A variety of researches have confirmed what many other less-sophisticated communities have known all along – that the female menstrual cycle has some appreciable and predictable effects on female social, psychological and even technical behaviour. Crime rates, industrial accident rates and incidence of illness, for example, have been correlated with the regular cycle.

A recent report by K. Dalton of the University College Hospital, London, reveals that young women writing examinations are affected by as much as 14 per cent by the time of their cycle at which they undertake some tests. The implications of this simple finding are of course enormous. For example, persons wishing to enter graduate school in the United States of America must take special examinations on a national standard. Should a woman write these during her low-performance time, she begins with virtually a second-class result and the work of her previous years in the educational system and her own personal qualities and skills may be severely devalued. In good part this is because she participates in a system which does not formally recognize her femininity, admitting that it may be the cause of changes in behaviour or performance of direct pertinence to the educational system which has so expensively provided her the opportunity for seeking graduate training.

The same effect will operate less dramatically but with persistent consequences throughout a young woman's educational career; it must also retain its impact during her functions in some job. This can become serious, not only for the individual woman's well-being and occupational success, but for the clients of her particular service or effort. The effect will be more clearly exposed as women increasingly perform tasks involving the exercise of technological judgement upon which depends the safety of other people. The relationship of the cycle to motor-car accidents has been pointed out; it seems inevitable that should women become airline pilots it will be necessary for their work schedules to conform to their biological rhythms – not because accidents are inevitable, but because they become somewhat more likely and hence a risk subject to control by conscientious managers.

Now, the writing of examinations and the flying of airliners are two rather extreme contrasts. But in both cases there is sufficient suggestion of the effects of the male-female difference on performance for sensitive communities to consider ways of mitigating the consequences of these for individual females and for the community at large. It seems likely that communities willing to take these factors into account will respond more suitably and immediately to situations involving real danger and the use of expensive artifacts such as aircraft and heavy industrial machinery.

The subtler and less tangible matter of scheduling educational, commercial, and other activities to take individual women's cycles into account seems much less likely to be implemented easily. Perhaps the two outstanding reasons for this are: (a) to do so would be to recognize formally and overtly real differences between males and females – something which communities have in a curiously successful and tenacious way managed to avoid doing, in the name of the ethic of equality, and (b) there appears to remain a widespread taboo of more or less severity against the formal statement by both men and women of the fact and occurrence of menstruation, though this depends on the attitude of the community involved, of course. It is not clear to what extent this is a function of females' desire to maintain some privacy in this respect, or of males' resentment and perhaps even fear of a process.

A study using the Human Relations Area Files by Young and Bacdayan describes an unusually bizarre correlation between political authoritarianism and the strength of taboos against various degrees of contact with menstruating females.[8] This suggests that there is considerable variation in this matter (as in any other) between communities and among individuals. It also portends that communities with relatively liberal political attitudes may be more likely to openly acknowledge the existence of the menstrual cycle and cope with its consequences. Yet, at the same time, in some such liberal communities – United Kingdom, the United States and Canada, for example – there is a particular reluctance to deal with such a female-specific matter as the cycle. Perhaps this is indeed for the first reason given above, namely that such an expression of difference might be construed as an expression of inferiority.

The particular reason for stressing the effect of menstrual cycling on work performance is that it is both relatively clear-cut and an excellent example of the general fact that the work patterns of industrial communities in particular are male-oriented. The 7-to-9-hour 5-to-6-day pattern of work of course represents the densest and apparently most efficient way of organizing the time of individual employees.

My point is very simple: were work adjusted to female propensities, as it is now to those of males, a more humane and effective "fit" between system and individual could follow.

The same principle applies to the relationship of working mothers and children. It is customary that those employees with full privileges are full-time employees. Again, this is defined in male terms. However, there is no special reason, beyond habit, inertia, the reluctance to face complexity, etc., why mothers cannot be treated as full-time employees with full privileges – in proportion to their contribution – even if they work only a few hours a week, or one or two days, or three half-days, or according to any other arrangement which permits them to spend as much time with their young as they regard to be necessary, while participating in the wider tasks available in paid employment.

At the moment, in technologically elaborate communities it is chiefly the unpaid volunteer lady worker who is permitted a flexible participation in the socio-economic network. If such arrangements can be made for volunteer employees, it is not inconceivable that they can be made, too, for paid employees. This is not to say that no part-time job possibilities are open to females, for of course there are. But such employment is always secondary in importance and individuals are discouraged from seeing their work as part of a continuing career; the very fact of their responsibility to their families permits them only a partial commitment to their employers.

The same comment applies to phasing of female work over years as well as days. The community spends great sums of money educating women to undertake jobs which the rigidity of its own structure makes it difficult for them to assume and maintain should they wish to bear children and spend considerable periods of time with them for several years or until they enter school. Similar problems of pension rights, seniority, retraining, and continuity arise for the mother whose working career is interrupted as for the part-time worker.

It is curious that at the same time as various organizations and governments claim difficulty in attracting and retraining committed and competent skilled employees, systematic and seemingly insurmountable barriers are placed in the way of the largest single pool of available personnel to fill these posts. And not only do the pertinent organizations themselves lose,

[8] F.W. Young and Albert A. Bacdayan, "Menstrual Taboo and Social Rigidity," *Ethnology*, Vol. 4, No. 2, April 1965.

but so does the community in general because of the resentment, confusion and conflict of loyalties between past training and present situation which more and more females experience simply because more and more women are being elaborately and carefully educated.

Male Bonds and Female Exclusion

Menstrual cycles and child-rearing are very obvious factors in any effort to assess the reality and possibility of female participation in what we can call the "macrostructures of society" – those involving large-scale organization and the major corporate enterprises. More subtle but perhaps even more significant influence are the essentially primate valences – bonding tendencies – to which I referred earlier. I mean by this that the tendency for males to form bonds in work, fighting, politics, etc., and the more obvious but equally pervasive propensity for males and females to form at least ephemeral bonds centering round the process of sexual titillation and consummation, may be as formidable barriers to egalitarian female employment as the obvious ones directly related to reproduction. And an additional factor which has been often overlooked both by ethnographers and students of contemporary politics and economics is the apparent difficulty females have in forming the bonds necessary in order to manage structures involving power and wealth. Let us explore these factors in turn.

It should by now be clear that the proposition here is that if males bond because it is "in the nature of the beast" to do so, then this places a considerable burden both on women seeking to join these bonds, and on those men willing to allow females into groups when this may signally affect the groups and the relations between group members.

One intriguing example of this phenomenon is the secret society; only exceptionally are these heterosexual. They are mostly all-male and when women do join them, this appears to mark the end of the society's particular drama and effect on its surroundings. This is, again, not to recommend a particular attitude or policy towards secret societies. But it is to suggest that they express certain effective propensities of the human male and that we can observe in these curious and unpredictable organizations a feature of male behaviour in the voluntary world which may find more formal expression in the more overt and legislated worlds of business, politics, etc.[9]

Team sport is another example of this phenomenon; with the exception of tennis and skating, team sports are overwhelmingly unisexual. Except for the most violent of sports, there is no reason why rules governing female participation could not be introduced and appropriate numbers of women join teams. But I suggest that team sports depend on the bonding process and that female participation is anathema to this and would severely curtail the enjoyment of both players and spectators because of the disturbance of the male sociodramaturgy of sport.

Perhaps both secret societies and sports will be seen by some as a wretched and retrogressive failure of males to embrace a modern and wholesome sexual egalitarianism. It may also be sensible to see them as rather complex projections of species-wide (and their incidence is species-wide) propensities for male bonding. In any event, they do exist (though secret societies are under considerable pressure, particularly in the United States) and represent a clue to what many men are concerned about and to the ways in which they willingly choose to spend their time.

[9] See Chapter Six, *Men in Groups,* op. cit., for a more detailed discussion of secret societies and male courtship and initiations.

If we apply the principles of these forms of association to other areas we can see that the rejection of female coworkers by males may stem from more than retrogressive pique, prejudice, lack of sympathy for females, or some other impetus regarded as malignant and uninformed. This means that coping with this rejection may involve dealing with subconscious which are possibly of ancient primate origin and which have for several million years and until just recently served very well for political and economic survival. What we may call "the anti-female tradition", has its origins, then, not only in belligerent male chauvinist ideology and, in economic exploitation of females, but in a genetic process which evolved because pre-hominids found they could survive and reproduce better if they excluded females from the processes of political dominance, with survival further aided by the exclusion of females from the hunting party.

The Disruptive Male Female Bond

One obvious reason for such exclusions has to do with the male-female bond. This linkage, we know, can be extremely passionate and vexatious and it is of endless interest to observers of both fictional and real stories of romance. What Desmond Morris has called "the pair bond" can be as ardently and obsessively maintained as any known among humans. While it can merge the lives of a male and female, it can also disturb class, religious and other social systems. Such is its strength that even the most persuasive rules of these systems are subject to disruption by a male and female caught in the bonding encounter. Yet while this bond can be intense it can also be ephemeral; this is the "crush," the "affair", the sporadic, problematic attraction of a male and female who presumably do not marry (though they may reproduce), some form of marriage being, presumably, the social end-product of this private valence. It is the ephemerality which is of interest to us here as we consider the role of females in male-dominated structures.

First of all, one may ask Desmond Morris: what about the "*au pair* bond", that common idyllic dilemma of the bourgeois male who finds his home occupied by a young woman with whom he is forbidden to engage sexually despite her proximity and possible attractiveness? After all, she was imported to solve his wife's problems, not add to them. While this is a frivolous example, it none the less underscores a particular feature of the male-female bond: that it is potentially disruptive, volatile, unreasonable, demanding, and very possibly hostile to the purposes which brought the male and female into contact in the first place. This is particularly so where the individuals concerned are relatively young and where they must spend extended periods of time together.

The implications of this for the conduct of socio-economic and political activities are obvious. One of the intriguing features of business organizations and similar structures in Europe and North America is the lack of formal recognition of the problem posed by the pair-bonding impctus. There is, of course, and undergrowth of awareness. Private taboos or understandings abound: teachers may not "pair-bond" with their student though students of other instructors may be more acceptable; and students at other institutions – why not?

The same rules of prohibited organizational incest are supposed to apply to relations between doctors and nurses, managers and secretaries, scientists and research assistants, and so on. That these are honoured in the breach as well as the observance suggests to some degree the difficulty of governing a highly charged primate pattern by the relatively fragile informal and even formal rules of organizations. And all this also suggests one further disadvantage which women face in their effort to secure comradeship with men in places of work – given the extant male-dominated structure.

The pair-bonding problem is soluble, with whatever strenuous efforts and accommodations that may be necessary, when the encounter is between unequal individuals. Doctors and nurses, actresses and producers, stars and starlets, can perhaps deal better with this disruption of formal relations than individuals in equal or near-equal positions.

Robin Fox has suggested that polygamy is not really about sex but about dominance, that sexual access is more important than sexual action.[10] His point is that polygamy may reflect human retention of the primate pattern where males dominate females and express this dominance in sexual activity. This point is, of course, contentious. But my argument remains that males and females of appropriate age and demeanour will seek some sexual *rapprochement* and that the conflict between formal colleagueship and differential sexual power may be disruptive in work and politics and other large-scale systems.

All this may seem far-fetched, and one dares broach this particular subject in this particular fashion only because of its obvious conflict with the demands of organizations for members to behave in formally sanctioned ways. Understandably, the introduction of this particular manifestation of sexual energy into organizations becomes more hazardous the higher in the hierarchy are the individuals concerned. Perhaps it is this combination – sex as a feature of male dominance, and sex as disruptive of relations between powerful equals – which augments the traditional cultural prejudice in very many communities against the installation of women in posts of power.

[10]Robin Fox, "The Evolution of Human Sexual Behavior", New York Times Magazine, 24 March 1968.

The Female-Female Bond

Before concluding, I must make several remarks about another kind of bond, the female-female bond. The groups which females form appear to occupy themselves less with macro-structures of society than with micro-structures. In general, they seem to be less persistent over time and involve fewer people in association having less organizational and technical complexity, both in humans and the other primates.

Recently there have been several sets of data published about primate groups in which female-managed "kinship structures" are significant for the continuity of primate bands across several generations, and even for the general guidance of the process of selection of male leaders.[11] Perhaps further research may reveal the female bonds that exist in the other primates, and that these serve evolutionary functions, particularly in the critical sphere of rearing the young and, through socialization, establishing the continuity of the future with the present. Be that as it may, the fact is that in the human case there is a highly significant division of food-gathering labour – as distinct from non-human primates, among whom all able individuals find their own food – and this suggests that the factor of male bonding among humans may well have been more important than among the other (vegetarian) primates.

Should we find that primate females specialize in social organization and the maintenance of kinship patterns, this could accord with human data about the centrality of the domestic arena for female

[11]Reynolds, "Kinship and the Family in Monkeys, Apes and Man", op. cit.

action. Necessarily, the facts of reproduction govern this at certain stages of the life cycle. Thus, there is the real possibility that selection has always favoured females who have surrounded themselves and their children with a group of kin and other females who provide information, security and the simple and necessary ease of social contact.

As much as solitary hunters face a difficult task, so do solitary mothers. In so far as biological processes can select for such gregarious characteristics, it is possible that a female bonding propensity does exist, with a focus on the relatively intimate matters of families and children rather than on the macro-structures involving war, hunting, defense, sport, religion and so on, which appear in virtually all cultures to obsess and stimulate human males. A host of present data about the reluctance of females to work for, vote for, or otherwise be associates of other females would then follow from this fact – if it should be proven to be a fact – that females in groups function best when occupied with tasks in the community consistent with those appropriate to their more limited familial ones.

Again, this is not to recommend this situation, nor to excuse the difficulty females have in penetrating male-dominated organizations or in contriving all-female dominated ones. My purpose is to note briefly that just as in pair-bonding or male-bonding, female-female bonding may involve biologically determined infrasocial processes. The factor of sexual competition for the attention and (implicitly) breeding potential of males is an additional restraint on the co-operation of females over extended periods of time and under various forms or degrees of social pressure.

I have tried to outline some of the parameters of biology within which efforts at social change may have to operate. Defining or describing a situation is not to excuse it, but presumably to provide some factual and theoretical basis for changing a system once it is understood. That the existing state of affairs is all involved with passionately felt prejudices, aspiration, fears and uncertainties makes it all the more necessary to ask why the position of women in society after society has remained unsatisfactory to idealistic and enterprising men and women.

The conspiracy theory of why this is so, and the class theory – curiously mixed as it is with an overtone of prejudice similar to that of racism – may be insufficient axes around which a discussion can revolve, for such discussion must also take into account what new biological and other data we have which may be pertinent.

Of course, there are limits to the utility of a biological model, too – it is too easy for some to say that what is true because of biology must always remain social reality; but this is just not so. Our human biology is, however, the fundamental foundation on which social reconstitution must perforce be based. There can be no other foundation. And the task of idealogists and makers of public policy must be to incorporate biological reality into the idealism of their programmes and the scope of their extensions of human possibilities.

VII. Reproduction

Professor Sir Alan S. Parkes, chairman of the Galton Foundation, London, has long had a special interest in the biology of human reproduction. He was professor of physiology of reproduction at Cambridge and for 30 years has been a staff member of the National Institute of Medical Research at Mill Hill, London. His essay is a straightforward, up-to-date account of the reproductive cycle in the human male and female. This somewhat pedestrian kind of information is valuable, we believe, in helping to overcome the sex misinformation that seems to persist in the minds of too many of us.

The second paper, ''The Mechanism of Bacterial Sexuality'' by Professor William Hayes, was included, despite its somewhat technical style for several reasons. Some new sexual systems that may play a role in human medicine are described. The study of bacterial reproduction has enlightened our understanding of the genetics of higher organisms. Bacterial research founded the Operon Theory (see an earlier reading, ''At the Controls of the Living Cell''). How useful a lowly, aberrant life form can be!

Professor Hayes has been director of the MRC Microbial Genetics Research Unit since 1957 and at present is designate professor of molecular genetics at Edinburgh University.

15.

The Reproductive Life Cycle

Sir Alan Parkes

In man there is a marked sex difference in the span of reproductive life. The end points are much more definite in the female than in the male. AUTHOR

Reproduction is characterized in all mammals by a series of cycles – of which the major one is the life cycle of immaturity, puberty, maturity, menopause and finally senility and death. The proportion of the life cycle occupied by the reproductive phase varies from species to species and from individual to individual of the same species; it is difficult to generalize, for not all humans – and probably few animals – survive to die of old age, or even to reach the end of their reproductive life.

In man, there is a marked sex difference in the span of reproductive life. The end points are much more definite in the female than in the male: the beginning and end of the female's menstrual cycle give a fairly clear indication of the maximum extent of her reproductive life. There are no such clear indications in the male; the onset of both puberty and of the climacteric – when sexual activity and potency decline – is less overt and more gradual. However, the male usually retains fecundity long after it has disappeared in females of similar age. One may well doubt reports of centenarians siring offspring, but there are well authenticated records of men of 70 to 80 years of age becoming fathers. By contrast, though women as a whole live longer than men, expectant mothers of 50 years of age are rare.

In mammals, the primary organs of reproduction known as the gonads (testes in the male and ovaries in the female) are essentially dual purpose organs of both external and internal secretion. The testes produce spermatozoa and the androgenic hormones, (mainly testosterone) which induce and maintain the attributes of maleness. The ovaries produce ova and the oestrogenic and progestational hormones, mainly oestradiol and progesterone, which are concerned with the development and maintenance of the attributes of femaleness, including sexual receptivity and the inception of pregnancy.

Associated with the gonads are accessory organs which play a part, mostly an essential role, in the reproductive processes. In the male, these consist of the scrotum which houses the testes, the epididymis which collects the spermatozoa from the testes, the vas deferens which conducts them to the urethra and thence to the exterior, the seminal vesicles and prostate glands, the secretions of which dilute the sperm mass, and the penis through which the spermatozoa pass to the female.

In the female, the accessory organs consist essentially of the female reproductive tract leading from the ovaries to the exterior – the fallopian tubes, uterus, cervix, vagina and vulva with its associated labia and clitoris. Usually essential also are the mammary glands (from which the class

Reprinted from the June 1970 issue of *Science Journal* by courtesy of *Science Journal*, London, England.

Mammalia takes its name) though the human breast in sophisticated societies has largely been rendered superfluous by artificial feeding and seems likely to degenerate into a mere psychosexual symbol. During the reproductive life cycle, the gonads and accessory organs show various changes.

Many animals – in addition to the essential reproductive organs, gonads and accessory organs – possess secondary sexual characters which, although mainly gonad dependent and showing highly significant changes during the life cycle, play no part (except perhaps as visual stimuli) in the reproductive processes. Where present, such characters are usually better developed in the male than in the female; in such cases, it is then possible to distinguish the two sexes without reference to the genitalia.

In man, for instance, the mature male can usually be distinguished from the female by the presence of facial hair, a different pattern of pubic hair, a deeper voice and a larger and more muscular body. All these are dependent on the androgenic substances produced by the testes, in the absence of which sexual differentiation of this kind does not appear. The female pattern of pubic hair, though not gonad dependent, being found in youths, castrated men and in ovariectomized women, is nevertheless dependent, as is the axillary hair, on maturation of the body at puberty. This neutral pattern of pubic and axillary hair appears to be dependent on androgenic substances from the adrenal glands.

J. M. Tanner of the Institute of Child Health, London, distinguishes six stages of pubic hair growth, of which the first five lead to the formation of the inverted triangle of coarse curly hair common to adolescents of both sexes. Stage six appears some years after stage five and is characterized by the upward extension of the hair to a point at the navel. According to Tanner, this stage is found in 80 per cent of mature men but in only 10 per cent of mature women.

Voice changes in the boy at puberty result in the pitch deepening from treble or alto to tenor or bass. The immediate cause is the enlargement of the larynx and lengthening of the vocal chords which accompanies the spurt in somatic and penis growth. The voice changes at puberty, being androgen dependent, tend to regress in old age.

The amount and distribution of head hair is of course greatly influenced by genetic factors; but baldness, which appears so often in early middle age, is not seen in eunuchs and evidently results from some action, probably on the scalp, of the male hormones. Skin changes at puberty are well known and congestion of the sebaceous glands caused by androgens, leading to a greater or lesser degree of acne, is frequently seen. At the other end of the life cycle thinning and wrinkling of the skin due to loss of elasticity and atrophy of subcutaneous tissue are well known effects of ageing.

As to body configuration, the smaller size of the female and the subcutaneous fat which causes her rounded contours are due at least in part to the bone ossifying and fat mobilizing effects of oestrogen and, together with the wider pelvis, constitute the chief positive secondary sexual characters of women – those that are caused by femaleness rather than the absence of maleness. By contrast, the greater musculature of the male is caused by the anabolic (tissue building) action of the androgens.

Five stages are recognized in the maturation of the external genitalia of boys: infantile condition, enlargement of scrotum, enlargement of penis, "sculpturing" and darkening of penis and lastly the adult condition. Two US workers have found that in a sample of American boys, stage two started on average about 12 years of age and that the time lag between stages four and five was much greater than between other stages — stage five being reached at about 17.5 years of age. The epididymis and accessory glands show appreciable activity at birth as a result of being subjected to maternal hormones during the later stages of foetal life. Postnatal regression then leads to a state of quiescence which lasts until puberty, when development starts at about 10 years of age and keeps pace with that of the testes.

The testes are normally fully descended early in life, but they increase little in size during the first decade. They then undergo relatively rapid growth which takes them to full size at the age of about 20 years. Further slight increases in size may take place during adulthood and there is little decrease even in extreme old age. These size changes are, of course, largely conditioned by the seminiferous tubules which produce the spermatozoa. The androgen secreting interstitial cells develop to a similar time table but contribute little to the overall bulk of the testis.

The interstitial tissue is probably responsible for the production of most, but not all, of the androgen which is excreted in the urine — the remainder being of adrenal origin. Total excreted androgen shows a very clear rise and fall during reproductive life (see Figure 1). This and similar curves have given rise to the notion that, androgenically, a man passes his prime at 30 years of age; functionally, however, androgen is adequate for much longer, as shown by the fact that the rise in the urinary secretion of gonad stimulating substance, indicating that the loss of androgen control of the pituitary gland starts only around 60 years of age.

It is not easy to determine the time when spermatozoa are first produced in any particular individual, especially as it may precede the capacity to ejaculate. However, according to Tanner, the tubules begin to show activity at about 10 years of age and active spermatogenesis starts some three years later. Boys are usually producing spermatozoa normal in numbers and quality at 14 to 15 years of age; such boys are potentially fecund and a few dozen fathers under the age of 15 (corresponding to fertile intercourse around 14 years) are recorded annually in the United States. From current observations it would seem that the United Kingdom is now emulating the United States in this respect. There is nothing obviously un-biological about such cases: the precocity involved is probably due to opportunity and outlook rather than to potential performance.

Once started, spermatogenesis normally proceeds vigorously right through adulthood and often into old age. This continuous production of vast numbers of germ cells by the male is in sharp contrast to the state of affairs in the female. Although always large in fecund men, there is naturally great variation in the number of spermatozoa produced by different individuals, and the number in one ejaculation will of course depend to some extent on the frequency of ejaculation. As to fecundity, 20 million spermatozoa in a single ejaculation is a very low count and marginal for fecundity; at the other end, 500 million is considered a very high count.

The physical signs of sexual capacity are erection, orgasm and ejaculation, but the three are not necessarily connected. Erection and orgasm can take place from an early age — even in infants. The capacity to ejaculate appears much later and may first be shown by nocturnal emission, often without erection. From the data given by A. C. Kinsey and his colleagues it seems that 50 per cent of his subjects had their first ejaculation at about 13 years and nine months. Much earlier and much later first ejaculations (8 to 21 years) have, of course, been recorded without any suggestion of physiological abnormality. At the other end of the life cycle, the capacity for erection and ejaculation is usually retained into old age; Kinsey's records show that about 75 per cent of American men were potent at 70 years of age, but

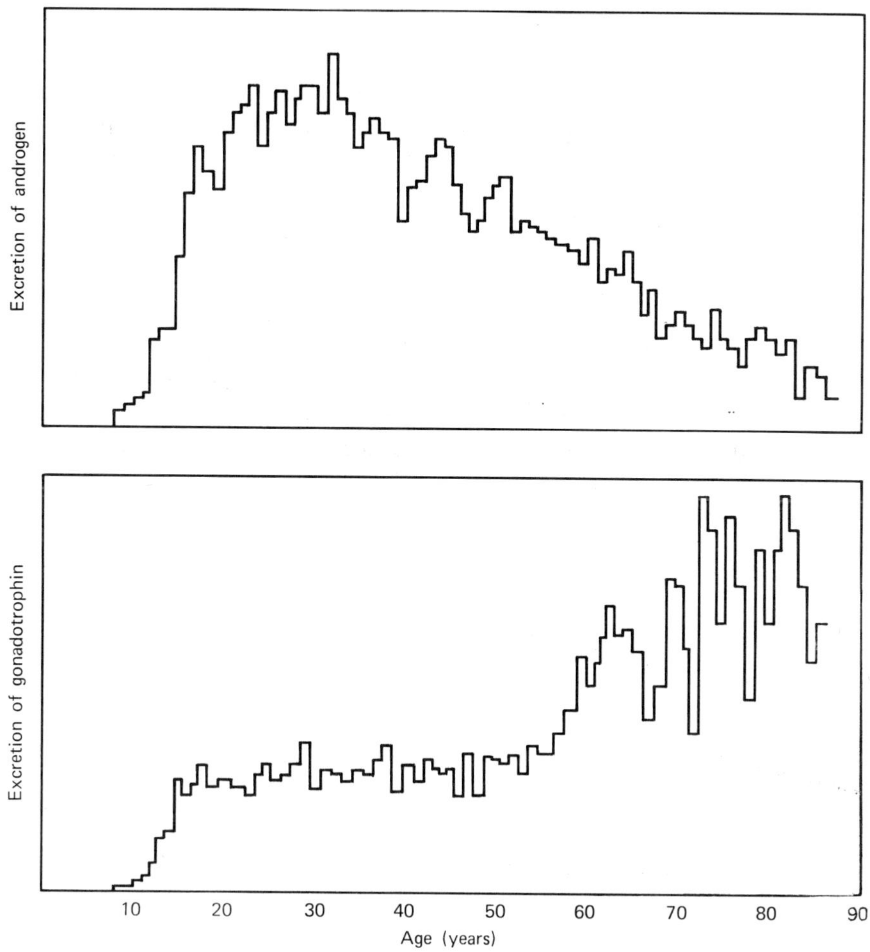

Figure 1 Man's peak, endocrinologically, is reached at about 30 years of age. The upper graph showing the daily excretion of androgen ("maleness" hormone) in the urine of normal boys and men indicates a clear rise and fall during the male's reproductive life span. Functionally, however, the androgen normally available is adequate for a much longer period, as shown in the lower graph by the rising daily excretion of the gonad stimulating substance gonadotrophin. Loss of androgen control by the pituitary gland starts around 60 years.

only 20 per cent of octogenarians. Ejaculation, however, is much less vigorous in old men.

The female's sexual maturation, like that of the male, is characterized by various somatic changes but more especially by development of the gonads and the accessory reproductive organs. Unlike the male, however, the process in the female has one clearly defined stage: menarche, or the appearance of the first menstruation. This has served as a reference point for almost all investigations into female puberty and, by the comparative ease with which it can be noted and recorded, has tended to obscure the importance of other events – such as the first ovulation, development of the breasts, changes in body configuration and so on.

The development of the breasts is an obvious feature of puberty in girls, but it is difficult to quantify. Five stages of development have been distinguished. The first stage involves elevation of the nipple. Stages two and three are signalled by an increasing elevation of breast and nipple to form a small mound, together with extension of the surrounding pigmented skin – the areola. In stage four the areola and nipple project to form a secondary mound above the level of the breast. By stage five only the nipple projects and maturity has arrived. Of these five stages, the first usually appears before the menarche and the last between 14 and 15 years of age. During the pubertal changes in the vagina and external genitalia, the vagina enlarges and its secretions become more acid owing to the formation of acid from the glycogen of the epithelium. The clitoris becomes erectile, Bartholin's glands secrete mucus and the labia enlarge.

Following slow enlargement during early childhood, the uterus starts on more rapid development around the age of 10 years and increases in size by many times in the next few years under the influence of ovarian oestrogens. Finally, with or without a preliminary ovulation, a sudden change in hormonal and neural stimuli causes breakdown of the endometrial lining of the uterus with the resulting effusion of blood. This drains to the exterior through the cervix and vagina, producing the first menstrual bleeding.

The age at which the first menstruation occurs varies considerably from individual to individual, but less so in different populations than is commonly supposed. Observations made on some 1500 Dutch girls, showed that menarche occurred between 13 and 15 years of age in about 60 per cent of the cases, with no less than 50 per cent having started to menstruate at between 13½ and 14 years. It is well known that the age of puberty in Europe and the US has decreased substantially in modern times. A figure of one year every 10 years is sometimes mentioned, but this alarming estimate is much exaggerated. More likely, as Tanner showed, the age of menarche has decreased by about only four years in the last century. This lowering in the age of onset of menstruation is presumably due to the improvement in environmental conditions; it is reminiscent of the observation that better physical development in boys is associated with earlier puberty.

Contrary to popular belief, the length of the menstrual cycle, even when well established, is very variable, both between individuals and from time to time in one individual. The late Carl Hartman published a diagram showing the frequency at which cycles of different length occurred in four subjects (in one of whom the cycle varied from 24 to 34 days in length). The median length in Hartman's four subjects ranged from 24 to 29 days. By contrast, the length of the cycle does not vary in any regular fashion during the major part of a woman's reproductive life, though it is often irregular in the years following menarche and during the menopause.

The menstrual cycle normally consists of two basic phases – the phase before ovulation (follicular) and the phase after ovulation (luteal). The first phase is characterized by the growth of an ovum bearing follicle in the ovary and regeneration and proliferation of the endometrium (lining) of the uterus. The luteal phase starts at the time when the ovum is discharged from the ovary at about mid-cycle and extends to the beginning of the next menstruation. During this stage the endometrium develops and becomes glandular under the influence of the hormone progesterone secreted by the corpus luteum (which develops from the shell of the ruptured ovarian follicle). This second phase of growth in the uterus serves to prepare it for the reception of the fertilized ovum, in the absence of which the corpus luteum starts to regress after about 10 days and about

14 days the lining of the uterus breaks down and menstruation occurs. Ovulation is thus associated physiologically not with the preceding menstruation but, through the corpus luteum, with the one that follows.

Like its overall duration, the two phases of the cycle are variable in length, though the luteal phase is thought to vary less than the follicular phase. Inevitably, therefore, there are many difficulties in trying to calculate the time of ovulation arithmetically. There is, in fact, no practicable method of detecting ovulation at the time it occurs; the best that can be done is to discover that it has occurred a couple of days or so previously, either by measurement of basal body temperature (BBT) which is very slightly elevated by the action of progesterone, or by determination of pregnanediol, the excretion product of progesterone.

If a fertilized egg is safely received by the uterus and implanted, the corpus luteum persists and menstruation is averted. The mechanism of this reaction is not known with certainty but it is obviously critical for the inception of pregnancy, and subject to a tight schedule. Such considerations have interesting corollaries. For instance, a cycle may be unusually short because one or both of the phases are truncated. If only the follicular phase is concerned there may be little effect; but if the luteal phase is reduced to, say, 10 days the cycle may necessarily be infertile because of inadequate build-up of the endometrium, or because of lack of time for the operation of the menstruation inhibiting mechanism. G. K. Döring referred to such cycles as incomplete.

A certain cause of an infertile cycle is failure of ovulation, resulting in an anovular cycle. Failure of ovulation, however, does not result in failure of menstruation, though the bleeding may be somewhat less; nor are anovular cycles necessarily shorter in length. It follows that neither the rhythm of the menstrual cycle, nor the mechanism which precipitates the breakdown of the endometrium, is dependent on the development and regression of a corpus luteum. Anovular cycles can be detected, therefore, only by BBT readings or estimations of urinary pregnanediol, neither of which in such cycles show the biphasic nature characteristic of the normal cycle. According to Döring, the incidences of both incomplete and anovular cycles show marked changes during the reproductive phase of the life-cycle and are frequent in teenagers.

The ovaries undergo far less absolute increase in size during sexual development that do the testes. According to published figures, they increase to a maximum size between 20 and 40 years of age and thereafter decrease to about half size. The most interesting fact about the ovaries, however, is that their oocyte population is laid down very early in life and is not added to thereafter. The size of the original population is not easy to assess but it may be around half a million in the human female. This number is greatly reduced by decay of follicles even before puberty and the process continues throughout reproductive life, so that by the time of the menopause the supply of ovarian eggs may be entirely exhausted. The loss of ovarian eggs by ovulation is negligible; even assuming the regular discharge of one egg each 28 days for 30 years, the total is only around 400 eggs.

The laying down of the ovarian ova very early in life has important practical implications: it means that during reproductive life the ova are anything between 15 and 45 years old – and that a woman aged 45 years who produces a child does so from an ovum of that age. This fact may perhaps account for the greater number of abnormal foetuses produced by older mothers – though there is the alternative, but less likely, explanation that the decreased frequency of intercourse among middle aged couples may result in the ovum or spermatozoa being "stale" at the time of fertilization and thus more likely to produce an abnormal embryo.

The release of a single ovum during each cycle is, of course, the rule for women; and, by analogy with other animals that give birth to single offspring, it is likely that the two ovaries usually ovulate alternately. However, a single egg may result in twins by splitting of the embryo at an early stage of development. Such monozygotic twins, commonly called identical twins because of their typically identical genetic make-up, form only a small proportion of all twins. The majority of twins result from the simultaneous, or near-simultaneous, ovulation of two eggs to produce dizygotic twins. These raise the total twinning rate to around one in 80 (1.25 per cent) of all births in western countries. From the point of view of the life cycle, the interesting thing is that the incidence of monozygotic twins shows no significant connection with the age of the mother, whereas the frequency of dizygotic twins shows a very marked correlation – rising by a factor of five or so to a peak in mothers between 30 and 40 years of age (see Figure 2). It is arguable therefore, that the ovaries reach their maximum biological activity, as well as their maximum size, comparatively late in reproductive life.

The menopause is very variable in its time of onset and has no definite end point – the cycle appearing to fade out rather than stop – and so the process is difficult to time accurately, even in a single individual. Many factors have been said to influence the age at menopause – including ethnic group, age at menarche, socioeconomic circumstances, occupation, urban or rural environment and marital and maternal history. At present one can only say that menopause usually occurs between 40 and 50 years of age, though earlier and later ones have often been recorded. It is difficult to know whether there is currently a trend towards later menopause as there is towards earlier menarche; certainly on the evidence available it seems possible.

Functionally, the menopause is characterized by increasing irregularity of menstruation, increasing failure of ovulation and increasing secretion of pituitary gonadotrophin (in response to the massive reduction in oestrogen secretion by the ovary). To what extent the onset of the menopause is determined by exhaustion of the eggs in the ovary, by failure of hormone production by the ovary, or by failure of the overall mechanism controlling reproductive function is uncertain.

In this article we have considered the development and regression of sexual organs and of sexual function during the life cycle in the two sexes as two separate issues. Fertility, however, in the sense of the production of viable young, involves the combined operation of coitus – the physiology, frequency and effectiveness of which change markedly during the life cycle. In the male, reaction time increases with increasing age, and the subjective characteristics of the ejaculatory process change gradually. In the female – in whom sexual response is more dependent on experience than it is in the male – interest in coitus and the capacity for orgasm may increase rather than decrease. The frequency of coitus, however, has been shown by surveys to decrease steadily with increasing age, possibly because it is largely determined by the male. Effectiveness also decreases: indeed the number of acts of coitus required to produce a pregnancy increases during the reproductive life.

At best, the probability that conception will result from a single act of coitus is small, and the advent of effective contraception has further dissociated sexuality and reproduction. Let us hope that this dissociation will not be completed by the development of ectogenesis – birth of a child outside the mother.

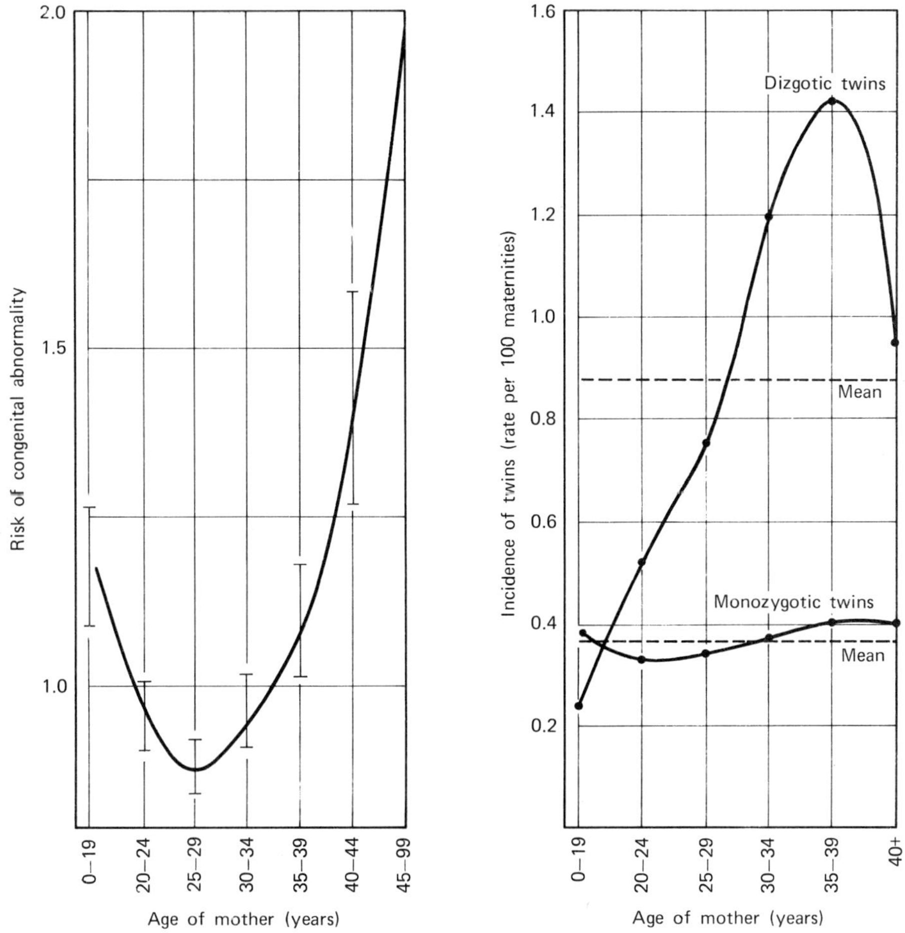

Figure 2 Relative risk of producing a congenitally abnormal child is lowest for mothers aged 25 to 29 years (diagram left). The laying down of the immature egg cells very early in the mother's life means that a child born to a 45 year-old mother comes from a 45-year-old ovum. This possibly accounts for the greater number of abnormal foetuses produced by older women. The graph on right shows that the frequency of monozygotic (identical) twins is independent of mother's age, while the frequency of dizygotic twins rises to a peak at 30 to 40 years of age.

Further Reading

Science and the Safe Period, by Carl Hartman (Bailliere, Tyndall and Cox, London, 1962)

Sexual Behaviour, by A. C. Kinsey and others (Saunders, London, 1948, 1953) "Ageing Processes in the Female Reproductive Tract," by P. L. Krohn *(Lectures on the Scientific Basis of Medicine,* 7, 285, 1958)

Multiple Births in Man, by A. S. Parkes (in *Journal of Reproduction & Fertility* Supplement 6, p. 105)

Growth at Adolescence, by J. M. Tanner (Blackwell Scientific Publications, London, 1962)

16.

The Mechanism of Bacterial Sexuality

William Hayes

The sex factor of E. coli is far from unique and . . . is only the prototype of a host of cytoplasmic genetic elements . . . under a variety of different disguises. Most important . . . [are those] which determine the synthesis of antibiotics . . . and the so-called resistance transfer factors. AUTHOR

If we look for a common denominator underlying the numerous expressions of sex that are found in nature, a rather simple central theme emerges. It is an alternating cycle, in the first phase of which two "haploid" cells – cells possessing only a single set of chromosomes and genes – pool their genetic material to form a "diploid" cell or zygote. In general, this zygote may be formed by the union, or conjugation, of the parental cells which then fuse to become one, or by the penetration of a female gametic cell (egg) by a male gamete (sperm). In either case the diploid zygote, or its descendents, subsequently completes the cycle by segregating haploid progeny cells, in which the inherited chromosomes or genes may be reassorted. The process whereby such reassortments arise is called "genetic recombination" and, when the parental cells differ in various characters, it leads to the inheritance of new character patterns by the offspring. Recombination is of fundamental evolutionary importance as a very efficient way of testing new character combinations for their fitness for the environment. This importance is shown by the common observation in man that, apart from twins from a single ovum, no offspring of the same parents ever appear the same.

Twenty years ago J. Lederberg and E. L. Tatum [1] discovered that even the humble colon bacillus *Escherichia coli* possesses a sexual mechanism that is mediated by conjugation. Hitherto, no overt sexuality had been recognized among the bacteria. This discovery, which at the time appeared to S. E. Luria [2] to be among the most fundamental advances in the whole history of bacterial science, was no accident, but the outcome of a carefully conceived experiment to test whether genetic recombination occurred in bacteria. *E. coli* is an organism of great synthetic ability which can build up all the amino acids and vitamins of the B group that it requires from glucose and an inorganic source of nitrogen. On the other hand, it is easy to obtain mutant strains of *E. coli* that have lost the capacity to synthesize one or more amino acids, and so are unable to grow unless these amino acids are present in the culture medium. Let us refer to these amino acids as *A, B, C, D,* and to the mutant strains that cannot synthesize one or more of them as A^-, B^-, and so on.

In the critical experiment of Lederberg and Tatum, the parental types were two mutant strains of a laboratory stock called K12, each unable to make at least two different amino acids; typical strains would be referred to as $A^-B^-C^+D^+$ and

Reprinted by permission from the January 1967 issue (No. 97) of *Endeavour*.

$A^+B^+C^-D^-$. Cultures of these parental bacteria are unable to grow on a synthetic agar jelly lacking amino acids, and in practice never produced colonies when spread alone on such a medium. Yet if a mixture of the cultures was spread on the same medium, colonies of $A^+B^+C^+D^+$ bacteria, possessing the inheritable capacity to synthesize all their amino acids, arose. They were in a proportion of about one for every million parental bacteria.

These bacteria were clearly genetic recombinants, resulting from pooling of the genetic material of the two reciprocally defective parents followed by selection of those recombinant progeny that inherited a non-defective reassortment of genes. The use of doubly-mutant parental strains excluded the possibility that they were mutational reversions, since the minimum probability of such an occurrence is of the order 10^{-14} to 10^{-16} per cell generation. Moreover, since recombinants never arose unless intact bacteria of both parental types were present, it was correctly assumed that the genetic transfer is mediated by cell to cell contact, that is, by conjugation.

Not long after the discovery of conjugation, the existence of gene linkage was clearly demonstrated. Recombinants, as described above, are selected on the basis of their inheritance of particular nutritional capabilities. If they were analysed for the concomitant inheritance of other characters in which the parents differed, and which were not selected by growth on synthetic medium, these characters were found to appear in differing but fixed proportions. This is compatible with a system in which the genes that determine these characters are arranged in a given order and at fixed distances from one another on a linear chromosome [3].

All of these results suggested the operation in *E. coli* of a more or less orthodox sexual system, in which a diploid zygote, formed by the fusion of two haploid parents and containing the full genetic complement of each, later produced haploid recombinants. This view was reinforced by the finding that crosses involving a particular variant of one parent produced a remarkably high proportion of relatively stable diploid bacteria that continued to form haploid recombinants. However, although the number of character differences – markers – available for study at that time was rather small, analysis of the diploids revealed a peculiar anomaly; a particular segment of chromosome from one of the parents was found consistently to be missing from the recombinants. This suggested that this segment was specifically eliminated from the zygotes. Moreover, with increase in the number of markers studied, other anomalies began to appear. For example, recombinants seemed to inherit most of their characters from one of the parents, and the progeny of crosses in which recombinant strains were used as parents often showed inherited character patterns very different from those expected. Thus the sexual system of *E. coli,* which at first had seemed rather simple and orthodox, took on such increasing complexity that it seemed likely that some important aspects of its true nature remained to be discovered.

Male and Female Bacteria

The first clue came from the observation that treating one of the two parental strains with streptomycin abolished the capacity of the cross to yield recombinants, while treatment of the other parent did not, although both were equally sensitive to the drug as judged by survival [4, 5]. From this it was deduced that there was a one-way transfer of genetic material from a donor (male) to a recipient (female) strain; the donor could be dispensed with once its function had been fulfilled, but survival of the recipient, in which the whole process of recombination and segregation took place, was essential. The fact that recombinants inherited most of their characters from the recipient parent further suggested that the genetic contribution of the donor bacteria to the zygotes is fractional. Then the remarkable fact came to light that the donor state is genetically determined, not, as one might expect, by a chromosomal gene or genes, but by an infectious agent called the sex factor or "F" (for "fertility"), which exists separately in the cytoplasm. This factor promotes conjugation between donor bacteria that harbour it (termed F+) and recipient bacteria (F−) that lack it, followed by its own efficient transfer to the recipients which are thus themselves converted into F+ donors [5, 6]. The efficiency of this process approaches 100 per cent under optimal conditions, despite the low frequency of recombinants for chromosomal genes, so that sex in *E. coli* can be said to be highly infectious.

More recent research has revealed that the sex factor consists of a deoxyribonucleic acid (DNA) molecule of about the same size as the chromosomal DNA of an average bacterial virus. One of its functions is to determine the formation of a new antigen at the surface of the male bacteria. This antigen appears to alter the surface charge so that the bacteria can now make intimate contact with females. However, the most unusual property of the sex factor is that it can exist in different states within its host cell. The key to this discovery was the accidental isolation, from an F+ strain, of a new type of male with quite novel behaviour. These were Hfr ("high frequency of recombinants") males, and they conjugate with females with the same efficiency as F+ males. However, in this case the unions are followed by transfer of the bacterial chromosome instead of the sex factor, with the result that recombinants containing chromosomal genes from the male arise about 1000 times more frequently than in crosses with F+ males; on the other hand, both the female population at large and the recombinants remain female, so that the sex factor has lost its infectious character. Nevertheless, it is clear that the sex factor itself is not lost, since Hfr males may revert to the F+ state [5, 6]. A further point of distinction lies in the effect of acridine orange. F+ males can readily be "cured" of their sex factor by the drug, to become females: it appears specifically to inhibit sex factor replication. Hfr males, on the other hand, are very refractory to attack by the drug.

The Features of Chromosome Transfer

The true nature of the differences between these two types of male was revealed by a brilliant series of experiments by E. L. Wollman and F. Jacob at the Pasteur Institute, Paris [7, 8]. Let us briefly look at two of these experiments, which have had a fundamental bearing on our knowledge of the states of the sex factor and of the nature of chromosome transfer. The first is the famous "interrupted mating" experiment. In this, samples of a mixture of Hfr male and female bacteria were removed at intervals during their mating, and violently agitated in a mixer to separate the mating individuals. The mixture was then diluted and spread on agar jelly to isolate recombinant colonies. After this, the recombinants issuing from each sample were analysed to see what genes they had inherited from the male bacteria. It turns out that for each gene there is a specific period after mating during which it is completely excluded from recombinants by interruption of the mating. After this, its incidence among recombinants rises until the level characteristic of uninterrupted crosses is attained. In other words a male bacterium transfers each of its genes to the female at a specific time after the commencement of mating. Moreover, the time sequence in which the various genes are transferred was found to coincide with the order of arrangement of the genes on the chromosome as determined by genetic analysis.

This means that a culture of Hfr male bacteria comprises a homogeneous population, of which all the bacteria transfer their chromosomes from the same point and with the same orientation. Transfer of the whole chromosome occupies rather more than 100 minutes which, remarkably, is more than four times as long as the generation time under optimal conditions.

It then turned out that recombinants selected for inheritance of genes located near the extremity of the male chromosome that is transferred last, frequently inherit the Hfr male character, although they comprise only a very small proportion of the total recombinants, due to a high tendency to breakage of the chromosome so that its distal end rarely enters the zygote. This implied that the sex factor in Hfr males, instead of being free in the cytoplasm as in F+ males, is an integral part of the bacterial chromosome. This conclusion is supported by the resistance of the sex factor in Hfr males to elimination by acridine orange, as well as by more recent experiments that show that its replication is under chromosomal control.

There was an important and novel outcome of the interrupted mating experiment. As it could be shown that the first half of the chromosome, at least, is transferred at a constant speed, the experiment enabled the distance between any two genes to be measured in the absolute terms of the time elapsing between their transfer under standard conditions. If the speed of transfer continued to be constant, the whole chromosome would be transferred in about 90 minutes so that the distance between genes can be expressed as a true proportion of the length of the chromosome as a whole. The observable nucleus of *E. coli*, of which there are normally several per bacterium, possesses only a single chromosome consisting of double-helical DNA, so that it is easy to estimate chemically the amount, and therefore the actual length, of DNA in the chromosome. This turns out to be about 1200 microns – about 500 times as long as the bacterium that contains it. Thus, if the assumptions are correct, the distances between genes, expressed in terms of transfer time, can be translated into the real physical terms of DNA length.

The Circular Chromosome

The second experiment that helped to define the relationship of the sex factor to its host cell had a number of important results. It led to the isolation of some independent Hfr strains from the same F+ male culture and it demonstrated that the ability of F+ populations to generate recombinants for chromosomal genes is probably due to the development of clones of "mutant" Hfr cells within it. Hfr clones certainly form, and the most interesting and important finding came from analysis of chromosome transfer by the different Hfr strains obtained. The strains were found to transfer their chromosome from different starting points and often in opposite directions [Figure 1C]. In addition, the pooled data from all the strains revealed the extraordinary and quite unexpected fact that whatever pair of genes one might postulate to lie near the extremities of a linear chromosome, an Hfr male strain could be found that transferred these two genes as adjacent and closely linked. In other words, the chromosome of the F+ male, from which these Hfr strains originated, must be a continuous or circular one, since it can be shown not to possess extremities [7, 8]. It has now been confirmed directly, by autoradiography, that the chromosomal DNA of all sexual types of *E. coli* is in the form of a loop [9]. Finally, in the case of any particular HFr strain, the male state and, therefore, the sex factor were shown to be inherited only in association with a gene transferred very late or terminally, even though this gene might be quite different in different strains (Figure 1C). Thus in the formation of Hfr strains, the sex factor may insert itself at any one of a number of sites around the circular chromosome of an F+ bacterium. Following conjugation, the chromosome opens up close to the site of insertion and is then transferred to the female as a linear structure with the sex factor at its tail.

Intermediate Males and Sex Factor Structure

The relationship between these three sexual types of *E. coli* – the female lacking the sex factor, the F+ male in which the sex factor exists and replicates autonomously in the cytoplasm, and the Hfr male in which the sex factor is inserted into the chromosome and is replicated as part of it – is shown in Figure 1. In addition, there is a third type of male bacterium which is

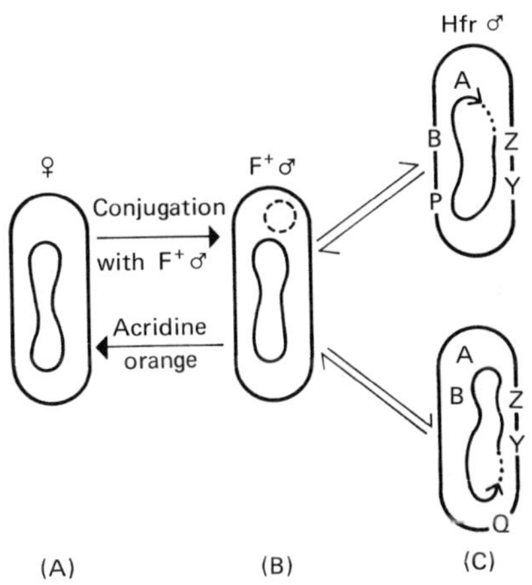

Figure 1 The main sexual types of *Escherichia coli*. The continuous, irregular line within each bacterium represents the bacterial chromosome; the broken line indicates the sex factor. The letters at (C) designate genetic loci (genes) on the bacterial chromosome. (A) No sex factor: the bacterium is female. (B) Cytoplasmic sex factor: the bacterium is an F+ male. (C) The sex factor is shown inserted into the bacterial chromosome at two different locations: the bacteria are Hfr males and can transfer their chromosomes from the point, and with the polarity, indicated by the arrowhead in the sex factor.

of practical importance as well as of theoretical interest. This originates from Hfr bacteria and is called an "intermediate male" because it displays the behaviour patterns of both F+ and Hfr males. That is, recombinants for chromosomal genes are generated at high frequency, and in addition the intermediate male state is itself highly infectious and spreads rapidly throughout the female population.

A major clue to the nature of intermediate males was the finding that the majority of intermediate male strains harbour a sex factor called F-prime (F′), which carries in its structure a recognizable chromosomal gene or genes originally located near the site of insertion of the sex factor. This suggested that the properties of intermediate males are conferred by a sex factor which has incorporated a fragment of bacterial chromosome. There is now much evidence that the contrasting behaviour of the various male types can be explained on the general hypothesis that the sex factor is a continuous DNA loop that can be inserted into and released from the chromosome by a single, reciprocal act of genetic exchange or recombination (Figure 2A) [10,11].

Genetic recombination is a two-stage phenomenon. It is first necessary for structurally similar or "allelic" regions of two chromosomes to "pair" or come into apposition; this is followed by breaking of the paired chromosomes at precisely corresponding points, and their cross-wise rejoining. At the level of molecular structure, it seems highly likely that the genetic similarity which determines pairing is the sequence of nucleic acid base pairs along the DNA double helix. When this sequence is near-perfect over long allelic regions, as happens when part of the chromosome of one *E. coli* K12 strain is transferred to another, the probability of recombination can be shown to approach unity. In contrast, insertion of the sex factor into the chromosome of an F+ cell to produce an Hfr bacterium is a rare event, of the order 10^{-4} to 10^{-5} per cell generation, so that pairing in this case probably results from very short or imperfect similarities of base sequence. *Pari passu* the Hfr state is a reasonably stable one, since re-establishment of pairing, leading to "recombination out" of the sex factor, may be expected to be equally rare. On the other hand, the situation will be drastically altered if a sex factor carrying a segment of bacterial chromosome is introduced into a female cell, for here the sex factor carries a region of near-perfect similarity to the corresponding region of the bacterial chromosome. As a result, the sex factor is continually being inserted into and released from the chromosome so that the bacteria alternate rapidly between the expression of Hfr behaviour and of F+ behaviour. In support of this hypothesis, if an F+ factor is introduced into a bacterial strain carrying a mutation which prevents genetic recombination, it fails to promote chromosome transfer, although it replicates and mediates conjugation in a normal manner.

A final question that may be asked about these relationships is how the sex factor acquires a fragment of bacterial chromosome in the first place. Figure 2 shows three kinds of pairing interaction between integrated sex factor and chromosome that could lead to recombination and liberation of the sex factor. The circles on the left represent the chromosome of an Hfr male in which the inserted sex factor is indicated by the dotted line. The radial arrows point to postulated regions of similarity that will tend to pair, and between which recombination will occur. The most common event is shown at A, where the paired regions that originally led to insertion of the sex factor are re-established. A normal sex factor is released and a normal chromosome left behind. In the second row (Figure 2B), pairing is assumed to occur between two different regions of similarity, existing perhaps by chance, one (2) within the sex factor, and the other (1) on the chromosome between genes *Y* and *Z*. In this case recombination leads to release of a sex factor carrying the *Z* region of chromosome, while the bacterial chromosome acquires a fragment of sex factor. If the sex factor is now eliminated by acridine orange, the "cured" bacterium should (and does) behave as a female. However if it is reinfected, by conjugation, with a normal sex factor it should behave, not as an F+ male, but as an intermediate male, because the sex factor fragment retained in the chromosome offers a good region of similarity with which the immigrant sex factor can pair and become inserted. Two cases of this kind have been reported.

A third possibility is also shown in Figure 2B. In this, two regions of the bacterial chromosome on either side of the sex factor pair. One (1) is between *Y* and *Z* and the other (3) is between *A* and *B* . In this case, recombination yields a sex factor that carries both genes *A* and *Z*. Two apparently identical sex factors of this sort have been obtained from independently isolated strains of the same Hfr male type [10, 11].

The Mechanism of Genetic Transfer

At the moment, the nature of the bridge connecting male and female bacteria, and the mechanism that the male bacterium employs to transfer genetic material across the bridge, are controversial. It used to be thought, and electron microphotographs appeared to reveal, that conjugating bacteria are directly united by contact between their cell walls. However, it has recently been discovered that the sex factor promotes the synthesis, by the bacteria that harbour it, of special, hair-like appendages called sex pili [12]. These can be distinguished from the numerous pili that are commonly produced by *E. coli* bacteria by the extraordinary fact that they act as the sites of adsorption for certain virulent

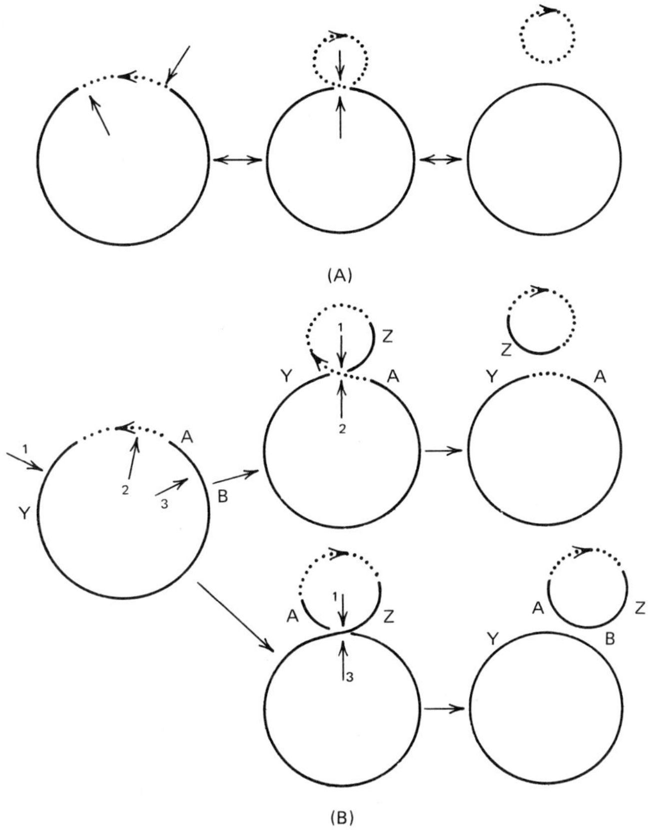

Figure 2 Genetic interactions between various regions of the chromosome of an Hfr male bacterium. The continuous and broken lines represent bacterial chromosome and sex factor respectively. The arrow heads indicate the initiation point and polarity of chromosome or sex factor transfer. The radial arrows point to regions of genetic similarity between which pairing and genetic exchange may occur. (A) shows the outcome of genetic exchange between the regions which initially led to insertion of the sex factor. (B) shows the outcome of similar exchanges between other regions of chromosome or sex factor. (From Hayes, 1966.)

viruses that infect only male bacteria. There is now little doubt that although common pili, which are hollow protein tubes, play no role in bacterial sexuality, the possession of sex pili is a necessary condition for conjugation. It has therefore been suggested that if the sex pili are also tubular structures they may represent the male sex organ of bacteria, through which the genetic material passes. As yet there is no evidence to confirm or refute this interesting hypothesis.

The problem of how the genetic material is transferred is a twofold one. First, in view of the very high efficiency with which genetic transfer may occur, how is it that the sex factor of F+ males, or a particular extremity of the chromosome of Hfr males, so easily finds the union joining the male and female bacteria? The most promising hypothesis is that the sex factor, whether free or inserted, is attached to the cell membrane and makes its antigen (or pilus) locally so that contact is established and the union made at the sex factor attachment site. Although there is as yet no evidence for this, there is good evidence [13] that the chromosome itself is connected to the membrane and that this plays a key role in chromosome replication.

The second problem concerns the way that the circular chromosome of either sex factor or bacterium is opened up and transferred as a linear structure to females. The most plausible hypothesis is that put forward by Jacob and Brenner [14]. This supposes that some metabolite, resulting from completion of the conjugation tube, initiates polarized replication of the DNA of the sex factor. The energy involved in the synthesis of the new DNA is then sufficient to drive one of the daughter DNA molecules through the tube into the female. If the sex factor is independent of the chromosome, it alone is transferred (Figure 3A). But if it is inserted into the bacterial chromosome, then replication initiated in the sex factor DNA will continue along the chromosomal DNA, which will therefore be transferred as well (Figure 3B). If the sex factor carries a fragment of bacterial chromosome, either the sex (F′) factor alone or the chromosome may be transferred. Which of these occurs depends on whether or not a genetic exchange has connected the F′ factor to the chromosome at the time when the replication point reaches the region of pairing (Figure 3C). In support of this ingenious hypothesis, there is now good evidence from autoradiographic studies that Hfr bacteria do not transfer the chromosome that pre-existed at the time of conjugation, but a newly synthesized replica of it [15].

An important feature of the sex factor is that it is widely infective. Thus it can be transferred from *E. coli* to many other species and genera of intestinal bacteria in which it propagates itself and also determines conjugation and its own transfer. However in only one other genus, *Salmonella*, which is closely related to *Escherichia*, can the sex factor generate Hfr males and promote chromosome transfer. Presumably the chromosomes of other genera do not possess sufficient genetic similarity to the sex factor to permit its insertion.

Other Cytoplasmic Factors

Within recent years it has become apparent that the sex factor of *E. coli* is far from unique and, in fact, is only the prototype of a host of cytoplasmic genetic elements of about the same size, which present themselves to us under a variety of different disguises. Most important among these are factors which determine the synthesis of antibiotics called colicins, and the so-called "resistance transfer factors". Only a few of the colicin factors are able to promote conjugation and their own transfer. In contrast, all the resistance transfer factors are, by definition, sex factors. Like the sex factor of *E. coli*, to which some seem to be related both genetically and functionally, they mediate conjugation and can spread epidemically through populations of intestinal bacteria which lack them. Some of these factors can also, rarely, initiate chromosome transfer, but the equivalent of Hfr bacteria have not yet been isolated from strains carrying these factors. Resistance transfer factors are of considerable importance in medicine because they have picked up and incorporated into a single transmissible structure the genetic determinants of bacterial resistance to a wide range of antibiotics in common clinical use. As many as seven such determinants have been reported to be carried by a single transfer factor. These factors can become extensively disseminated among the normal flora of the intestine in human and animal populations, and thence be transferred by conjugation to a wide range of dangerous intestinal pathogens to initiate epidemics which cannot be treated effectively [16, 17].

It is interesting to speculate on the phy-

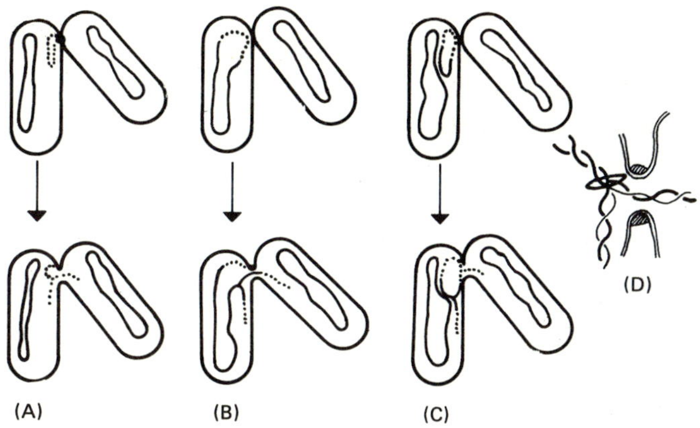

Figure 3 Diagram of the probable mechanism of conjugation and genetic transfer in *E. coli*. In each pair of bacteria the male is shown on the left and the female on the right. The bacterial chromosome is represented by the continuous line, the sex factor by the broken line. The black blob at the point of bacterial contact represents the conjugal antigen (whatever its nature) which is synthesized under sex-factor control. (A) The sex factor is extra-chromosomal. (B) The sex factor is inserted into the chromosome. (C) The sex factor carries a chromosomal fragment. (D) A magnified representation of the transfer process. The ring represents a molecule of replicating enzyme attached to the cell membrane, through which the DNA passes as it is replicated, one replica entering the female cell. The old, parental DNA strands are shown as heavy lines, and the newly synthesized strands as light ones. (From Hayes, 1966.)

logeny of sex factors. In the pattern of its relationship to its host cell, the sex factor of *E. coli* closely resembles the genetic material of those bacterial viruses that can be carried as provirus, that is, their DNA can be incorporated in the bacterial chromosome and the virus is not virulent. Both are transmissible agents that are able to replicate autonomously in the cytoplasm, to insert themselves into the chromosome, and to pick up and incorporate fragments of host chromosome into their structure. Moreover, sex factors as a class fulfil all the criteria whereby A. Lwoff [18] defined viruses. They are infective, they depend for their metabolism on the biochemical machinery of the host cell, and they possess only one type of nucleic acid. It therefore seems logical to regard sex factors as viruses that ensure their efficient spread to a wide range of host cells by the novel method of mediating conjugation between them. In this way they avoid the risks attendant on exposure to the environment and the need to elaborate a protein coat to protect their genetic material from it. According to this idea, the role of sex factors in promoting transfer of the bacterial chromosome, though possibly of some evolutionary advantage to the bacterium, would be merely incidental to their function as viruses. It should be remembered that although sex factors appear to be common and widely distributed, at least among intestinal bacteria, chromosome transfer mediated by them is a rare phenomenon under natural conditions.

References

[1] Lederberg, J. and Tatum E. L. *Nature, Lond.,* **158**, 558, 1946,

[2] Luria, S. E. *Bact. Rev.,* **11**, 1, 1947,

[3] Lederberg, J. *Genetics, Princeton,* **32**, 505, 1947.

[4] Hayes, W. *Nature, Lond.,* **169**, 118, 1952.

[5] *Idem. Cold Spring. Harb. Symp. quant. Biol.,* **18**, 75, 1953.

[6] Lederberg, J., Cavalli, L. L., and Lederberg, E. M. *Genetics, Princeton,* **37**, 720, 1952.

[7] Jacob, F. and Wollman, E. L. *Scient. Am.,* **204**, 93, 1961.

[8] *Idem. 'Sexuality and the Genetics of Bacteria.'* Academic Press, New York, 1961.

[9] Cairns, J. *Endeavor,* **22**, 141, 1963.

[10] Broda, P., Beckwith, J. R., and Scaife, J. *Genet. Res., Camb.,* **5**, 489, 1964.

[11] Hayes, W. *Proc. R. Soc., B.,* **164**, 230, 1966.

[12] Brinton, C. C. Jnr., Gemski, P. Jnr., and Carnahan, J. *Proc. natn. Acad. Sci. U.S.A.,* **52**, 776, 1964.

[13] Jacob, F., Ryter, A., and Cuzin, F. *Proc. R. Soc., B.,* **164**, 267, 1966.

[14] Jacob, F. and Brenner, S., *C. r. hebd. Séanc. Acad. Sci., Paris,* **256**, 298, 1963.

[15] Gross, J. D. and Caro, L., *Science,* **150**, 1679, 1965.

[16] Watanabe, T. *Bact. Rev.,* **27**, 87, 1963.

[17] Anderson, E. S., *Brit. med. J.,* **2**, 1289, 1965.

[18] Lwoff, A. In 'The Viruses', Vol. 2, p. 189, edited by Burnet, F. M. and Stanley, W. M. Academic Press, New York, 1959.

VIII. Heredity

A rather diverse group of papers comprises this unit although all of them focus on aspects of human genetics. Two of them consider the most critical stage in life, embryonic development. With conception a new individual begins. The success or failure of the new combination of parental genes is measured largely by the development of a normal embryo, a congenital abnormality, or a miscarriage. The latter are natural concerns of parents but must be understood for what they are. In "Heredity and the Embryo," David H. Carr, professor of anatomy at McMaster University, Hamilton, Ontario, deals mainly with the causes of abnormalities leading to natural abortions. In many cases, prospective parents, especially those with family histories of birth defects, will find consultations with a genetic counselor in order. In "Will the Baby Be Normal?" Robert W. Stock, staffer for *The New York Times Magazine*, examines many aspects of that recent outgrowth of modern genetics, genetic counseling. Both papers raise profound ethical questions for an informed public to consider.

So few people know that genetic counseling services exist that we have included a list of United States clinics compiled by Henry T. Lynch of the Creighton University School of Medicine.

The fourth reading by Nobel laureate Joshua Lederberg, world-renowned geneticist of Stanford, considers the mechanics, difficulties, and ethics of controlling man's genes. This unit ends with an interesting problem in population genetics by Charles M. Woolf and Frank C. Dukepoo of Arizona State University. In local isolates of Hopi Indians the high frequency of the gene for albinism presents an application of the Hardy-Weinberg law and the kinds of factors that may disturb a genetic equilibrium.

17.

Heredity and the Embryo

D. H. Carr

Spontaneous abortion is nature's usual way of limiting the population of abnormal individuals. Although nature does a very efficient job it has been the history of medicine to observe and improve on nature's lead. AUTHOR

As you sit reading this article you may congratulate yourself on having survived the most hazardous phase of your life. The nine month period between fertilization and birth presents more hazards than any similar span up to the ninth decade. These hazards include those due to abnormalities within the sperm and ovum as well as numerous environmental threats associated with illness of the mother, but I am mainly concerned to review abnormalities which arise from defects in the male and female germ cells themselves.

Each newly conceived embryo, or conceptus, is genetically different from the mother's tissues in which it grows. It contains genes derived from the father which lead to the production of proteins foreign to the host mother. For some reason which is still not completely understood the uterus does not reject this foreign material. It is believed that some form of insulating barrier exists between the mother and offspring which allows her body to tolerate the genetically different tissue. On the other hand, if the conceptus is in some way abnormal the uterus rejects it with great reliability. The means for recognizing the abnormality in the embryo is again an unknown one, but it is highly efficient. For every abnormal child born alive at least five are lost as spontaneous abortions or stillbirths.

Miscarriage is the familiar term for loss of pregnancy before the baby is capable of an independent existence but, medically, all such losses are abortions. If the pregnancy terminates naturally it is a spontaneous abortion, the medical synonym of "miscarriage". At some time the foetus becomes capable of independent existence. This point used to be considered to be the 28th week of pregnancy, but many foetuses born before that date now survive. Pregnancy lasts about 40 weeks. At 20 weeks the foetus weighs about 400 grammes, and I know of no foetus under that weight that has survived. Most hospitals regard the loss of a pregnancy before 20 weeks as a miscarriage and after as a stillbirth.

The estimated frequency of spontaneous abortion depends on the source of the figures. Most patients who know they have had a miscarriage do not go to hospital; some do not even seek medical advice. Hospital records of miscarriages are therefore of little value. The most consistent and reliable estimates of the incidence of spontaneous abortion are based on personal interviews. From such studies it appears that about 15 per cent of all pregnancies are lost before the foetus is viable at 20 weeks. Stillbirth is only about one tenth as common, 1-2 per cent of

Reprinted from the June 1970 issue of *Science Journal* by courtesy of *Science Journal*, London, England.

pregnancies being lost between weeks 20 and 40. And these figures are not the whole story. Many fertilized eggs are probably lost either before they can become attached to the wall of the uterus or very soon after they start to grow there. In these cases the woman may not ever have been aware that she was pregnant. Menstruation may be unaffected or it may be rather late and unusually heavy. Studies of very early pregnancies in humans and other primates suggest that conceptuses lost in this way are probably twice as common as recognized miscarriages. So 30 per cent of all fertilized ova are lost without patients being aware of their pregnancy, 15 per cent as recognized miscarriages and 1.5 per cent as stillbirths.

Very early losses appear to be abnormal in many ways and there may be no actual embryo formed. I will not discuss these abortions further as we know nothing of their cause. I have mentioned them only to emphasize that, of all eggs fertilized, about half fail to produce a living child. But in recent years a great deal has been learned about abnormalities in later development.

Inherited characteristics are transmitted from parent to offspring through the chromosomes. These are contained in the nucleus of the cell but are visible as individual structures only when the cell is dividing. The number of chromosomes is characteristic for the species. It is only 14 years since the number for man was correctly designated as 46, and since 1960 it has been found that there are many departures from the normal number and structure. These abnormalities arise either in the male or female germ cells or in the ovum at fertilization.

The chromosomes undergo spatial rearrangement in the germ cells. At the time of ovulation in the female the chromosomes in the ovum are reduced in number from 46 to 23, the other 23 being shed in a small cell called the first polar body. Abnormalities may arise at this time. But the final parcelling of chromosome material in the ovum does not occur until the moment of fertilization. The 23 chromosomes in the ovum are double stranded and the DNA content is therefore twice the required amount. These double stranded chromosomes consist of 22 nonsex chromosomes or autosomes and one sex chromosome (X). These elements are lined up on the spindle awaiting the stimulus to separate into two single strands.

The male sperm also contains 22 autosomes but the sex chromosome may be an X or a Y. Fertilization of the ovum by an X-bearing sperm leads to the development of a female embryo but a Y-bearing sperm results in a male. The entry of the sperm into the ovum normally signals several other events. All other sperms are prevented from entering the ovum. The two strands of the chromosomes separate from one another, one strand remaining in the ovum and the other entering the second polar body. The full completment of 44 autosomes and two sex chromosomes is restored. Then the ovum starts to divide.

The cause of abnormalities in the inherited material may be disorders of the genes or of the chromosomes as a whole. Any departure from normal in the genetic information is called a mutation. If this mutation occurs in the genetic matrix (DNA) of the chromosome it will not be visible under any of today's microscopes. Gene changes can be recognized only by their effects which may change the chemical material within the cell. Some of these abnormal chemicals produced by mutated genes may be demonstrated under the microscope by using special staining techniques which show that the cells look different. Even then it is not the gene mutation itself which is visible but only its effect on the metabolism of the cells. In many cases, such substances can be demonstrated only by techniques which allow the separation of normal and variant proteins; and often gene mutations cannot be recognized at all in single cells but only in whole organisms.

Changes at the gene level can physically alter the whole organism in a way that even the casual observer can notice. Among such changes are extra fingers or toes, a cleft palate or a club foot. Changes in genetic information may produce abnormalities involving several systems in the body. But at present all we can observe is the effect of genetic mutation and not the actual alteration of molecular structure. And the recognition of genetic effects is further complicated by the fact that many of the abnormalities which they produce may have other causes: cleft palate, for example, may be the result of environmental agents.

In contrast, mutations or abnormalities at the chromosome level can generally be observed directly, because changes in chromosome number or structure are visible under the highest power of the microscope. Even then our present relatively crude techniques may miss small changes but they give valuable information and are being improved.

In order to see chromosomes it is necessary to have dividing cells. Few cells in the body divide frequently enough to provide us with chromosome preparations so it is necessary to grow cells outside the body in artificial conditions *(in vitro)*. There are many techniques but in all of them the principle is that pieces of tissue or free cells are grown in an artificial medium in a sterile container at body temperature. At a suitable time the cells are arrested during division and treated with a solution more dilute than the cellular fluid. This "hypotonic" solution flows into the cells causing them to swell and allowing the chromosomes to spread. The cells are then fixed and stained.

When magnified about 1000 times human chromosomes in a cell about to divide into two can be seen to be double stranded. Chromosomes are often obtained from a cell grown from tissue of a spontaneous abortion. Although the embryo itself has usually been dead for some time before a miscarriage occurs, the sac surrounding it is living and in about 70 per cent of specimens of such tissue it is possible to grow cells and obtain chromosome preparations. These chromosomes are from general tissue cells and, as often explained elsewhere in this issue, contain 46 elements, 23 from each parent, comprising 23 pairs of chromosomes, 22 pairs of autosomes and one pair of sex chromosomes, XX in the female and XY in the male.

As our present methods are relatively crude it is impossible to identify all the chromosomes in the cell. Even in this computer age the best way we know of detecting abnormalities is to cut out each chromosome from a photomicrograph and pair them up like a jigsaw. This array, known as a karyotype, is constructed according to an international convention which improves the likelihood of noticing not only if there are too many or too few chromosomes but also if there are extra long or missing segments. But it is only experience that tells us that particular differences between chromosome pairs are normal variants or technical artefacts.

Genetic mutation is a known cause of abnormalities in the infant. The gene defect may be carried on an autosome or a sex chromosome. It may behave in a "dominant" or a "recessive" manner, though these terms are not absolutes. Generally a dominant gene produces its abnormal effects in the person who carries it, so that the defect is manifest. If the effect on development is severe the affected individual may die before reproductive age or fail to find a suitable mate. On the other hand, recessive genes carried on autosomes rarely produce obvious effects in their carriers. The normal chromosome which pairs with the one carrying the recessive mutant gene prevents the expression of the latter. If two carriers of the same recessive autosomal gene were to marry, any of their offspring who received one chromosome with the recessive gene from each parent (statistically, one child in four) would have the disease. A familiar example of such an abnormality is cystic fibrosis.

Recessive mutant genes may be carried on the X-chromosome. If the carrier of this abnormal gene is a male he has only one X-chromosome and the effect of the gene is inevitably manifest. In contrast, a female with a mutant gene on one X-chromosome appears perfectly normal because the dominant normal gene on her other X-chromosome suppresses it. Familiar examples of X-linked inheritance are haemophilia and the commonest muscular dystrophy.

From studies of mammals other than man it is known that certain gene mutations are so serious that they cause death of the embryo, abnormalities and abortion. Such mutations are called genetic lethals. If these are carried as recessive genes in man, the marriage of two carriers should produce an increase in spontaneous abortions. Blood relatives have more genes in common and the closer their relationship the more genes they share. The rate of spontaneous abortion has therefore been studied in the families of married relatives (usually cousins). There was no increase in the number of miscarriages in these families, but this does not necessarily mean that recessive lethal genes do not occur in man. It may mean that this technique of detection is too crude; and the mutant genes may be so lethal that pregnancies have been lost too early to be obvious.

There is statistical evidence that certain types of mating are associated with increased pregnancy loss and sterility. This conclusion is based on studies of marriages between people of certain A, B and O blood groups. But it is not possible to pick out particular families with high risk of pregnancy loss since the results are based on population findings rather than on the testing of individuals. Although A, B, O incompatible matings appear to be associated with an increased risk of spontaneous abortion, Rh blood group incompatibility is not. Rh haemolytic disease is a cause of late foetal loss (stillbirth) but not of abortion. In view of recent advances in immunology it is probable that this disease will soon be eliminated.

Work carried out in Liverpool, New York and elsewhere during the past nine years has focused attention on the prevention of Rh haemolytic disease in the newborn (previously the main effort had been directed towards treatment of the affected infant). The foetus can become affected because some Rh negative mothers produce antibodies to blood cells of their Rh positive infants. In future pregnancies these antibodies rise higher if the woman carries another Rh positive infant. The antibodies then cross through the placenta to the infant's blood and destroy its red cells, causing severe anaemia and often death.

The new treatment of Rh haemolytic disease aims at prevention of antibody formation in the mother's blood. If the mother has not previously developed antibodies the main danger of her becoming sensitive to Rh positive blood cells arises during the delivery of her baby. To prevent this she is injected with antibodies to destroy the baby's Rh positive cells which have entered her circulation. If the Rh negative mother already has antibodies from previous delivery of an Rh positive baby she is not a candidate for this treatment. However, it offers hope of almost eliminating Rh haemolytic disease among previously unaffected women.

Chromosome abnormalities are found in about one in every 200 liveborn infants. The commonest type of such abnormality in the living infant is that associated with mongolism or Down's syndrome. This disease occurs in one in every 600 or 700 infants and is associated with the presence of an extra small chromosome in group G.

Most chromosome abnormalities are found in spontaneous abortions. About 15 per cent of all pregnancies end in abortion and perhaps one third of these have a chromosome abnormality. Rough calculations show that ten such abnormal offspring are aborted for every one born alive. In addition to those with chromosome abnormalities, many other malformed embryos are lost in this way. So, although a miscarriage may be a disappointing experience for a woman, it is usually nature's way of removing abnormal individuals from the population.

There are three main categories of chromosome abnormality in abortions. The largest group is associated with the presence of one extra chromosome, giving a count of 47 instead of the normal 46. This usually arises as an abnormality of cell division during the formation of the first or second polar body by the ovum. It may also arise in the male germ cells, when the extra chromosome is carried in the sperm. The understanding of the events leading to this disorder is based on observations made in lower animals and plants over the past 50 years.

When the double stranded chromosomes in the ovum separate into two single strands following fertilization, one strand remains in the ovum while the other enters the second polar body. But if one chromosome fails to divide both strands will remain in the ovum. The polar body will lack one of these strands and contain only 22 elements whereas the ovum will receive an extra chromosome to give a count of 24. When this ovum is fertilized by a normal sperm containing 23 chromosomes the total count will be 47 instead of 46.

The extra element may be any of the autosomes or a sex chromosome. An extra sex chromosome rarely leads to abortion and has a much less severe effect on development; males or females so endowed may appear quite normal though some are mentally retarded. But if the extra element is an autosome the conceptus will most likely be aborted before the end of the third month of pregnancy. In many cases, the embryo is resorbed and only the membraneous sac remains. The effect of an extra chromosome varies according to which one is in excess: generally, the smaller the extra chromosome, the less severe the effect on development.

The second major type of abnormality in spontaneous abortion arises from whole extra sets of chromosomes. This means that there are exact multiples of the number found in the germ cells (23), the commonest multiple being three: possession of 69 chromosomes is known as triploidy. The double stranded chromosomes may fail to divide when the sperm enters the ovum, so no polar body is shed and the ovum contains 46 elements. With the 23 chromosomes of the sperm the count then becomes 69. On the other hand the ovum may be normal but the first sperm may fail to trigger the mechanism which shuts out all others. A second sperm enters and this dispermy brings the count to 69.

Triploidy is quite common in man but such embryos nearly always abort, even though they may appear quite normal. Many embryos have developed to later stages and a very few triploid infants have been born alive, although dying within hours or days. The abnormalities of these infants may not appear severe. Death and abortion are more likely due to the chemical imbalance resulting from such a major chromosome disorder. In triploidy, each element is represented three times instead of twice. As two sets of chromosomes come from one parent and one from the other the sex chromosome complement is variable; XXY is commoner than XXX or XYY.

The chromosome number may be doubled, causing tetraploidy. This abnormality probably arises at the first cell division of the fertilized ovum. Normally the 46 chromosomes each divide into two and then the cell divides. If after the chromosome divided the cell did not, the count would become 92 instead of 46. A tetraploid ovum is capable of a little further development and abortuses have been found with 92 chromosomes in all their cells. But tetraploidy appears much more serious than triploidy. Tetraploid abortuses are lost at an earlier stage than triploid ones, and usually the only evidence of development is the empty membraneous sac which normally surrounds the embryo. Tetraploidy appears to be very serious in mammals other than to man, yet in lower animal forms both triploidy and tetraploidy are compatible with normal development. In horticulture polyploid strains are deliberately developed for their hardiness or large size.

In about one in five of all miscarriages with a chromosome disorder one sex chromosome is absent, giving a count of 45. The disorder is apparently due to simple loss of a sex chromosome during development of a germ cell, the sperm being the offender in about 75 per cent of such cases. Embryos and foetuses with only one X-chromosome may be quite normal or grossly abnormal. It is estimated that 97 per cent of all embryos with only one X-chromosome are aborted. The remaining three per cent become living infants, many being relatively normal girls of small stature who are almost always infertile due to the death of the germ cells in the ovary.

All these three types of chromosome disorder arise as chance events. The parents are almost always normal and recurrence of a chromosome anomaly in a subsequent pregnancy is rare.

Abnormalities in a high percentage of abortuses are obvious simply by inspection. About half of all structually abnormal conceptuses have abnormal chromosomes, but the cause of abnormality in the other half is uncertain. Their chromosomes may have small defects which are not detectable by present techniques. I have already described the difficulty of detecting abnormalities at the molecular level. And there may be environmental factors, locally in the uterus or in the body generally, which affect embryological development.

Recurrent abortions are rarely the result of a chromosome disorder, although a few are associated with a chromosome abnormality in an apparently normal parent. Some women spontaneously abort one pregnancy after another, even though they are physically normal and their embryos usually normal. In many instances these women seem intolerant to their pregnancies, perhaps as a result of a nervous overreaction by the uterus, but even after repeated failures such women can usually be treated and carry pregnancies to term.

About 2.5 per cent of liveborn infants have abnormalities which are detectable at, or soon after, birth and which interfere with normal life or require surgical correction. Many genetic abnormalities do not become manifest until later in life. These include conditions associated with multiple abnormal genes or due to an interaction between genetic defects and the environment. If all diseases with underlying hereditary defects are classified together they undoubtedly form the major challenge to preventive medicine today. It has been estimated that there are in the United States about 15 million cases of congenital defects and 6 million mentally retarded individuals. The cause of most of these is unknown. The first step is usually to detect the underlying abnormality. Then it is necessary to find factors which predispose to this disorder.

For instance mongolism and some other abnormalities are commoner in children born to women in their later reproductive years. It has been estimated that if no woman over 37 years of age had a child, mongolism would be reduced by one-third. At the population level this is a substantial contribution. But what of a woman who marries when she is 38 and passionately wants to have a child? Should she take a chance? Now we're at the personal and not the population level, and the answer I would give is "yes, of course". But a compromise between the individual and the population can be realized. It is my personal opinion that, if a woman has an established family of two or three by the time she is 35, she should not have more children.

If an abnormal child is born, neither parent is "to blame". The problem is not in those terms. But it is a fact of nature that, while no new female germ cells are formed from before birth to the menopause, this makes them susceptible to certain changes which appear to apply particularly to the chromosomes. Conversely, the rapidly renewing sperms arising from the male germ cells are more likely to suffer genetic damage and there is some evidence relating gene defects to increased age of the father. Many abnormalities are unrelated to age, while others are commoner among very young mothers. Fewest abnormal offspring are found in mothers aged 25-29.

Radiation exposure is known to increase the risk of a woman having a mongol child later in life. But, as with maternal age, we must be realistic. If an X-ray is needed for diagnosis and treatment it should be taken. But X-rays should not be taken unnecessarily and the ovaries and testes should be shielded if possible.

Some abnormalities can be diagnosed before birth. Sometimes an abnormality not compatible with live birth can be diagnosed by X-ray. Recently it has become possible to diagnose a wide range of abnormalities, most of them rare, in families with a high risk due to abnormal genes or chromosomes in the parents. Cells from the foetus are obtained by withdrawing a sample of the amniotic fluid which surrounds the foetus and contains loose cells from it. This fluid is contained in the sac surrounding the embryo from early pregnancy. From a practical point of view it is unlikely that fluid can be obtained before the 12th or 13th week of pregnancy, when the womb can be felt just above the pubic bone. After local anaesthetic a fine needle can be inserted through the abdominal wall into the amniotic cavity. The fluid extracted contains a small number of cells from the embryo and these can be processed in various ways according to the abnormality which is suspected. The cells may fail to grow outside the body and the procedure may have to be repeated. It is not a test to be undertaken lightly and the techniques are too complex to be applied to every pregnancy. Where justified, the parents are told the findings. If an abnormal condition is detected the woman may choose to have the pregnancy terminated where this is legal.

Historically, infections have been the dreaded enemy of man and have repeatedly threatened his existence on this planet. The discovery of vaccination in 1798 and antibiotics in 1928 were milestones which changed the course of history. Today congenital defects have become a major cause of human suffering. Spontaneous abortion is nature's usual way of limiting the population of abnormal individuals. Although nature does a very efficient job it has been the history of medicine to observe and improve on nature's lead.

But doctors are now faced with a dilemma: whether to save the life of a deformed child, creating a painful and tortured existence for it, or to allow it to die as it would naturally without the benefit of modern medicine. Spina bifida children, born with deformed backbone and spinal cord, are a case in point. Without an operation they generally die within days or weeks of birth but with the benefit of modern surgery they may survive as incontinent paraplegics needing constant nursing care.

Who should decide whether such a child is allowed to live: the doctor or the parents or society as a whole? Not only do we need to learn more and offer more to reduce suffering, we are faced by pressing ethical problems which need wide and informed public discussions.

FURTHER READING

Spontaneous and Habitual Abortion, by C. T. Javert (McGraw-Hill, New York, 1957)

"Chromosomal Studies of Spontaneous Human Abortions," by S. L. Inhorn (in *Advances in Teratology II,* Academic Press, New York, 1967)

"Chromosomal Errors and Development," by D. H. Carr (in *American Journal of Obstetrics and Gynecology,* 104, 327, 1969)

18.

Will the Baby Be Normal?

Robert W. Stock

. . .[Perhaps] five percent of the American population needs or could benefit from genetic counseling, but only the barest fraction of that number is even aware that such a service exists. AUTHOR

The process of human reproduction, whatever else it may be, is a lottery – a game of chance stacked in our favor. For all their impossible, miraculous complexity, the genes carried by randomly assorting eggs and sperm are remarkably dependable. The species survives.

Yet there are losers. Every year some 200,000 American children are born with an inheritance of disabling defects: deaf, mentally retarded, physically deformed. It has always been so, a fact of the facts-of-life – to be feared, to be accepted. Until now.

The biological-genetic revolution of the past decade has enabled scientists to probe the depths of the human cell, to understand and even predict the workings of the genes. Suddenly there have appeared the first concrete glimmerings of hope that man may be able to transcend his hereditary limitations, to turn back the inevitable, to control the lottery.

The man whose task it is to translate this revolution into human terms is the genetic counselor, member of a small but growing group of medical experts that is virtually unknown to the public at large. His patients are frightened men and women – married couples whose first child was abnormal, engaged couples with a family history of birth defects. They want to know: "Can we have normal children?"

Until now, the counselor's reply has had to be generalized: "Given your family background and the nature of the defect, the odds are 3 to 1 that your child will be healthy." But the rules are changing. Many defects can today be detected in the cells of potential parents, even in the cells of a child in the womb. Treatments have been devised to correct or modify the effects of some abnormal genes in the newborn. The genetic revolution is giving the counselor new, precise facts and options – and forcing grave ethical decisions upon his patients and, in the last resort, upon the society as a whole.

It has been estimated that 5 per cent of the American population needs or could benefit from genetic counseling, but only the barest fraction of that number is even aware that such a service exists. Which may be just as well, in a way, since there are nothing like enough counselors to serve them.

For counseling today is something of a stepchild. It is not a formally recognized medical specialty. It does not attract Government research grants. There are in this country some 114 hospital and laboratory facilities that are classified as genetic counseling "centers," a six-fold increase since 1955, but counseling is very much a sideline for most of them.

The typical counselor is a physician with an interest in genetics; he may have a traditional medical specialty (often pediatrics or gynecology) or he may be a laboratory scientist. In either case, he is primarily interested in counseling patients whose genetic problems fall within his particular area of competence, or will help advance his particular research projects. It is not that he lacks compassion for those he must turn away; he lacks time.

The preponderance of scientists in the ranks of the counselors is a new development, an outgrowth of the genetic revolution. The researcher studies hereditary defects because they are, as one geneticist puts it, "experiments of nature that present unique opportunities for expanding our knowledge of many basic biological processes." Along with the knowledge have come the patients, eager to take advantage of the scientists' new discoveries. "If we even began to take on all the people who call us," says one scientist-counselor, "there'd be no discoveries." Where can a person in trouble turn? "To his family doctor. He can refer a patient to the right counselor if necessary. More likely, he'll be able to provide the essential information himself."

The fact is that the new methods of detecting hereditary defects apply thus far only to a few dozen of the 1,500 or so recognized gene abnormalities. For the rest, the counselor's basic tool remains what it has always been – a series of mathematical formulations laid down a century ago by a Bohemian monk named Gregor Mendel.

As Mendel demonstrated, the physical traits that make us what we are come in two varieties – dominant and recessive. We all have a double dose of the machinery for producing each and every trait – one set of directions from our fathers, the second from our mothers. (Inevitably, there is an exception: a single dose of the male-determining gene is enough to make us a male.)

Whether we will have brown eyes or blue is settled at our conception: If the genetic directions, the genes, governing eye color are both programed for blue, we'll have blue eyes; but if either is programed for brown, our eyes will be brown. We say that brown eyes are dominant over blue; blue eyes are a recessive trait.

The genetic counselor follows this same line of reasoning. A man and his fiancée seek help. The man's mother was born deaf as was a maternal uncle; the man himself has normal hearing, as does his wife-to-be. Can they have normal children?

The counselor studies their family histories, constructing diagrams called pedigrees. He finds no history of deafness in the woman's family. He finds that the deaf members of the man's family suffer from a type of deafness which is inherited as a recessive defect. And he arrives at the following conclusions:

In this case, as is happily most often true in matters genetic, normality is dominant over abnormality. Since both man and woman have normal hearing, both must have at least one of the dominant normal-hearing genes. Furthermore, both the woman's genes are probably normal, there being no deafness in her family and the gene being rare. Thus any child they have will almost certainly have normal hearing, since he will necessarily receive one dominant, normal-hearing gene from his mother.

The bane of the genetic counselor's existence is that word "almost." He must hedge his prediction because it is always possible that somewhere in the woman's family tree a serpent lurks in the form of that same recessive deafness gene. She may have inherited it – in which case she has one normal and one abnormal gene. Her fiance has probably inherited the abnormal gene from his family. The result: any child they have would stand a 25 per cent chance of receiving a double dose of the undesirable gene, and such a child would be born deaf. Unless and until the counselor can actually detect such recessive genes in the cells of his patients' bodies, he can never be sure.

This version of the doctor's dilemma is central to modern genetics – and to the controversy over abortion and eugenics. According to Dr. Kurt Hirschhorn, president of the American Society of Human Genetics and professor of pediatrics and genetics at Manhattan's Mount Sinai School of Medicine, all of us carry from three to eight potentially lethal genes. If our mates carried one of these same genes, our children would have a 1 in 4 chance of inheriting a fatal disease.

A majority of the recognized inborn defects, including some of the most common, are recessive. One such is cystic fibrosis, with an incidence of 1 in 1,600 births. If we were willing to abort all fetuses known to be carriers of or affected by the disease over the next 40 years, says Dr. Hirschhorn, we would "eradicate cystic fibrosis from the American scene." He is not enthralled by the notion, but he points out that scientists can already detect fetuses that are affected by or carriers of that defect.

As such eugenic programs become more and more feasible, the genetic counselor is likely to find himself involved in guiding national policy on the matter. As researcher and counselor, he is capable of understanding not only the scientific aspects but the human implications as well. The fear of begetting and bearing an abnormal child strikes at the very roots of the men and women who come to his office – at their self-esteem, at their personal relationship, at their basic instincts of racial survival.

When the mysterious phenomenon that maintained the species misfired, the ancients reacted with hate and foreboding. The birth of a defective child was a punishment of the gods, an omen of ill fortune; such children were slain, their parents ostracized. Even as man's knowledge of the reproductive process increased, the abnormal birth remained invested with feelings of guilt and shame. A Scandinavian woman who gave birth to a deformed child in 1683 was burned alive as a witch. An American woman who bears such a child in 1969 is allowed to live, but she – and her husband – carry invisible, indelible scars.

There are some signs that such attitudes are changing. "Most of the people who seek genetic counseling today," says Dr. James German, director of the Laboratory of Human Genetics at the New York Blood Center in Manhattan, "are more enlightened. For example, they have a better understanding of the mechanics of fertilization, the fact that egg and sperm play equal roles." Dr. German credits the "admirable advances" in the teaching of high school biology.

Yet the old fear and ignorance still persist, particularly among the poorly educated. A counselor describes a case in point:

"The wife was pregnant, and she had borne one defective child. The husband was angry. He said he just wanted to know what to expect, but he kept his eyes trained on his wife's protruding abdomen. He kept talking about the abnormal child he thought was in there – 'she' had failed him on the first pregnancy, 'she' might do it again. What he really seemed to want was evidence that whatever happened, it would be her responsibility, her fault, her guilt."

This heritage of shame accompanying the birth of a defective child is but a part of the emotional tangle through which the genetic counselor must find his way. There is, for example, no way of measuring the anguish of the mother of such a child, or its effect on her.

Dr. C. F. Fraser, director of the Department of Medical Genetics at the Montreal Children's Hospital, tells of a patient who had given birth to a child with a cleft lip. The chances that she would have a second child with the same defect were 1 in 20 – a figure, he says, which most people find reassuring. But this woman found such a risk so formidable, after the experience with her first child, that she determined not to have any more children; and when she did accidentally become pregnant, she had an illegal abortion. "It seems," Dr. Fraser admits, "we failed to convey a realistic view of the risk in this case."

Because of their desire to protect their patients' privacy, counselors generally refuse to allow themselves to be identified in detailed descriptions of particular cases – unless the details are grossly altered. Thus, in the undoctored description of the case that follows, the New York geneticist is not named.

The young couple who came to the counselor's office 10 months ago were intelligent, well-educated, seemingly in excellent health in every way. They had been referred by a physician who was treating the husband's mother for a disease with symptoms that suggested it might be Huntington's chorea.

Huntington's is one of the most insidious of all known genetic diseases, a dominant gene defect which does not usually begin to take effect until age 35 or later – well into or after the child-bearing age. The afflicted person gradually loses his mental capacity; he becomes subject to uncontrollable jerking and flailing of his body and arms; he falls into an irritable premature senility.

The first step in the genetic counseling procedure is to establish the diagnosis. The geneticist found that the mother's disease seemed to check out as Huntington's. Then he took the husband's family history, seeking other instances of the disease.

He traced the family tree painstakingly, relative by relative, asking detailed specific questions. How old was an aunt when she died? What were her symptoms? Was she alert up to the end? "If you ask in a general way whether a relative is normal," he says, "a patient is apt to answer yes without thinking. Families tend to forget the unpleasant."

The preparation of the pedigree is often a lengthy process taking hours of knee-to-knee conversation between patients and counselor as together they construct the family tree. In addition to providing the basis for Mendelian prediction, it also gives the counselor an opportunity to learn about his patients – their social position, their intelligence and education, their emotional problems. "You get clues about how to speak with them later," the New York counselor explains, "after you have all the facts. You find out, for instance, that maybe the man would leave his wife if she refused to have children because of a genetic risk."

The counselor discovered that the young husband's grandmother had died in a mental hospital in a distant state. His patients knew nothing about her condition; with their permission, he contacted the hospital and had her records examined. The symptoms matched those of Huntington's. He was also able to determine that two of the husband's uncles had died of the disease.

The next step was painful. The counselor explained the genetics of the matter to the couple: "I give people a lecture, with diagrams and pictures." Some counselors use dice to get across the risk figures, but he says, "I'm no good at that sort of thing." What counts is that the patient understands the odds and remembers them. Montreal's Dr. Fraser recently advised a group of his colleagues that they should write down the odds and send a copy to the family physician – he has had patients who have reversed the odds he gave them to fit their real wishes.

What the New York counselor had to tell the couple was this:

The evidence was overwhelming that the husband's mother, uncles and grandmother had died of Huntington's. Since that gene defect is dominant, the odds were 50 per cent that the husband had the abnormal gene – though he was too young to exhibit the symptoms. And if he died there was a 50 per cent chance that any child he sired would be afflicted with Huntington's.

"That's a high risk for such a serious disease," the counselor comments. "The husband said he wouldn't take it, but the wife insisted that they should. So now she refuses to use contraceptives, and he only infrequently dares to make love to her. Their marriage is in danger of being ruined."

What can be done? "The husband might ask to be sterilized," the geneticist suggests. "They could adopt a child. They could have a child by artificial insemination. At the moment, there's little that medical science can do. Based on the facts now available to them, they have to work it out for themselves."

There is an oft-stated tradition among genetic counselors that the patient should make the decision whether or not to have a child, possibly in consultation with his family doctor; the counselor simply presents the odds. In fact, however, as most counselors will privately admit, their own attitudes often influence the decision.

"It makes a difference," one says, "how you state the odds. Do you tell a couple they have a 75 per cent chance of having a healthy child or a 25 per cent chance of having an abnormal child? Do you let them get to the point of saying, 'Doctor, what would you do?' or do you steer them away from it? When a couple is utterly confused, unable to make a sober judgment, you have to help them. Does that mean I'm playing at being God? I don't think so – I hope not – I don't know what else I can do."

Approximately 250,000 of the 4 million infants born in America each year have important birth defects, but only 20 per cent of them can be attributed specifically to inherited gene defects. According to the National Foundation-March of Dimes, which specializes in the birth-defect field, some 20 per cent are caused by environmental factors during pregnancy – drugs such as the tranquilizer Thalidomide, diseases such as rubella (commonly known as German measles) and radiation from X-rays. Some 60 per cent of abnormal births are thought to be the result of interaction between environment and heredity. A knowledge of mutagens – the substances that cause gene mutations – is an integral part of the genetic counselor's armamentarium.

It has long been known, for example, that X-rays administered in excess are capable of causing mutations. Sometimes the radiation alters only a single gene; sometimes it is sufficient to cause actual breaks and rearrangements of the embryo's chromosomes – the rod-shaped objects, 46 in each cell, that contain the genes. In either case, the event is on an impossibly tiny scale, since all of the genes in all of the cells of the body weigh less than a fifth of an ounce. The effect on a developing fetus, however, can be momentous. Genes control the production of enzymes – catalysts that govern the chemical reactions which make life possible. In the embryo, a damaged or defective gene translates into a defective enzyme which may translate into a defective child. When the parents of an abnormal child come to a genetic counselor, he must consider the possibility that mutagens have been at work on the fetus.

Mutations seem to have relatively little immediate impact on the fully formed human, with two possible exceptions. Their role in relation to leukemia and cancer is under study. And they may have a profound effect on future generations. If the parents of a child-to-be have broken, rearranged chromosomes or gene defects in the germ cells of testes or ovaries, his conception may be ill-starred; the mutations may be passed on to him in his mother's egg or in his father's sperm.

The crucial question is this: Do human germ cells undergo the same or similar changes as the other cells of the body when a mutagenic substance is operating? Most geneticists answer in the affirmative, and there is evidence from animal experiments that supports this conclusion. Definitive proof in human beings, however, remains elusive.

One New York genetic counselor cites a case in point. The patient's first child had a tumor and required X-ray treatment; during the treatment, she held the child on her lap – an unorthodox procedure. Within two hours of the X-ray, a second child was conceived – and it was born with multiple defects and died in the first year. The woman wanted more children; would they be normal?

The counselor ordered a chromosome study of the second child and of the parents. The child's cells contained broken chromosomes; the cells of the parents appeared to be normal. The counselor was able to tell the parents that the chances of their having another child with defective chromosomes was minimal, but he suggested that the mother stay away from X-rays. He couldn't prove that the radiation had caused mutations in the mother's eggs and thus been responsible for the defective child, but the circumstantial evidence was strong.

The significance of such speculation is heightened by the degree to which modern man is exposed to potentially mutagenic factors. In recent years, scientists have compiled a long list of substances that are suspected of causing gene alterations, and the Federal Food and Drug Administration is initiating a broad study of the field. For among the suspects are: the artificial sweetener, cyclamate, 8,000 tons of which are consumed by Americans each year in soft drinks, candies and other foods . . . fungicides and pesticides . . . caffeine . . . LSD.

For the genetic counselor, one of the most exciting frontiers of medicine today is the diagnosis of gene and chromosome disorders in the developing fetus. It brings "a new dimension and precision" to counseling, says Dr. Henry L. Nadler, head of the Division of Genetics at the Children's Hospital in Chicago and a pioneer in the field. It is now possible, in the case of 30 or more defects, to spot the specific enzyme deficiency by examining samples of cells from amniotic fluid – the liquid in which the developing fetus floats. (The 30 include cystic fibrosis and galactosemia, the inability to metabolize a substance found in milk, leading to mental retardation.) Chromosome abnormalities can also be observed in these cells. "I got into this area," Dr. Nadler says, "because I found it frustrating to have to tell parents that they had, say, a 1 in 4 chance of having an abnormal child. Now, at least for a group of diseases, we can be precise – the risk is either 100 per cent or zero."

The procedure for obtaining the fluid, usually performed between the 12th and 14th week of pregnancy, is called amniocentesis. A hollow needle is introduced into the womb through the abdomen and a sample withdrawn. There is a small risk of damage to the mother or child, but amniocentesis is only performed when the counselor has reason to suspect a defect. Typical reasons: the mother has borne earlier defective children; the parents are in their 40's or 50's.

The major risk facing older parents is the birth of a child with Down's syndrome, also commonly known as mongolism; typical symptoms include severe mental retardation, an average life expectancy of 10 years and an altered, "Oriental" facial structure (hence the name mongolism). About 1 in 600 children are born afflicted with the disease. The odds that a 25-year-old woman will bear such a child are 1 in 2,000; for a 45-year-old woman, the odds soar to 1 in 50.

No one knows for certain why this is so. One theory is that human eggs, like those of other species, deteriorate over a period of time; a woman's eggs have been in her ovary since before her birth, and the longer they remain, the better the chance they will become "stale." When such an egg is fertilized, it may carry the extra chromosome characteristic of the most frequent kind of mongolism. Another possible explanation has been propounded by the New York Blood Center's Dr. German, though it has recently come under attack. When the egg is released from the ovary, he points out, it remains fertile for about 24 hours, toward the end of which time it deteriorates. Sperm remains capable of fertilizing an egg for perhaps two days after deposition in the womb. Since older couples tend to have intercourse less frequently than their juniors, there is a greater chance among older couples as a group that fertilization will take place toward the end of the egg's cycle, with the same adverse effect on the child conceived.

The ultimate value of amniocentesis, of being able to detect birth defects in the unborn child, will depend on the ability of scientists to correct these genetic mistakes. Today they lack that ability; there is a heartbreaking time lag between detection and treatment. Amniocentesis can relieve a parent's mind, or confirm his worst fears. And if it is the latter, he faces an agonizing decision – to allow the abnormal fetus to come to full term, with all the attendant emotional and financial problems, or to have it aborted.

Abortions based on amniocentesic diagnosis have been and are being performed in some American hospitals, but because of the legal dangers posed by rigid, antiquated abortion laws, physicians are not inclined to speak of them publicly. One man who does is Dr. Carlo Valenti, a genetic counselor who is associate professor of obstetrics and gynecology at Downstate Medical Center in Brooklyn.

Dr. Valenti tells of a young mother who had borne one mongoloid and, pregnant again, feared she would have another. Amniocentesis proved that her fears were justified. She requested an abortion, which was authorized by his hospital's abortion committee to protect the mother from "unjustified psychiatric trauma." The operation was performed successfully; examination of the aborted fetus indicated that it would have developed into a mongoloid; the mother plans to try again for a healthy child, provided amniocentesis is again available to her.

In the case of rubella, Dr. Valenti points out, amniocentesis may soon be a life-*saving* procedure. "Right now," he says, "if a mother has rubella, there's a strong chance her child will be deaf, have cataracts, etc., so we often abort – but this means we're also eliminating all those babies who are healthy. When it becomes possible to identify the virus of rubella in cells of the fetus, and there's been some recent progress in this direction, those healthy babies will be rescued. Why destroy – I guess that's the word – normal babies?"

In advising his patients, a genetic counselor must also weigh other medical and scientific developments. Many birth defects, for example, once crippling or fatal, can now be treated with considerable success. Hemophiliacs can lead relatively normal lives with the use of a new blood plasma product. Surgical techniques and speech therapy overcome the effects of cleft lip and cleft palate. Missing limbs are replaced by prosthetic devices. A hormone isolated from the human pituitary is adding inches to congenital dwarfs. Such treatments must be part of the equation when a high-risk couple decide whether to chance a child.

For those birth defects that affect the metabolism of the child, it is often vital that treatment begin soon after birth – sometimes even before the obvious symptoms of the disease appear. Tests that can detect a few of these diseases have been devised, and more are on the way. By using such tests to screen all newborn children, the affected child can sometimes be spotted quickly and saved.

The prime example of this kind of

screening concerns the disease called phenylketonuria (PKU) – an inherited enzyme deficiency that causes the body to accumulate phenylalanine, producing severe mental retardation. Approximately 1 in 10,000 newborns is affected. Some years ago it was found that PKU mental retardation can be halted, though not reversed, by feeding the child a special diet low in phenylalanine. The sooner the child is placed on the diet, the more normal he will be. The nation's medical establishment ground into high gear, and today 38 states (including New York) have laws requiring that all newborn babies by given the so-called Guthrie test for PKU.

Such mandatory screening for defects is most likely the wave of the future, but the history of the PKU program indicates some of the pitfalls of screening. Recently, for example, scientists have found many atypical cases of PKU, including infants whose mental development is not retarded by the disease but who have been automatically put on the special diet. The diet is low in protein, and there is a suspicion that these PKU children, who would have had normal intelligence, now are mentally retarded because of malnutrition.

In the world of tomorrow, screening procedures may not be limited to finding infants who bear a double dose of a defective gene and thus exhibit the symptoms of the disease. As indicated earlier, most inborn defects are recessive, and we all carry about with us a single dose of a few dangerous defects. Recent research has made it possible to identify carriers in the case of 50 or more recessive gene defects including one type of muscular dystrophy and PKU. Thus it may one day be possible to screen all newborn infants and identify the carriers of these diseases.

The year is 1975, and Linda is a few hours old. Samples of her blood, urine and skin and placental tissue are going through the hospital's computerized disease scanner. Three red lights flicker on the console, and the fact that she is a carrier of three specific defects is noted on her Universal Health Card, which is forwarded to Washington, D.C. Eighteen years later, when she and childhood sweetheart Mark apply for a marriage license, the cards are compared. It turns out that Linda and Mark have a defect in common – any child they have faces a 25 per cent chance of being mentally retarded.

As genetic counselors today look toward the future, they are haunted by their own versions of the Linda-and-Mark story. For it represents that point in man's history when he will be able, consciously, to control his own heredity and evolution. Given the facts, what will all the Lindas and Marks do?

In the first place, will they have a choice? The concept of eugenics is not particularly popular today, but it was hailed by many Americans of a few decades ago. Nobel laureates Sir Peter Medawar and Dr. Linus Pauling have taken the position that two such carriers should be prevented from marrying each other. "I have advised, not entirely joking," says Dr. Pauling, "that individuals should have tattooed on their foreheads symbols for the defective genes they carry . . ." Under pressure of the population explosion, such selective breeding theories will inevitably gain adherents. And thus it may well be that Linda and Mark will be refused permission to marry – or forced to undergo sterilization if they insist.

On the other hand, the ability to detect inborn defects in the fetus may be fully developed by that time. In that event, they might be allowed to marry, with a single proviso: should Linda become pregnant, she will have to submit to amniocentesis and agree to an abortion if the fetus has a double dose of the defect.

By the time Linda and Mark face their crisis, another option may be open to them. Scientists are rapidly gaining insight into the complex workings of various enzyme systems within the cells of the body.

Since enzymes are the end product of the genes, it may be possible to bypass a defective gene by artificially supplying the missing enzyme. This year researchers at Rockefeller University and at the drug firm of Merck, Sharp and Dohme announced the first synthesis of an enzyme.

It may also be possible, according to Dr. W. French Anderson of the National Institutes of Health in Bethesda, Md., to make it easier for a defective enzyme to do its job. The enzyme sparks chemical reactions; if these chemicals were artificially changed to allow for the enzyme's deficiencies, the reactions could go forward on schedule.

Another possibility may be available to a child born to Linda and Mark: the actual correction of the gene defect itself. Such an achievement might not only cure the child's ailment but also assure that it would not be passed on to *his* children. Most scientists feel that such genetic engineering will not be available to patients until well into the 21st century, if then, but there are already some straws in the wind.

Scientists have observed the fact that genetic material can flow between cells and recently such a transfer was seen from a normal to an abnormal human cell. Thus one day scientists may be able to re-educate defective cells by removing them from a child and letting them grow next to normal cells; once the original cells have learned their lesson, they could be reimplanted. Experiments with viruses have raised another possibility: a specially trained virus could be loaded with the normal gene for a particular defect and sent through the body to infiltrate the cells and take over from the defective gene.

The ethical and social implications of these and other such upcoming scientific wonders are of major concern to many genetic counselors, and they are actively seeking to promote public knowledge and discussion of the issues. They frequently make reference to the parallels between their situation and that of the nuclear physicists during the forties. Of more immediate concern, however, is a matter generally kept within the family: the shortage of educational programs, personnel and facilities to meet the growing public demand for counseling services.

At a recent symposium sponsored by the National Foundation, Dr. Orlando J. Miller, associate professor of obstetrics and gynecology at New York's Columbia University-Presbyterian Medical Center, told his colleagues: "We're not concerned with eugenics," but rather "with treating individual patients." And Dr. Valenti, of the Downstate Medical Center, insists that, as a counselor, it is not his business to "try to change the rate of mutation in man. When Mrs. Smith comes to me, she's the whole universe for Mrs. Smith."

Because so many counseling units have developed as adjuncts of research programs, there has been a natural tendency among the scientist-counselors to seek out and treat the unusual cases – those that will contribute to the progress of the program. (A side benefit of this approach is that many counseling patients are treated without charge, as part of the research budget of the hospital or laboratory. In general, the cost of counseling varies with the amount of time and testing required and the patient's ability to pay.) Foundation grants and Federal funds are provided almost entirely for research, not for counseling. "In terms of money input," says a Federal official in Washington, "a number of people here feel that, on the service level, there is not a very high-level need for genetic counseling. They're not worried about it now." Most Federal support of counseling is through the programs of the U. S. Children's Bureau for research to aid retarded and otherwise handicapped children; the counseling aspect is secondary.

There are a handful of service-oriented counseling centers. One such is at the University Hospitals of Cleveland, where any newborn suspected of bearing gene defects is given a chromosome test; if the test is positive, his parents are referred to counselors. According to Dr. M. Neil McIntyre, professor of anatomy and pediatrics at Western Reserve University, the center also welcomes referrals from private physicians. "We do research as well," he says, "but the emphasis, the reason for the unit, is service."

In 1964, New York State underwrote the cost of establishing a genetic counseling clinic at Albany Medical Center; it was specifically set up as a service for patients, with research benefits, secondary. And in 1967, the Legislature passed a law calling for the creation of a birth defects institute at the Albany center; its director, Dr. Ian H. Porter, says his organization will soon be giving grants to several institutions to set up counseling clinics.

The first major clinical center specifically designed to detect inherited defects in the fetus was recently established at The Johns Hopkins Hospital in Baltimore under a grant from the National Foundation. Genetic counseling will be available.

Such programs, however, cannot begin to meet the need. Most people are not even aware that counselors exist, but already, New York's Dr. Miller admits, he receives many more requests for service than he can handle. "We need to meet community needs," he says. "We all recognize they are not being met now."

One solution offered is a drastic improvement of the training in genetics given medical students. Some geneticists feel that the bulk of counseling duties in the future will necessarily be handled by the family doctor. Others expect that men and women specially trained as genetic counselors or counselors' aides will take up the slack. Sarah Lawrence College has plans to establish a program to train such aides. And Dr. McIntyre envisions a special university curriculum to develop counselors – "expert geneticists with some medical training and sufficient knowledge in psychological approach."

It's a tall order, as Dr. McIntyre admits, but not unreasonable. The genetic counselor can be nothing less.

19.

Genetic Counseling Units in the United States

Henry T. Lynch

State	City	Unit	Department
Alabama	Birmingham	Univ. of Alabama Medical Center	Laboratory of Medical Genetics
Arizona	Tempe	Arizona State Univ.	Zoology
California	Berkeley	Univ. of California	Zoology, Genetics
	Los Angeles	Children's Hospital	Pediatrics, Biochemistry
	Los Angeles	Univ. of Calif. School of Medicine	Pediatrics, Medicine
	Martinez	Genetics Consultation and Counseling Service	Health Department
	Oakland	Children's Hospital Medical Center	Birth Defects
	Palo Alto	Stanford Univ. School of Medicine	Medicine
	Palo Alto	Palo Alto Stanford Hospital	Pediatrics
	San Bernardino	St. Bernadine Hospital	Pediatrics
	San Francisco	Univ. of California	Pediatrics
	San Francisco	Univ. of Calif. Medical Center	Pediatrics
Colorado	Denver	Univ. of Colo. Medical Center	Pediatrics, Biophysics
Connecticut	Hartford	Connecticut Twin Registry	Health
	New Haven	Yale Univ. School of Medicine	Medicine
	New Haven	Yale Univ. School of Medicine	Pediatrics
	Ridgefield	New England Inst. for Medical Res.	Cytogenetics
Washington, D.C.		Children's Hospital	Neurology
		Georgetown Univ. Hospital	Obstetrics, Pediatrics
		Howard Univ. School of Medicine	Medical Genetics
Florida	Coconut Grove	Univ. of Miami Child Development Ctr.	Pediatrics
Georgia	Atlanta	Georgia Mental Health Institute	Psychiatry
	Augusta	Medical College of Georgia	Endocrinology
Illinois	Chicago	Children's Hospital	Biochemistry
	Chicago	Illinois State Psychiatric Inst.	Mental Health, Pediatrics
	Chicago	Medicine Blood Center	Pediatrics, Cytogenetics
	Evanston	Evanston Hospital	Research
	Springfield	Dept. of Public Health	Preventive Medicine
Kansas	Kansas City	Kansas Univ. Medical Center	Medicine
Kentucky	Lexington	Univ. of Kentucky Medical Center	Pediatrics
	Louisville	Child Evaluation Center	Pediatrics
	Louisville	Univ. of Louisville School of Med.	Pediatrics

State	City	Unit	Department
Louisiana	New Orleans	Genetic Counseling Center	Anatomy
Maryland	Baltimore	Johns Hopkins Hospital	Medicine
	Baltimore	Sinai Hospital	Pediatrics
Massachusetts	Boston	Birth Defects Center	Pediatrics
	Boston	Children's Hospital Medical Center	Clinical Genetics
	Boston	Massachusetts General Hospital	Medicine
	Boston	Massachusetts General Hospital	Pediatrics
Michigan	Detroit	Univ. of Detroit	Biology
	Lansing	Michigan State Univ.	Zoology
	Northville	Plymouth St. Home & Training School	Mental Health
Minnesota	Minneapolis	Minnesota Dept. of Health	Human Genetics
	Minneapolis	Univ. of Minnesota	Dentistry
	Minneapolis	Univ. of Minnesota	Genetics
Missouri	Columbia	Univ. of Missouri Medical Center	Pediatrics
	St. Louis	Children's Hospital	Pediatrics
	St. Louis	St. Louis Univ.; Cardinal Glennon Memorial Hospital for Children	Pediatrics
	St. Louis	Washington Univ. Medical School	Medicine
Nebraska	Omaha	Children's Memorial Hospital	Birth Defects
	Omaha	Creighton University School of Med.	Preventive Medicine
New Hampshire	Hanover	Dartmouth Medical School	Pathology, Medicine
New Jersey	Newark	New Jersey College of Medicine	Pediatrics
New Mexico	Albuquerque	Univ. of New Mexico School of Med.	Pathology
New York	Albany	Birth Defects Institute	Health Dept.
	Buffalo	Buffalo General Hospital	Medicine
	Buffalo	Roswell Park Memorial Institute	Pediatrics
	Buffalo	State Univ. of New York at Buffalo	Pediatrics
	Jamaica	Creedmoor State Hospital	Psychobiologic
	New York (Bronx)	Albert Einstein College of Med.	Medicine
	New York	Albert Einstein College of Med. of Yeshiva Univ.	Genetics
	New York	Cornell Univ. Medical College	Pediatrics
	New York	Mount Sinai School of Medicine	Pediatrics
	New York	New York State Psychiatric Inst.	Medical Genetics
North Carolina	Durham	Duke Medical Center	Obstetrics, Gynecology
	Morganton	Western Carolina Center	Birth Defects Evaluation Clinic
Ohio	Cincinnati	Children's Hospital Research Fdn.	Pediatrics
	Cleveland	Cleveland Metropolitan General Hosp.	Pediat. Neurology
	Cleveland	Cleveland Psychiatric Institute	Medical Genetics
	Columbus	Univ. Hospital	Medicine
	Dayton	Barney Children's Medical Center	Birth Defects Evaluation Center

State	City	Unit	Department
Oregon	Beaverton	Oregon Reginal Primate Center	Genetics
	Eugene	Sacred Heart Hospital	Pediatrics
	Portland	Univ. of Oregon Medical School; Crippled Children's Division	Genetics Clinic
	Salem	Fairview Hospital and Training Ctr.	Medical Research
Pennsylvania	Philadelphia	Children's Hospital	Pediatrics
	Philadelphia	Hahnemann Medical College	Anatomy
	Philadelphia	Jefferson Hospital	Medicine, Obstetrics, Gynecology
Rhode Island	Providence	Rhode Island Hospital	Pediatrics, Pathology
Tennessee	Knoxville	Univ. of Tenn. Memorial Res. Ctr. and Hospital	Birth Defects Evaluation Center
	Nashville	Vanderbilt Hospital	
Texas	Austin	The Univ. of Texas The Genetics Foundation	Human Genetics Res.
	Ft. Sam Houston	Brooke Army Medical Center	Pediatrics
Utah	Logan	Utah State Univ.	Zoology
	Salt Lake City	Primary Children's Hospital	Birth Defects Clinic
	Salt Lake City	Univ. Medical Center	Internal Medicine
Vermont	Burlington	Mary Fletcher Hospital	Pediatrics
Virginia	Charlottesville	Univ. of Virginia Hospital	Preventive Medicine, Internal Medicine
	Charlottesville	Univ. of Virginia School of Med.	Chromosome Res. Lab.
	Richmond	Medical College of Virginia	Biology, Genetics
Washington	Seattle	Mason Clinic	Pathology
	Seattle	Univ. of Washington	Medicine
West Virginia	Morgantown	West Virginia Univ. Hospital	Pediatrics
Wisconsin	Madison	Univ. of Wisconsin Medical School	Medical Genetics

Reprinted from the January 1968 issue of *Today's Health* as adapted from Lynch, Henry T. *Dynamic Genetic Counseling for Clinicians*, 1968. Courtesy of Charles C. Thomas, Publisher, Springfield, Illinois.

20.

Genetic Engineering: Controlling Man's Building Blocks

Joshua Lederberg

When we approach functions as complex as human intelligence and sympathy, we must be quite humble about our capacity to unravel the components of heredity and environment. AUTHOR

Humanics is the scientific study of human nature. It tells something of our past traditions that the word is an unfamiliar one; but the subject it describes is inevitably caught up in the recent rush of progress in experimental biology. In particular, dramatic advances in our knowledge of the biochemistry of deoxyribonucleic acid (DNA) and of its function as the material basis of heredity have provoked much new speculation about the application of this new knowledge to man and his problems. As a result of these advances, we anticipate better tools to mitigate disease, to improve agriculture, and to exploit microorganisms in industry.

We must also visualize, however, the impact of genetic engineering on humanics, which includes the possible modification of human nature toward previously unattainable ideals. Phrases like "genetic programming" or "genetic engineering" may conjure up the Frankensteinian image of a mad scientist or a technocratic dictator pushing the buttons that will control an assembly line of babies produced to order for service as infantrymen or storm troopers or docile subjects. Some may even imagine that their own genes may somehow be subjected to alteration at someone else's command or that they will have unlimited options to create any manner of offspring they wish – perhaps a child who might grow up as an athletic prodigy with an IQ of 350 and a head of hair that automatically shears itself at regular intervals.

Actually, our present knowledge of genetic science is not the obvious limiting factor for the furtherance of such aims. Rather, we lack the necessary insight into the essential biochemistry, developmental biology, psychology, and social dynamics of these phenomena. And indeed, were we to gain such insight, genetic engineering would probably be a redundant tool in competition with many other ways of influencing human development and behavior. To avoid the distorted view of genetic engineering that is all too prevalent in contemporary journalism, the topic must be examined within a broader view of man's evolutionary history and of the impact of established institutions on human biology.

Genetic engineering has been an important element in human cultural progress. The beginnings of agriculture depended on the remarkable insight that the seeds of a given plant would beget others like it. Early agricultural man, in his development

---

Reprinted from the November 1969 issue of *Today's Health* as condensed from the *1970 Britannica Yearbook of Science and the Future*.

of crops like Indian corn and wheat, accomplished technical miracles that have still to be surpassed by contemporary plant science. This kind of "biological engineering" – to produce reliable food crops from wild grasses – achieved a phenomenal result without the benefit of profound insight into the mechanism of heredity or the chemistry of DNA.

We have no way of knowing whether prehistoric man consciously applied similar principles to guiding his own evolution by selective breeding. In many subhuman primates the social hierachy does give a dominant male privileged access to receptive females during their intervals of maximum fertility. With the development of democratic ideals, however, the very concept of compulsory selective breeding as a method of engineering human improvement has been discredited as a violation of elementary human rights.

The principal utility of genetics in modern medicine is in diagnosis, now applicable to many specific genetic diseases with great precision, mainly by the use of biochemical and microscopic methods. It is often possible to counsel parents in a family where a rare disease has cropped up about the prospects of a similar anomaly occurring again in future children. Since many parents will respond to discouraging advice by not taking chances, this kind of genetic counseling is a *de facto* form of selective breeding. Its principal benefit, however, is intended to promote the integrity of the family and to prevent the conception of children likely to suffer from a serious defect. A by-product of genetic counseling in this situation is to reduce the frequency of defective genes in the next generation.

Because certain deleterious genes also can be detected in the hybrid carrier state, some individuals may use this information in their selection of mates. As yet, we have no reliable statistical information on the subject, but it is doubtful that any significant number of people take genetic factors into account when they fall in love and marry.

From a population-genetic point of view, selective mating does not help to eliminate a deleterious gene from the population; it merely postpones the overt occurrence of defective offspring. Since future generations may well be better equipped to repair a genetic defect than we are at the present time, selective mating can hardly be called an imprudent policy.

The genetic diseases to which such considerations apply are each quite rate, but there are enough of them to warrant the spreading practice of genetic diagnosis in order to furnish significant information to a considerable proportion of the population. The average human being carries the equivalent of eight or 10 potentially harmful genetic defects, all of which are usually masked in the hybrid condition. Although most of these defects are not now recognizable by biochemical analysis, studies of DNA specificity and of cell fusion are beginning to revolutionize our approach to these problems.

Cell cultures from specimens of blood or from tiny fragments of skin have enhanced the diagnosis of many genetic diseases. More recently, this technique has also been applied to cells obtained from the amniotic fluid that surrounds the developing fetus. By this method the occurrence of a serious genetic disease in a young fetus can be diagnosed, and the mother may request a therapeutic abortion to avoid bearing a severly crippled or retarded child. [In many states, this request cannot be granted legally.]

A number of genetic diseases can now be detected prenatally. Of these cystic fibrosis, a metabolic disorder in children, is the most prevalent. Its incidence, however, is too rare to recommend the routine examination of amniotic fluid in every pregnancy. On the other hand, the carrier state for cystic fibrosis can also be deter-

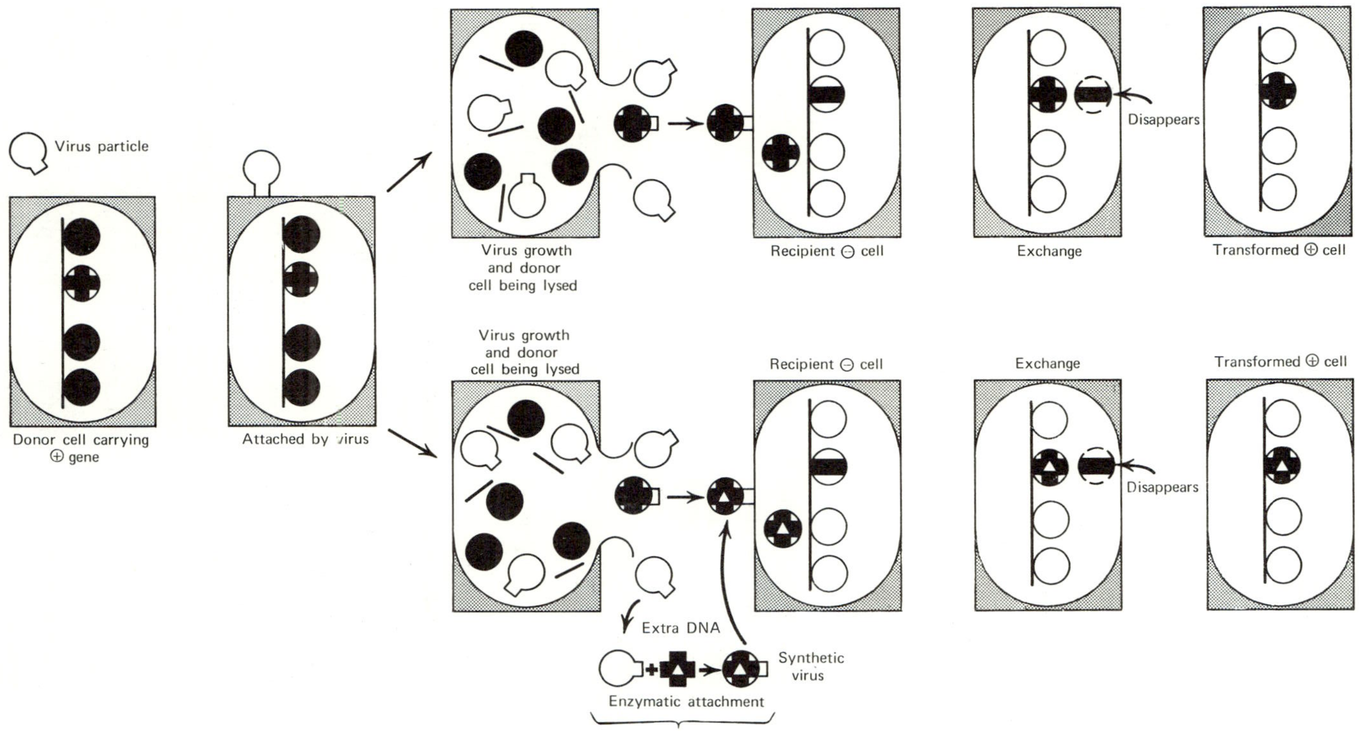

Figure 1 Diagram depicts man's laboratory accomplishments in genetic engineering. Following nature's lead molecular biologists can transplant a segment of DNA into a bacterium and change its genetic constitution. The process of transduction with a natural virus (top row) occurs commonly in certain kinds of bacteria. It requires the aid of a bacterial virus, which displaces certain genes in bacterial DNA. In the laboratory, transduction with a synthetic virus is accomplished by using enzyme to attach desired bit of DNA to special carrier virus. Resulting synthetic virus then infects a bacterium, transferring previously incorporated DNA into the virus (bottom row).

mined in the parents, and fetal examination would be indicated if there is already one chance in four that the fetus may be diseased.

While the elimination of fetuses having this serious genetic disease may appear to be a negative approach, this procedure should be weighed against the assurance that can be given parents of being able to nurture a child free from disease on future attempts. Eventually, a better understanding of the biochemistry of cystic fibrosis may lead to methods of treatment so effective that the disease would no longer be a serious burden to the young child.

In many respects, mongolism, or Down's syndrome, is more serious than cystic fibrosis because of its severe mental retardation. Prenatal examination can reveal the extra chromosome that causes the condition. Down's syndrome occurs in one of about 600 births, but certain individuals have a chromosome pattern that predisposes them to a much higher frequency of afflicted progeny. For such mothers, and mothers with pregnancies at advanced ages, a prenatal examination of fetal cells is especially indicated.

Paradoxically, additional conceptions undertaken to compensate for an eliminated fetus will tend to increase the frequency of the deleterious gene in the population. For example, the child with cystic fibrosis was, until recently, not likely to survive until reproductive age and, therefore, did not contribute to the gene pool of the following generation. Also, the diseased child tended to displace a potential sibling whose odds are two in three of being a carrier for the gene and who would eventually contribute to the gene pool.

If, however, our objective in this kind of medicine is to alleviate unnecessary human distress, then we should focus our attention on the reduction of the disease, rather than on the elimination of the gene for it. In spite of the obvious natural selection against it, the gene's very capacity to survive in the human population indicates that it might also carry some still unexplained and even beneficial function in human fitness.

Another characteristic that can be diagnosed by prenatal examination is the sex of the fetus. Improved determination of fetal sex at an early age and the development of drugs to induce a voluntary abortion may make individual control of the sex of offspring technically plausible. The use of abortion for this purpose, however, would probably be repugnant to most people.

Germinal choice, which is another approach to selective breeding, has been advocated strongly by Julian Huxley and the late [geneticist] Hermann J. Muller. Their scheme would provide for the banking of sperm from preferred men in cold storage for later voluntary use in artificial insemination. Real problems arise, of course, in the identification of preferred males.

However bizarre these schemes for selective breeding may appear, the present world does exhibit a wide disparity in the number of offspring produced by different parents. In some sense, our other social policies establish the pattern for these discrepancies. Yet, we know too little of human genetics to sustain an informed criticism (or approval) of that pattern. It is much more difficult not to be alarmed at some examples of negative family planning in relation to parents' ability to provide each child with the parental care that should be his birthright. On the other hand, we have still to devise compulsory schemes that can discourage overbreeding where it demeans a child's rights without at the same time creating an unacceptable invasion of the personal freedoms of the parents.

It is not easy to set clear boundaries for the subject of genetic intervention. Since one could postulate that the right set of genes could forestall the possibility of a

PROGRESS IN CELLULAR ENGINEERING

While genetic engineering someday may have revolutionary effects on medicine—and on mankind—another type of "biological engineering" is already saving lives.

Called cellular engineering, it is a technique in which the bone marrow of one person is transplanted into another. Today it is being used to treat patients with certain hereditary defects, such as Wiskott-Aldrich syndrome and lymphopenic agammaglobulinemia. Infants with one of these rare immunologic deficiency diseases (the inability to resist infections) have been injected with marrow cells "borrowed" from members of their families. When the transplanted bone marrow (the organ which manufactures disease-fighting corpuscles) "takes," it provides their bodies with the ability to protect themselves.

In cellular engineering, whole cells are transplanted. In genetic programming, on the other hand, scientists must deal with the DNA molecule within the chromosome within the nucleus of the cell. Partly because of this, many experts feel that cellular engineering may produce more immediate results in humans than genetic engineering.

But its future is by no means assured. Even with recently developed tissue-typing methods, there have been marrow transplant failures because of rejection. Further research into cell compatibility is needed. So far, cells from immediate family members have been used in most successful transplants. Research continues using donors outside the family.

In addition, no one knows what the long-term effects of cellular engineering will be. Only about a dozen of these operations have been successful. The longest living bone marrow transplant patients have survived slightly more than one year. But the early results seem encouraging.

"We are at the threshold of cellular engineering," observes Dr. Robert A. Good, professor of pediatrics and microbiology, University of Minnesota. "If we can cross this threshold, we will be able to attack many other diseases."

In addition to various birth defects, future targets of cellular engineering include certain types of cancer—chiefly leukemia—with associated immunologic deficiencies, liver and kidney diseases, and some types of inborn errors of metabolism. Some limited successes in leukemia research already have been achieved.

"There are many problems that must be solved first," Doctor Good acknowledges. "But the challenges of the future are exciting."

disease developing in a person, any disease can then be said to have a genetic aspect. Preventive measures, such as the Sabin vaccine for polio, can be regarded as artificial replacement of the human genes that are unequal to this one of life's challenges. We do not know, however, whether any living human beings already contain genetic factors for resistance to polio virus. If they did, the comparison of resistant and sensitive individuals would put the genetic aspect of this kind of medicine into sharper form.

The brain, in its growth, must be subject to some explicit regulation from external stimuli. This is an urgent item on the humanicist's agenda. We need powerful tools to deal, on the one hand, with obvious defects that cry out to be corrected and, on the other, with the possible enhancement of human intellectual ability. The closest parallel to this in present practice is the care that physicians take to be sure that pregnant mothers do not suffer from thyroid deficiency.

Another approach to the modification of an established genetic makeup is the transplantation of the tissue or organ from another individual. When the indication for such a transplant is a failing heart or kidney, obviously the operation is not a compensation for a genetic defect. The message is clearer, however, when the indication is a metabolically insufficient pancreas – say diabetes, though the primary lesion may be elsewhere – or a congenital deficiency in some other endocrine gland.

Tissue transplantation is still seriously impeded by two factors: the phenomenon of tissue rejection based on genetic incompatibility of different individuals and the serious difficulty of obtaining viable organs for transplant. Fundamental genetic studies on the determination of the protein structure of antibodies and of tissue antigens may be expected to eliminate the first obstacle. As for organ supply, a thorough understanding of tissue rejection may make it possible to use animal organs for transplant purposes. One approach that would allow the use of animal organs would be the early inoculation of infants with purified pooled antigens representing the tissue specificities of potential future organ donors.

Transplantation, in the sense of mixing cells of different origin into one organism, can be done experimentally at very early stages of embryological development. Already, as many as four different mouse eggs, representing eight different parents, have been fused to form a single embryo that matured into a single adult mouse. This procedure has great theoretical interest because of its potential for incorporating complementary advantages from a variety of different parental strains.

Except for a calculated choice of parentage, intelligent design plays a limited role in controlling the genetic makeup of an individual. In microorganisms, however, it is now possible to introduce specific new genetic information in a much more controlled fashion. But such experiments still have considerable random components, and usually it is not possible to instruct one particular cell to adopt a specific new genotype. Instead, a large number of cells are exposed to DNA that has been contrived to have the desired characteristics. One out of the many cells may incorporate the foreign DNA and with it some new characteristic. The occasional cell that responds in the appropriate fashion can then be separated from the other cells.

The great force of recent work in molecular biology stems from the use of relatively simple experimental materials, such as viruses and bacteria. The direct manipulation of individual genes within the chromosomes in cells of higher organisms, however, present formidable and possibly insuperable technical difficulties. Nevertheless, we can foresee the use of viruses to mediate the transmission of specific genetic information.

Contemporary work with animal viruses by a number of scientists now strongly suggests that these viruses may also be capable of introducing genetic information in the cells that they infect. For example, the SV-40 virus of monkeys, which as far as is known is harmless in man, leaves a number of copies of its DNA sequences in the chromosomes of cells infected in tissue culture. This suggests that viral DNA can be engineered and that synthetic viruses can be used for the modification of genetic defects.

For instance, it should soon be technically feasible to attach the genetic DNA that codes for the enzyme phenylalanine hydroxylase, which functions in the liver of normal men and animals, to the DNA of SV-40 virus. The inoculation of an infant with such a hybrid virus would be expected to alleviate the disease phenylketonuria (PKU).

Man has, in fact, been practicing a similar form of genetic engineering for quite some time. When Edward Jenner discovered the vaccination against smallpox, he introduced the use of a variant virus to compensate for a "genetic defect" shared by all mankind – namely, our inherent sluggishness in producing antibodies against the smallpox virus. The projected design for the use of more carefully engineered viruses to generate specific enzymes shows an obvious parallel to this long-established medical procedure.

This approach to genetic engineering also has the advantage that, in all likelihood the genetic information carried by such viruses is not incoporated in the sex cells for transmission to the next generation. This is a purely empirical observation. To be sure of keeping future options open, the limitation of virogenetic effect to somatic tissues must be carefully verified in every case.

At this writing, only one important technical difficulty remains – that of attaching specific segments of DNA from totally unrelated sources. However, enzymes discovered in 1968 for rejoining DNA molecules broken in just one strand are already being used as essential reagents in research.

The theoretical possibility of virogenic attachment to chromosomes and propagation to further generations cannot be completely determined without empirical study. If we do not keep a vigilant lookout on the effects of viruses – whether used for vaccination against disease, for genetic repair, or as infectious agents in our environment – we may be in for some unpleasant surprises.

Because the use of viruses for vaccination purposes has not been generally associated with the alarm of "genetic engineering," these agents do not receive the close attention they deserve in view of their biological potentialities. At the very least, viruses used for vaccination should be chemically purified and identified as having only the one desired species of DNA or RNA (ribonucleic acid). This standard has not yet been adopted by the pharmaceutical industry, nor has it been included in the regulations enforced by governments.

The extraordinary specificity of pairing by the two strands of a DNA molecule has opened the way for studying biological specificity, mainly by molecular hybridization. The specific reagent in these experiments is a solution of DNA single strands prepared from reference material. For some purposes, this DNA may be incorporated into a culture medium of solid agar or attached to the surface of filter membranes. When exactly complementary strands of DNA, or sometimes RNA, are added, conventional double-stranded DNA structures will be re-formed and can be detected by a variety of different methods. Some of these methods are so sensitive that it may be possible to discern even single nucleotide differences between a reference and an unknown in a sample.

These procedures will, undoubtedly, be instrumental for the isolation of specific-gene DNA, an objective which has already been achieved to a limited extent. Molecular hybridization also furnishes a method of distinguishing from each other the messenger-RNAs produced by different cells.

Other advances in cell biology have opened up some additional technical possibilities for the evasion of genetic scrambling that now invariably accompanies sexual reproduction. The propagation of new plants from cuttings is a familiar experience in horticulture. In lower animals such as earthworms, vegetative reproduction is a common occurrence; missing organs can regenerate spontaneously in small fragments cut from the previous individual. Deeper insights into the mechanisms of this embryological development could lead to similar phenomena, even in man. But these are remote prospects indeed.

Groups of individuals derived by vegetative propagation and having identical genetic constitutions are called "clones." The prospect of producing genetically homogeneous groups of individuals presents some interesting issues. In addition, it is a way of propagating a genotype already tested in one generation for further trial in a second. We already have a foretaste of the properties of a clone in the behavior of identical twins.

Clonal propagation would afford an otherwise unavailable opportunity for certain humanic experiments. Without such tests it is unlikely that we will ever be able to know the extent to which the performance of acknowledged geniuses or athletic stars are manifestations of unusual genetic endowment.

The technical limitation to human cloning is mainly the much smaller size of mammalian eggs when compared with the egg of a frog, but this is not an insuperable difficulty. There may be, however, other obstacles based on differences in the biology of the frog egg and that of the human that are not yet known.

Within the last few years, it has been discovered that tissue cells can be made to fuse with one another in the presence of certain virus-derived particles. These cells thus form "vegetative hybrids" that can originate from such widely distinct species as fish and human. The technique has already become quite important in the analysis of the genetic functions carried by different human chromosomes, which can be tested for their ability to make up for known defects in other animal cells.

Vegetative hybridization and the use of the hybrid cells to renucleate an egg open the door to another form of genetic engineering – hybrid plants or animals containing some or many chromosomes from distant species.

When we approach functions as complex as human intelligence and sympathy, we must be quite humble about our capacity to unravel the components of heredity and environment. Certainly, there is no gene that can ensure the ideal development of a child's brain without reference to tender care and inspired teaching. The paths to intelligence can be deviated at many points. For example, the child born deaf was, for all practical purposes, an idiot until we learned the special techniques needed to teach him.

These considerations suggest that the main role of genetic science may be to sharpen perceptions of how to engineer the environment for the optimum development of existing genetic types. When we have reached some mastery of this challenge, we can more reasonably advocate the extension of genetic engineering beyond the repair of the most obvious and urgent forms of genetic defect.

21.

Hopi Indians, Inbreeding, and Albinism

Charles M. Woolf and Frank C. Dukepoo

Although selection for the heterozygote or genetic drift may account for the high frequency of the albino gene in the Hopi population, the most apparent explanation is acquisition of the gene by migration and cultural selection . . . . The frequency of albinism among the Hopis will decrease rapidly with the decline of their culture. AUTHORS

Albinism results from an inborn error of metabolism that involves the conversion of tyrosine to a brown or black insoluble polymer. Melanogenesis takes place in melanocytes, a group of specialized cells. The first step, the hydroxylation of tyrosine to dihydroxyphenylalanine (DOPA), is catalyzed by tyrosinase. The second step, the oxidation of DOPA to the quinone, is also catalyzed by tyrosinase. The rest of the pathway leading to the production of the polymer (melanin) may be nonenzymatic (*1*).

Although albinism in man is most often caused by a recessive autosomal gene, genetic heterogeneity may be present. The Trevor-Roper pedigree (*2*) of two albino parents producing normally pigmented offspring is suggestive evidence for at least two different types of recessive albinism in man. Supportive evidence comes from the experiments of Witkop, Van Scott, and Jacoby (*3*) who placed unfixed albino hair bulbs into a solution of L tyrosine (50 milligrams per 100 milliliters, *p*H 6.8). They observed that the hair from some albinos formed pigment, but the hair from others did not. Genetic heterogeneity is not unexpected since a mutation in the structural gene, a controlling gene involved in the production of tyrosinase, or at some other locus resulting in the presence in the melanocyte of some enzyme-binding substance, would all disrupt melanogenesis.

Albinism occurs at different frequencies in various human populations. In Europe the overall frequency is about 1 in 20,000, with estimates ranging from 1 in 10,000 in Norway to about 1 in 30,000, or less, in southern Europe (*4*). In marked contrast are Indian populations in Central and North America. The Cuna Indians of San Blas Province, Lower Panama, show a frequency of about 1 in 200 (*5*). Frequency values of a similar magnitude characterize the Hopi Indians of Arizona and the Jemez and Zuni Indians of New Mexico (*6*). These North American Indians are not well known as having a high frequency of albinism, mainly because of the reduced population. A frequency of about 1 in 200 results in a conspicuous number of albinos in a total population of about 20,000 Cuna Indians; the impact of this frequency is not as great in populations of less than 6000 individuals.

The detrimental nature of albinism has been well established. Albinos are prone to skin cancer. Myopia and lateral nystagmus are usually part of the genetic syndrome,

Reprinted from *Science*, Vol. 164, pp. 30-37, April 4, 1969.

and albinos are extremely sensitive to sunlight. Neel *et al.* *(7)* conclude that, in present-day European and Japanese populations, the relative reproductive (Darwinian) fitness *(f)* of albinos is about 0.7 to 0.8, but in past generations it was more like 0.4 to 0.5. A relative fitness of 0.4 implies that albinos leave 40 functional offspring for every 100 left by nonalbinos. When a deleterious trait is due to an autosomal recessive gene *(c)* and when carriers *(Cc)* and noncarriers *(CC)* have the same relative fitness value ($f=1$), the frequency of affected individuals *(cc)* in a large population is a function of both the mutation rate (μ) and the selection coefficient ($s=1-f$) against affected individuals. The frequency at equilibrium is $R=\mu/s$. The mutation rate at the albino locus is unknown, but if a value of 1/40,000 is assumed (that is, there is one gamete containing the mutant gene among 40,000 produced), and the relative fitness of albinos is given the value $f=0.4$, the frequency of albinos at equilibrium in the population would be $R=\mu/s=(1/40{,}000)/(0.6)=1/24{,}000$. For this mutation rate, which is an acceptable value for man, the frequency of albinos at equilibrium would be relatively low, even if selection against them were less severe. For example, if the fitness of albinos were 0.9, then the frequency of albinos at equilibrium would be $R=(1/40{,}000)/(0.1)=1/4{,}000$. The occurrence of Indian populations with an albino frequency of 1 in 200 indicates that some force other than mutation and selection against albinos has been operating in past generations.

Hopi Indians and Albinism

During the spring of 1961, one of us (C.M.W.) became interested in albinism among the Hopi Indians because an acquaintance remarked that she had observed an albino child in one of the Hopi villages. A trader than told her that albinos are common on the reservation, and the Hopis consider it a "good luck charm" to have one or more residing in their village. Hrdlička recounted a visit to the Hopis in 1900 and described 11 albinos from ten sibships (*8*). Based on this number of albinos and an estimated population size of 2000, the frequency in 1900 was 1 in 182.

The Hopi reservation is located in northeastern Arizona, where it is completely surrounded by the Navajo reservation. The Hopi villages with one exception are located on the top or at the base of three mesas, known as First Mesa, Second Mesa, and Third Mesa, which are fingers off Black Mesa (Fig. 1). The villages on the mesas are adjacent or separated by only a few kilometers. The villages of Hano, Sichomovi, and Walpi are on First Mesa. Polacca, which is part of the First Mesa population, is located at the base of the mesa. About 19 kilometers westward is Second Mesa with the villages of Mishongnovi, Shipaulovi, and Shongopovi. Another 16 kilometers westward is Third Mesa with Bacavi, Hotevilla, and Oraibi on the top, and New Oraibi at the base. Oraibi, occupied since about A.D. 1150, is the oldest continuously inhabited community in the continental United States (*9*). Moencopi, founded originally by members of Oraibi, is located 64 kilometers farther west on the western range of the Navajo reservation. The Hopi inhabitants of Moencopi are included in the Third Mesa population.

The village of Hano on First Mesa was established by Tewa emigrants from the Rio Grande area in New Mexico. Seeking refuge after the Spanish revolt, they arrived in about 1700. They were allowed to remain on First Mesa on the condition that they help protect the Hopis from hostile nomads (*10*). The Hopis and Tewas on First Mesa have intermarried extensively in recent generations. In this paper, unless specified otherwise, the Tewas are considered as part of the Hopi population.

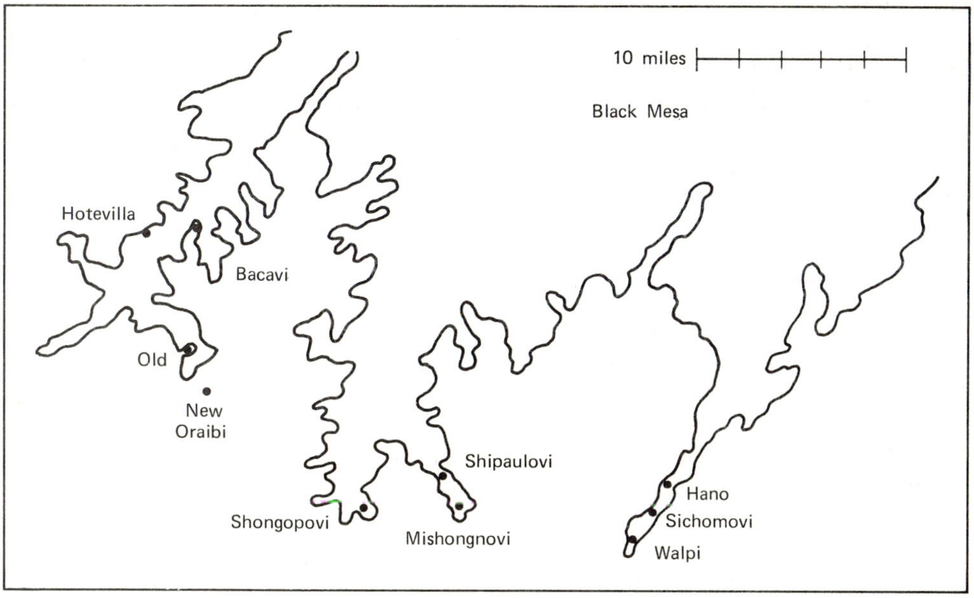

Figure 1 Location of the Hopi villages on or at the base of First Mesa, Second Mesa, and Third Mesa.

Accurate census data are not available for the Hopi population. The Hopi Indian agency, Bureau of Indian Affairs, Keams Canyon, Arizona, estimates (1968) that there are about 6000 Hopi Indians living on or off the reservation. The Hopi-Tewa population on First Mesa numbers about 1000.

Objectives of Study

A study was initiated with the following objectives: (i) to determine the frequency of albinos in the Hopi population; (ii) to construct pedigrees using living albinos and those described by Hrdlička as reference points; (iii) to estimate the mean population coefficient of inbreeding for the various village and mesa populations in order to determine the influence of inbreeding on the frequency of albinos; and (iv) to determine, if possible, what forces account for the high frequency of the detrimental gene in this population.

A pilot study (*11*) verified Hrdlička's conclusion that albinos are common in the Hopi population. The observed frequency in 1962 was 1 in 227. It was soon noted that generalized albinism occurs only in the Second Mesa and Third Mesa populations (*6*). Two male siblings with piebaldness (partial albinism) and subtotal nerve deafness are from First Mesa. These males are striking because of their uniformity in depigmented pattern even though they differ in age by 4 years (*12*). The gene responsible for this syndrome is different from the one causing albinism in the Second and Third Mesa populations.

It was a formidable and sometimes impossible task to construct Hopi pedigrees. Hopis tend to be suspicious of anyone, especially a white person, who attempts to obtain information about their families or culture. They fear he is going to "write a book" and expose Hopi secrets. The inhabitants of Hotevilla and Bacavi are especially suspicious. There are no written records and traditional Hopis wish not to mention the name of deceased relatives. "Don't speak his name, let him rest in peace." It is not uncommon to find a person who has never heard the given name of a deceased grandparent. Among traditional Hopis, pre- and extramarital relations are accepted as part of the Hopi way of life. Divorce is easy to obtain, and new arrangements are accepted by the villagers. These activities introduce errors into pedigree studies; however, Hopis talk about these activities and are often aware of true paternities. The matrilineal clan system and occurrence of ceremonial and adopted relationships present a problem of communication in a genetics stydy. Hopis inherit membership in the clan of their mother. Men and women of the same clan and same generation are considered as brothers and sisters. When a young Hopi is initiated into a society he acquires a ceremonial father. Furthermore, an older man who befriends a Hopi in a special way may become an adopted father. In a conversation with a white person, a Hopi may not make a distinction between a biological, clan, ceremonial, or adopted relationship. He may refer to several different women as his mother, name distant relatives as brothers and sisters, or state that a given man is his father, then later refer to the same man as his older brother or son (*13*). Rclationships are bewildering to white persons and even to young Hopis, as evidenced by an account in the autobiography of Sun Chief (*14*).

"So this old man came in the morning for peaches and collected peach stones to plant. He would walk past me with his stick and say, 'Good morning and good luck to you, my father.' I did not like to be called father (ina'a) and showed that I was offended. One day my mother said to me, 'Don't treat your son that way, Chuka. He used to have a father who belonged to our Sun Clan. His father was your great-great uncle; that makes me his aunt and you his father. Try to treat your son better.' After a while I grew used to having him call me father"

Definitive progress was made in the collection of pedigrees when one of us (F.C.D., a Hopi Indian from First Mesa) became associated with the study. He obtained pedigree data (*15*) which would have otherwise been most inaccessible.

In this matrilineal society, all property is owned by females, and even though exceptions do occur, it is customary for an individual to identify with the place of residence of his mother, which is usually his place of birth and rearing. Even if a Hopi is born and reared off the reservation, he will identify, with his mother's mesa. If his mother is not a Hopi, an individual may be adopted into a clan of his choosing and may identify with the place of residence of his father. With these few exceptions, the matrilineal system was used in our study to denote village and mesa membership.

In the total population of 6000, there are 26 albinos, or 1 in 231. As in 1962, all cases of albinism observed occur at Second Mesa and Third Mesa. Apart from First Mesa, the frequency is 26 in 5000, or 1 in 192. The albinos are affiliated with the following villages: Mishongnovi, nine; Shipaulovi, four; Shongopovi, one; Bacavi, one; Hotevilla, five; New Oraibi, one; and Moencopi, five. Three albinos living in 1900 were not described by Hrdlička – two from Oraibi and one from Shipaulovi.

Abbreviated pedigrees showing living and related albinos are given in Fig. 2. Also shown in these pedigrees are the 11 albinos observed by Hrdlička in 1900. These latter albinos are designated by the letter *H*.

Inbreeding in the Hopi Population

The frequency of a recessive trait in a large population is a function of the frequency (q) of the recessive gene and the degree of inbreeding among the members of that population. This is expressed by Wright's equilibrium law (*16*) where the

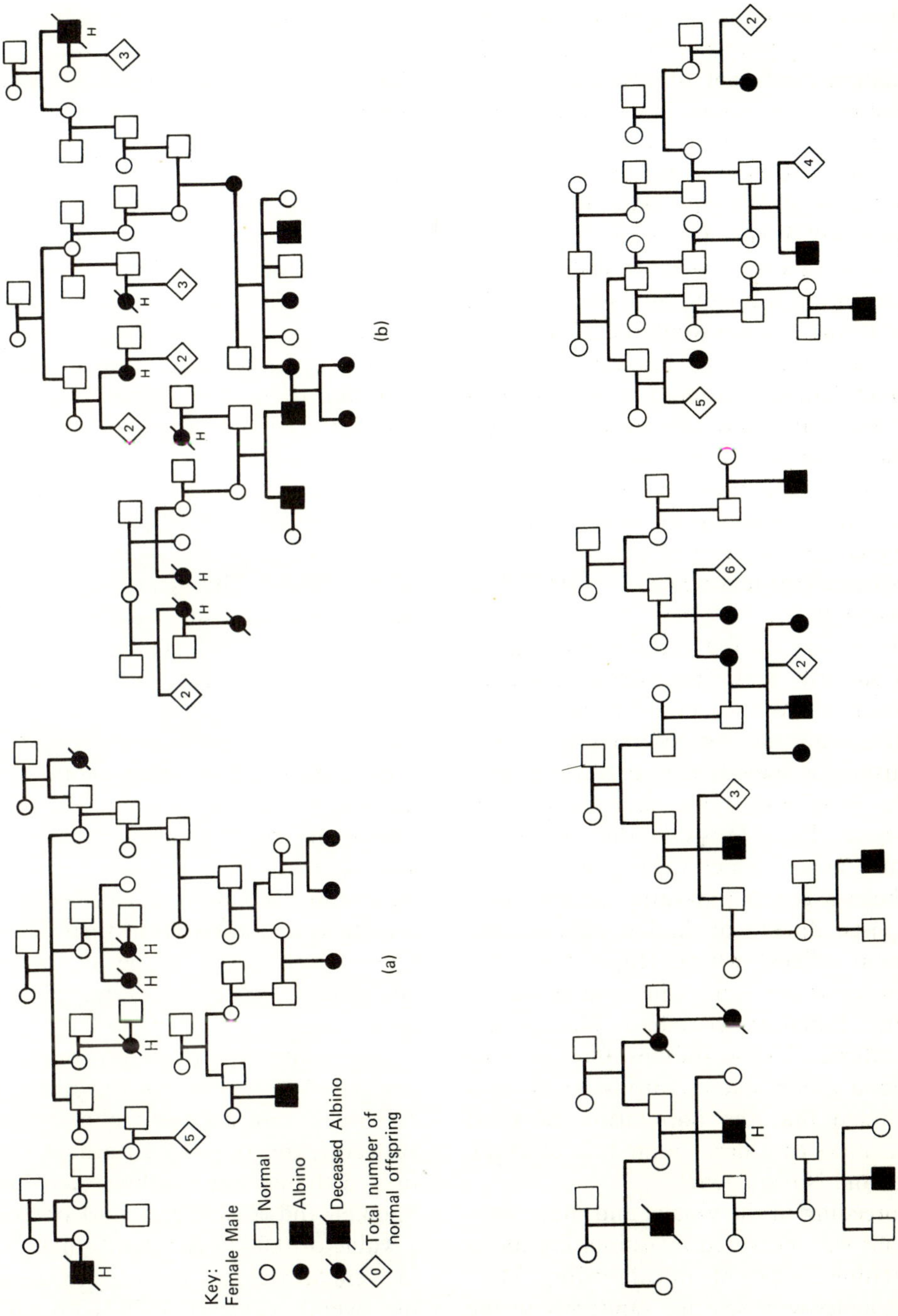

Figure 2 (a)–(e) Pedigree Nos. 1-5 of albinism. The albinos symbolized by *H* in these pedigrees were observed by Hrdlička in 1900 (*8*).

frequencies of *CC, Cc,* and *cc* individuals in a population at equilibrium are given by $D = p^2 + \alpha pq$, $H = 2pq(1 - \alpha)$, and $R = q^2 + \alpha pq$, respectively. The mean population coefficient of inbreeding symbolized by α can range, theoretically, from 0 (random mating) to 1 (complete inbreeding). The Hardy-Weinberg law is a special case of Wright's equilibrium law, where $\alpha = 0$. When, for example, the frequency of a recessive gene is $q = 0.001$, the frequency of those homozygous for the gene ($R = q^2 + \alpha pq$) would be 20.98 times more frequent in a population where $\alpha = 0.02$ than in a random mating population. Inbreeding has relatively little effect on the frequency of individuals homozygous for a common recessive gene. If $q = 0.1$, the frequency of individuals homozygous for the gene would only be 1.18 times more frequent when $\alpha = 0.02$ than when $\alpha = 0$.

The history of the three mesa populations suggests that First Mesa should be the least inbred and should have the lowest frequency of the recessive gene for albinism. An obvious reason is the mixing of the Hopi and Tewa gene pools. Furthermore, First Mesa is the closest to Keams Canyon, which during recent times has been a center (boarding school, trading posts, Bureau of Indian Affairs government offices) where Hopis have encountered white Americans and members of other Indian tribes, mainly Navajo and New Mexico Pueblo Indians. On the basis of blood group studies, Brown *et al.* (*17*) concluded that the population of First Mesa shows considerable mixture of Hopi, Tewa, and Navajo genes.

Inbreeding in the village and mesa populations was measured by estimating mean population coefficients of inbreeding. Married couples were selected randomly in the various villages. For each couple, an inbreeding coefficient was calculated from the formula $F = (\frac{1}{2})^N [1 + F_A]$, where N is the number of individuals along the chain from one individual to another and F is the inbreeding coefficient of a common ancestor (*18*). The F represents the inbreeding coefficient of a child born to each couple; therefore, this method assumes that each married couple has the same reproductive potential. The number of married couples sampled was 388 (First Mesa, 131; Second Mesa, 133; and Third Mesa, 124). The F values were averaged to obtain an estimate of the mean population coefficient of inbreeding. Moencopi was not studied because it is not on the Hopi reservation, and it was considered more informative, for the time available, to concentrate on the villages geographically associated with the three mesas.

The data show that Second Mesa is the most inbred ($\alpha = 0.01365$), Third Mesa is intermediate ($\alpha = 0.00702$), and First Mesa, as expected, is the least inbred ($\alpha = 0.00250$). The mean population coefficient of inbreeding based on all 388 F values is 0.00797. These estimates are lower than the true values since it was difficult to trace pedigrees back for more than about six generations. Furthermore, even a Hopi Indian from another mesa is unable to obtain cooperation from all inhabitants of Hotevilla and Bacavi. Consultants from these villages helped collect data, and even they had difficulty in obtaining cooperation from their fellow villagers. Refusal to cooperate introduces additional bias in the estimates of α for these villages.

The Hopis are relatively inbred when compared with various European, white American, and Japanese populations (*19*). However, the Hopis are perhaps no more inbred than other southwestern Indians. Spuhler and Kluckhohn (*20*) obtained an α value of 0.0080 for the Ramah Navajos in New Mexico, a value almost identical to the overall value (0.00797) obtained for the Hopis.

Before the coming of roads and modern means of transportation, the mesa populations were semi-isolates. The distances involved, the rigors of a trip during the wintertime, and the threat of attack by Navajos and Utes during certain periods of time discouraged the casual exchange of visitors among mesas. These factors promoted mesa endogamy in the biological sense. Even today cultural activities are strongly mesa-oriented, resulting in a certain degree of rivalry among the mesas. Although the physical location of the villages has strongly favored biological endogamy, cultural regulations forbidding certain types of consanguineous matings have reduced the inbreeding potential. Closely related clans are grouped into a single phratry, a brotherhood or fraternity of clans. An individual is discouraged from marrying or having a sexual relationship with a member of his own clan or phratry, or with a member of his father's clan or phratry. Such restrictions eliminate certain relatives as spouses. However, it is of interest that this cultural regulation allows for marriage between half-sibs when they have the same father and their mothers belong to unrelated clans. Such marriages occur.

The frequency of the albino gene may be used as an indicator of the "purity" of the Hopi gene pool. Second Mesans tend to boast that they marry non-Hopis less frequently, and that they are the "most Hopi" of all. They also boast the largest number of albinos. No Hopi now living recalls the birth of an albino at First Mesa; however, the albino gene is present in this population. Two males from First Mesa married to females from Mishongnovi and Moencopi are the fathers of albino children. Each male had grandparents from Second Mesa. The mixing of the First Mesa gene pool has greatly reduced the frequency of the albino gene. Mixing is also found in a section of the Third Mesa population.

At the turn of the century, a serious dispute arose among the inhabitants of Oraibi, then the only village on Third Mesa. Some (the "friendlies") wished to support the government by enrolling their children in schools. Others (the "hostiles") wished to maintain a traditional Hopi way of life. As a result, the hostiles moved from Oraibi in 1906 and established the village of Hotevilla, and later the village of Bacavi. New Oraibi was established at the foot of the mesa by friendlies so that they could live in close proximity to the government school. Some of the friendlies from New Oraibi selected spouses who were not Hopi. The split has led almost to the total decline of Oraibi. Only a few families live there. The residents of Hotevilla and Bacavi soon found themselves in the dilemma of having too few clans. They sought marriage partners from Moencopi rather than from Oraibi or Second Mesa. Moencopi, located 64 kilometers away on the Navajo reservation, consists of Hopi and Navajo inhabitants. It is divided into an upper and lower village. The lower village consists mostly of traditional Hopis; the upper village is composed of Hopis and Navajos. Mixing of Hopi and Navajo genes has occurred at Moencopi, especially in recent years. These marriage patterns are reflected in the data on inbreeding (Table 1) and the distribution of albinos. The 26 albinos tend to be of pure Hopi lineage. The five albinos from Moencopi are from the lower village. A major exception occurs in the village of Shipaulovi on Second Mesa. The mother of two albinos is a Navajo. The albino children have been adopted into a Hopi clan, identify with the village of their father, and are considered Hopis by the Hopis. Albinism is not uncommon to the Navajo, especially in the vicinity of the Hopi reservation (*6*). This is best explained by the Moencopi influence and the fact that for centuries the Navajos have been raiding the Hopi villages for the purpose of obtaining food, women, livestock, and other booty.

Table 1. Inbreeding Coefficients in Different Segments of the Hopi Population.

Village	Number of marriages	1/8	1/16	1/32	1/64	1/128	Other	α	
				First Mesa					
Hano	50	0	0	1	1	1	0	0.00109	
Sichomovi	51	0	1	0	0	1	0	0.00137	
Walpi	30	1	0	1	0	0	1	0.00677	0.00250
				Second Mesa					
Mishongnovi	43	1	2	2	2	2	3	0.01226	
Shipaulovi	28	0	5	3	3	1	3	0.02176	
Shongopovi	62	0	6	5	5	4	5	0.01222	0.01365
				Third Mesa					
New (Old) Oraibi	59	0	2	1	2	1	1	0.00463	
Hotevilla	37	0	4	2	1	3	1	0.00950	
Bacavi	28	0	3	1	1	1	0	0.00864	0.00702
Total	388	2	23	16	15	14	14	0.00797	
		(0.5%)	(5.9%)	(4.1%)	(3.9%)	(3.6%)	(3.6%)		

Estimate of Gene Frequency

The frequency of albinos and the inbreeding coefficient can be used to estimate the gene frequency in the Second and Third Mesa population. Based on an estimated population size of 5000, the frequency of albinos becomes $R = 26/5000 = 0.0052$. If $\alpha = 0.01088$ for Second and Third Mesa, and if we assume equilibrium, and solve for q in the equation $R = q^2 + \alpha pq$, we obtain a gene frequency of $q = 0.067$. The frequency of heterozygotes in this population becomes $H = 2pq(1 - \alpha) = 0.124$. The high frequency (12.4 percent) of carriers of the gene for albinism is reflected in the pedigrees shown in Fig. 2. The frequency of the gene for albinism is so high that even if the population resorted to random mating, the frequency of albinos would be reduced only slightly ($R = 1/223$ as compared with the present value of $R = 1/192$). The intriguing problem is to account for the high gene frequency in this population, as well as the Cuna, Jemez, and Zuni Indians.

Selection for the Heterozygote

A deleterious recessive gene is maintained at a relatively high frequency in a population when the heterozygote has a reproductive advantage. For example, if the relative reproductive fitness values for heterozygotes and albinos are set at 1.00 and 0.5, respectively, it can be shown that the relative fitness of homozygous persons would have to be only 0.962 in order to maintain an albino frequency of $R = 1/200$ in a large population. This slight advantage of the heterozygote, which would be difficult to measure in a practical situation, might manifest itself by increased fertility, vigor, maturation rate, or resistance to a disease. The implication is that, in the presence of a specific homozygous background produced by genetic drift and many generations of inbreeding, heterozygosity for certain mutant genes increases reproductive fitness. The corollary is that the advantage ceases with the advent of genetic variability. The presence of albinism at a high frequency in some populations and not others would imply that the latter lacked the proper genetic background or the allele had not been introduced by migration or mutation. This form of selection cannot be ruled out as the explanation for the high frequency of albinism in various Indian populations.

Genetic Drift

Genetic drift implies that a gene could be common in a population if one or more of the few early settlers carried the gene (founder principle), or because of chance segregation and recombination during a time when the population was small. Indian populations in the southwest have been plagued by epidemics, periodic starvation conditions, and raids by hostile groups. These populations have often consisted of a small number of individuals. If the albino gene were present in one of these small populations, the frequency would fluctuate and might rise by chance to a relatively high value. Furthermore, it is likely that many, if not all, of the present southwest Indian populations were founded by a small group of migrants. The frequency of the albino gene could be high in a present population if one or more of the founders were heterozygous for the gene.

If the high frequency of the albino gene in the Hopi population is accounted for by a chance process occurring in past generations, what is the origin of the mutant gene? The albino gene in the Hopi, Jemez, and Zuni Indians is likely "identical by descent" since gene flow occurred among the Pueblo Indians of Arizona and New Mexico in historic and prehistoric times. The presence of albinism in the remote Cuna Indians suggests further that the mutant gene is not of recent origin, and draws attention to Central America as the possible site of origin. There were many albinos in Montezuma's royal household at the time of the conquest of Mexico. Albinos along with malformed individuals were considered by the Aztecs as suitable wards of the state (*6*). The extent of this practice in Central America before the coming of Cortez is unknown. The protection of albinos in early American cultures would have favored the propagation of the albino gene in certain segments of the population. Groups migrating northward could then have carried the albino gene to the southwestern part of the United States. Climate would influence greatly the destiny of the albino gene acquired at a high frequency by migration. Natural selection against the gene would be more effective in the hot desert populations of southern Arizona than in the more temperate climates of northern Arizona and New Mexico. Pueblo dwellings which afford protection from the sun would lessen the effectiveness of natural selection.

The hypothesis of genetic drift may be questioned because of the existence of albinism at a high frequency in three different southwest Indian populations. It is not likely that a deleterious gene would reach a high frequency in all three populations by only a chance process. Some selective mechanism has likely been operating in one or more of these populations.

Role of Cultural Selection

A high frequency of a gene in a population is determined by processes operating in previous generations. We believe that the most acceptable hypothesis to explain albinism in the Hopis is the acquisition of the gene by migration (or mutation) and maintenance at a high frequency by cultural selection. Albinos have been protected in the Hopi society.

Traditional Hopis believe that albinism, along with other congenital defects, is the result of some specific action in the life of a parent or close relative. Typical retrospective explanations are as follows.

"When ——— was a young man, he owned a white donkey. Everywhere he went he rode his donkey. He liked it so much two of his granddaughters were born white."

"———liked to portray Eototo (white kachina). Therefore [his daughter] was born white."

"——— worked with white sand while his son was being formed."

"His mother [father] slept with a white man [woman]."

There is no recognition of the heredity phenomenon. An albino child born to an albino parent is explained merely by the existence of "whiteness" in the parent. Because of this logic it is assumed that albino parents may produce albino children.

Younger Hopis who have learned that albinism is a genetic defect may show a certain degree of discrimination against albinos; however, comments made by traditional Hopis reflect only a positive attitude toward them. Albinos are accepted completely in the Hopi society. They perform in ceremonies along with other Hopis. Several have been influential chiefs and priests. One, who was chief at Oraibi about 100 years ago, is legendary as a rainmaker. "During a plague of prairie dogs one fall he is said to have called for a Masau Katcina dance, which caused a veritable cloudburst that drowned all the prairie dogs" (*13*). Comments by traditional Hopis illustrate the acceptance of albinos in the Hopi society.

"Albinos are smart, clean, nice and pretty. There is nothing wrong with them."

"Albino girls are very pretty. I like their nice color. I would like to marry one, but they probably wouldn't have me."

"I would like to have an albino baby. There were albinos when I was a child, so they are not new to me. In fact, I am related to one. He's my brother [clan affiliate]."

"I know lots of them, and they are good Hopis."

Here [on Second Mesa] we have lots of them and we are very proud of them. There are more here than at the other mesas. We take good care of them."

"It is lucky to have albinos [in the village]. They are very special."

Albinos are considered as part of the Hopi heritage on Second and Third Mesa. This was evident in the revealing answer to the question asked of a member of Third Mesa as to why there are no albinos on First Mesa. "Oh, they are Tewas."

Marriage is important in the life of a Hopi. It is extremely rare to find an adult, especially a woman, who has never been married. For this reason, the fact that many albinos have remained unmarried contradicts the attitude of complete acceptance in the traditional Hopi society. One albino male stated he was unable to work in the fields and therefore could not grow corn for a wife and family. Although the inability to provide adequately for a family is the excuse usually given for males, the reason is not obvious for females. Answers to questions as to why specific albino females did not marry usually implied withdrawal of the female. The difficulty of being an albino in a dark-skinned population seems to have such a repressive action that suitors are rejected. This point needs further investigation. The acceptance by Hopis of this Hopi heritage apparently stops short of marriage. Females should reject males who cannot work in the fields. Titiev (*13*) quotes a discriminating comment regarding the albino chief from Oraibi who had the reputation of being a good rainmaker: "Nakwaiyamptiwa never married, 'because in those days ladies did not like albinos.' "

However, any real or subconscious bias towards marrying an albino apparently does not manifest itself in the selection of a sex partner. For this reason albino males in past years had an advantage in regard to sexual activity. Hopis for centuries have made their living by farming. Men and boys would leave the mesa tops and go to the fields, often many kilometers from the mesas. It is commonly known, among the inhabitants of Second and Third Mesa, that albino males were not expected to labor in the fields. They remained in the villages and performed tasks such as weaving. In the villages they were protected from the sun and hostile nomads, and most importantly, they had ample time and opportunity to engage in sexual activity. One albino is especially legendary for this activity.

"I knew that old man. Some say he had about twelve kids; others say about fifteen."

"He never married."

"They say he was always around trading with the ladies. He would make babies with them."

"He was real funny, and knew a lot of good stories. Everyone liked him."

In a small population, a sexually active male will have an imposing effect on gene frequencies of the next generation.

The acquisition of the albino gene by migration (or mutation) and a form of cultural selection whereby albino males are given a slight sexual advantage would explain the frequency of the albino gene in the Hopi populations. Whether a similar hypothesis is acceptable for the high gene frequency among the Cuna, Jemez, and Zuni Indians remains to be determined. Indeed there is no reason to conclude that the high frequency in all four of these Indian populations is due to the same force or forces. There appears to be cultural selection against albinos in the Cuna population (*5*).

Albinos are not considered as "good luck charms" for a village. The word lucky is used frequently in the English vocabulary of a Hopi. The meanings of the expressions "He is lucky with corn" or "He is lucky with girls," are apparent. However, the statement "It is lucky to have albinos," can be given the wrong interpretation by an outsider. The statement implies that it is difficult to produce albinos because they do not appear in all families or in all villages. There is no evidence that any special religious significance was ever attached to albinism that would predispose to the present positive attitude toward them. This point was investigated thoroughly because Hopi tradition would allow for a religious symbol being attached to albinism.

Hopis share the legend permeating Central American Indian cultures that someday a white leader will return and lead them to a better life. Just as Cortez was welcomed by the Aztecs as their white leader, Pedro de Tovar, the first white man to visit the Hopis, was welcomed by the Hopis as their "lost white brother" (*21*). Tovar and his men quickly subdued the Hopis. The lost white brother has yet to arrive. If the birth of an albino was ever considered as a symbol or reminder of their lost white brother, it is unknown to any living Hopi. The intriguing problem remains, however, as to why albinos, especially albino children, are considered so affectionately in all villages. The admiration of their whiteness is clearly not an identification with white Americans; it represents instead an association of whiteness with cleanliness, goodness, and purity; attributes honored by traditional Hopis. Rain is essential for existence in the barren mesa country. White clouds symbolize rain and are a constant reminder of the necessity of living a pure life. "If Hopis live a clean, good and harmonious life, there will be plenty of rain and an abundance of food for the children to eat."

Conclusion

Although selection for the heterozygote or genetic drift may account for the high frequency of the albino gene in the Hopi population, the most apparent explanation is aquisition of the gene by migration and cultural selection of the type described here. A study of the Hopi people also indicates that time will soon erase albinism as a Hopi heritage on Second and Third Mesas. Paved roads now link the Hopi mesas so that only minutes are required to travel from one mesa to another by truck or automobile. These modern transportation facilities, schools on and off the reservation where members of different ethnic groups associate, and a growing population forcing young people to seek employment off the reservation all promote outbreeding, reduce the frequency of the albino gene, and decrease the probability of homozygosity for this gene. The eclipse of agriculture as a way of life negates any reproductive advantage held by albino males in past generations. The frequency of albinism among the Hopis will decrease rapidly with the decline of their culture.

References and Notes

1. T.B.Fitzpatrick,in *The Metabolic Basis of Inherited Disease,* J. B. Stanbury, J. B. Wyngaarden, D. S. Frederickson, Eds. (McGraw-Hill, New York, 1960).
2. P. D. Trevor-Roper, *Brit. J. Ophthalmol.* **36** 107 (1952).
3. C. J. Witkop, Jr. , E. J. Van Scott, Jr. , G. A. Jacoby, *Excerpta Med. Sect. 1 Anat. Anthropol. Embryol. Histol.* **F-169**, Abstr. 381 (1961).
4. K. Pearson, E. Nettleship, C. H. Usher, *A Monograph on Albinism* (Draper Company Research Memoirs, Biometric Series, Dolan, London, 1911-1913), vols. 6 and 9.
5. C. E. Keeler, *J. Hered.* **55**, 115 (1964).
6. C. M. Woolf, *Amer. J. Hum. Genet.* **17**, 23 (1965).
7. J. V. Neel, M. Kodani, R. Brewer, R. C. Anderson, *ibid.* **1**, 156 (1949).
8. A. Hrdlička, "Physiological and medical observations among thc Indians of southwestern United States and northern Mexico, " *Bull. Bur. Amer. Ethol.* No. 34(1908).
9. L. L. Hargrave, *Mus. N. Ariz. Mus. Notes* **4**, No. 7 (1932).
10. E. P. Dozier, "The Hopi-Tewa of Arizona" *Univ. Calif. Publ. Amer. Archeol. Ethnol.* **44**, No. 3 (1954).
11. C. M. Woolf and R. B. Grant, *Amer. J. Hum. Genet.* **14,** 391 (1962).
12. C. M. Woolf, D. A. Dolowitz, H. E. Aldous, *Arch. Otolaryngol.* **82,** 244 (1965).
13. M. Titiev, *Old Oraibi, A Study of the Hopi Indians of Third Mesa* (Peabody Museum Papers, Harvard Univ., Cambridge, Mass., 1944), vol. 22, No. 1.
14. Sun Chief, *The Autobiography of a Hopi Indian,* L. W. Simmons, Ed. (Yale Univ. Press, New Haven, 1942), p. 68.
15. F. C. Dukepoo, thesis, Arizona State University, Tempe (1968).
16. Ching-tS'ung Li, *Population Genetics* (Univ. of Chicago Press, Chicago, Ill., 1955).
17. K. S. Brown, B. L. Hanna, A. A. Dahlberg, H. H. Strandskov, *Amer. J. Hum. Genet.* **10**, 175 (1958).

18. C. Stern, *Principles Human Genetics* (Freeman, San Francisco, ed. **2** 1960).
19. C. M. Woolf, F. E. Stephens, D. D. Mulaik, R. E. Gilbert, Amer. J. *Hum. Genet.* **8**, 236 (1956).
20. J. N. Spuhler and C. Kluckhohn, *Hum. Biol.* **25,** 295 (1953).
21. F. Waters, *Book of the Hopi* (Viking, New York, 1963).
22. Supported by AEC contracts AT(11-1)-1415 and AT(11-1)-2013-01 to Arizona State University.

IX. Development

The fascination that man has long had with the size of living things, from the minuscule to the gigantic, is reflected in an essay by John Tyler Bonner, plant morphologist and professor of biology, Princeton.

He correlates size with organismic aspects such as generation time, structural complexity, division of labor, stages of development, adulthood and movement and coordination. Among the phenomena considered are the commonality in size that most organisms have, the reproductive cell, and the disparities which result from its development.

22.

The Size of Life

John Tyler Bonner

Increased size brings with it the complicated problems of servicing the organism so that it can operate properly. AUTHOR

In the full sweep of all the sciences, there are at one end of the scale, galaxies of inconceivable size, and at the other, objects as minute as atoms and elementary particles. A common characteristic of these extremes is that they are hard to see. Astronomical bodies are too far away and can only be studied with vast telescopes. Atoms and their internal composition are so small that even a microscope is no help. Furthermore, both travel about at speeds beyond our powers of observation, almost beyond our powers of imagination.

Living organisms are in a range somewhere in between. Most plants and animals are of a size we can easily see. It is true that bacteria and other microorganisms may be exceedingly small, but with the help of a microscope, they too can be observed. And the rate of movement for organisms of all sizes – large and small – is such that they can be seen. Even their life histories are within our comprehension. At the size level of living organisms we are closest to our direct experience.

If living organisms are arranged in a scale according to size, the smallest will be bacteria, a few microns in length; the largest will be giant sequoias, some more than 350 feet long, and in the animal kingdom, blue whales, some as much as 100 feet long. When we compare the characteristics of different-sized organisms, a rough direct proportion evolves: the larger the organism, the longer it takes to grow to maturity. It is easiest to consider the period of growth as the time from fertilization – or for asexual organisms, the moment of fission – to the moment the organism is sufficiently mature to be able to reproduce itself. This so-called generation time is a more consistent measure than life-span, or the time that an organism survives before it dies; the latter seems to vary greatly and shows only a crude correlation with size. Generation time, on the other hand, is directly related to size.

Looking at the size extremes again, we find that bacteria divide every half hour or so, while giant sequoia begin producing cones only after 60 years of growth, when they have reached a height of some 260 feet. The reason for this correlation between an organism's size and generation time is obvious. Since each generation is mainly a period of growth, a period of construction, it simply takes longer to build something large than something small, just as it takes longer to build a skyscraper than a one-room shack. There are more operations; more steps that are required.

There is another interesting correlation with size: the larger the organism, the more complex it is. This principle applies to buildings as well as to living constructions, and again the reason is obvious. Increased size brings with it the compli-

---

Reprinted by permission from the January 1969 issue of *Natural History*. Copyright 1969 by *Natural History*.

cated problems of servicing the organism so that it can operate properly. To give a simple example: if an organism is small, it can get oxygen directly from the environment by diffusion; but if it is large, it is too thick for the oxygen to diffuse inward. Therefore, if combustion is to take place at all in a large animal, there must be lungs or gills for gas exchange and a circulatory system to carry the oxygen inside the body. These structures are highly elaborate: the cost of increased size is considerable in terms of complexity and division of labor.

But why is it necessary to do so much building? We always hear that nature is efficient, even parsimonious, and never does anything in vain. Why, then, do we find any animal or plant larger than a bacterium? As with all such questions the biologist immediately falls back on the notion of natural selection. He argues that there must have been adaptive advantages to being larger and, therefore, selective pressure for size increase. Nothing could be easier to rationalize; for instance, a larger organism can more effectively chase and eat a smaller one. Or in the case of a plant, a larger one can reach above a smaller one and take first chance at the sun's rays for photosynthesis.

There is another aspect to the question of why large organisms require a long period of building. In bacteria and other unicellular organisms, the cell divides in half, and each half grows until it reaches a mature size ready for the next division. This involves some construction, but of a relatively simple nature, for only half the organism is regenerated. But large trees and mammals, or for that matter all reasonably sized animals or plants, do something far more radical than split in half. They start from scratch, from a single cell, and build a large and complex structure each generation.

Again the harder road seems to have been followed, and the reason comes from the genetic system. Recombination of genes produces variation required for natural selection and evolutionary change. The genes from the parents are carried in the chromosomes – one set from the father and one from the mother – and these are brought together by fertilization. Each fertilized egg can vary depending upon how its genes are combined, and it is only possible to make this kind of arrangement with single cells. To make the point clear, if an elephant split in two lengthwise, both halves would be the same, and would not offer the variation necessary for natural selection to take place. In bacteria this is not a problem because they are always single cells, and they vary primarily by mutation, by a sudden change in their characteristics. That is, if one cell possesses a gene that has been altered by mutation, it can pass it on directly to its offspring. With large organisms, however, the mechanism of inheritance and variation depends upon their beginning life as single cells. It is not surprising, therefore, that the single-cell stage has been retained, even in the case of the gigantic plants and animals. In fact, it would be fair to say that these larger animals never could have evolved if they had not remained minute single cells at one stage in their life history.

Now, we come to some of the burning questions of modern biology. How do organisms make these size changes, how does a single cell grow into a whale or a tree? What are the rules that make this construction sufficiently precise so that it is consistent from generation to generation?

It is known that a great many of the messages, the instructions that govern development, come from the genes in the nucleus. In single-celled organisms these instructions are carried out directly in the cytoplasm, the living substance immediately surrounding the nucleus. There is considerable evidence that a set sequence of events, consisting of a series of reactions, alternately affects the nucleus and the cytoplasm. Gene products from the nucleus enter the cytoplasm, which is modified and in turn affects the nucleus, thereby activating a new gene, which makes new products. As one might expect, the biochemist has been busy trying to identify these different substances and to analyze these steps on a biochemical level.

The kind of events that occur in development can be classified in a number of ways, but it is convenient to think of them as three constructive processes: growth, morphogenetic (or form-causing) movement, and differentiation. By growth is meant synthesis of new protoplasm, and in multicellular organisms this is usually accompanied by cell and nuclear division. Morphogenetic movements are found primarily in the embryos of animals where the cells move and produce a new shape. The best known such movement is that of gastrulation, in which an embryo folds upon itself to become multilayered. In the process of differentiation, different parts of an organism take on a different composition and structure. This is really equivalent to a division of labor, for the parts are associated with different functions. For example, in one body some cells are muscle, some nerve, some cartilage, and these different cell types are organized in well-formed organs and tissues that divide the labor.

One of the problems the developmental biologist wishes to solve is the nature of these three processes themselves, even without considering their over-all role in development. In the case of growth this has turned out to be primarily a problem of chemical synthesis, and many of the biochemical pathways are known. The mechanism of morphogenetic movement is far less understood, largely because the mechanism of ameboid motion (which is the way the embryonic cells move) is not known. Many kinds of differentiation are also being studied by the biochemist, who seeks to follow the chemical reactions that lead to the altered structure.

There are other problems, at a higher level that bring us to the heart of developmental biology. We want to know how these three processes – growth, morphogenetic movement, and differentiation – are controlled so that the end result is the perfect and constant form of the adult. If growth is not held in check, then some parts will grow way out of proportion to others; precisely what appears to occur in the growth of cancers. If morphogenetic movement is not carefully guided, the result is formless chaos. If differentiation is not controlled the delicate balance of proportions between the different cell and tissue types, as well as their proper positioning in space, is lost. Therefore, we not only want to know how each process occurs but also how it is controlled so that everything comes out to perfection.

We think of each of the processes of development as producing an internal condition that inevitably leads to the next stage. It is like a series of steps in a set sequence; the end result is the inevitable one, given the initial reactants and their initial organization. This is not a new idea; it goes back at least to Aristotle. The problem we face now is to solve the riddle of this sequence of steps.

The approach today is to come back to consider the reactants in terms of gene products and changing cytoplasmic environments. We have already mentioned this for unicellular organisms, but the problem becomes more complex in multicellular ones. Here a nucleus divides repeatedly by mitosis, so that each cell apparently contains the complete genetic complement of the organism. How is it possible, in this population of identical nuclei, to obtain differences in parts, that is, differentiation?

If the cells are firmly stuck to one another, as they are in plants with rigid cell walls, then regions of the organism will be different in some way. This may be due to the environment; some parts, for instance, may be more exposed and, therefore, have greater gas exchange than others, or more radiation. This may affect the cytoplasm of those regions, which affects the nuclei, so that nuclei in the various parts receive different signals from the cytoplasm. There are also internal ways in which differences in parts may arise. One of the most interesting, yet puzzling, ways is by the polar movement of substances. Chemicals, such as hormones in the higher plants, are produced by the cells, but move in one direction more than in another. The result is an accumulation of the substance at one end of the organism, again causing a differential stimulus on the nuclei at that end.

If the cells are mobile and can undergo morphogenetic movement, they can enter areas that differ in chemical environment, thereby producing the necessary stimuli to begin the events leading to differentiation. In animals there is usually a combination of growth and morphogenetic movement that, with the help of polar movement of substances and other spacing behavior, produces the desired and consistent adult shape.

The development biologist seeks to understand all these steps so that he can explain the wonderful process of how a fertilized egg turns into a tree or an animal. The difficulty has been that there are so many steps that, if they are considered in terms of chemical reactions, to catalogue them all would be well-nigh impossible. He has, therefore, attempted to find those steps that are particularly significant in the developmental process. Progress has been slow, but hope for the future is great. In the meantime there has been some impatience, for genetics has made such great strides forward, while developmental biology has not yet shown such flowering.

We enter into two of biology's older domains, physiology and anatomy, when we consider the functioning of the adult, the main period of size equilibrium. I will begin with plant anatomy and physiology, because plants are far simpler than animals, but even more important, because most plants do not have a mature stage that corresponds with anything we find among animals. Large trees (as well as many smaller plants) keep growing, they never cease their period of size increase. The botanist refers to this property as continuing embryology. It means that the growth zones at the apex of the plant and at the apexes of all the shoot and root branches continue to grow. So does the cambium, which is responsible for the thickening of the tree's branches and trunk. If the tree seems to have a size limit, it is only that, eventually, through age and decay, the wind will fell it. Other plants, such as annuals, do reach a growth limit, but in this case all the cells, except for those of the seed, die. As a hay field turns brown, or a wheat field turns to light gold, the cells of the plant wither and dry and only a dead stalk is left to hold the seed so it may scatter in the winds.

Yet even though plants do not have an adult stage that corresponds to that of animals, they certainly have an anatomy and a physiology. But plant physiology is largely, for the very reasons we have given, the physiology of development. The concern is with growth hormones, oriented growth movements with respect to light and gravity, and related problems, all of which are completely within the outlook of the developmental biologist. Plants have no nervous system, no muscles, no means of locomotion comparable to that of animals. (For simplicity's sake I am confining my remarks to higher animals and plants, for indeed the distinction between the two is rather fuzzy in more primitive forms.) Among higher plants there are a few exceptions, such as the sensitive plant that lowers its leaf suddenly at a touch, and certain fly-catching plants that move relatively rapidly. But these are physiological curiosities and hardly in the main line of significant problems in biology.

If we turn to animals, some aspects of adult physiology hold pressing problems for the future, and here again we are confronted with the matter of size level. Physiological activities, for example, can be studied at the size level of the entire organism – the organ, the tissue, the cells; even the biochemical events that take place within the cells.

Let us briefly consider some physiological functions of the whole animal. The assimilation of energy in the form of food is a prime essential. It involves organs to grasp the food, a mouth to begin the food processing, and an alimentary canal to digest the food and absorb it into the blood and thence to the tissues. Another function is locomotion. Swimming, flying, or running involves the contraction of muscle situated in key positions, and all of these processes must be properly co-ordinated, so that the animal can always make the appropriate responses to its changing environment. Of these three primary physiological activities characteristic of living organisms (energy assimilation, locomotion, and co-ordination) the first two are fairly well understood, at least in their broad outline. This is especially true of assimilation. In the case of movement there remain some outstanding questions, especially on the biochemical and cellular level where the answer to the molecular basis of muscle contraction is being sought. There has, however, been much exciting progress along these lines in recent years.

One pertinent point concerning the movement of animals and size should be mentioned. If we group animals according to size categories, and then compare the rates of speed of the fastest animal in each group, we find that the larger the animal, the faster it can move. This relation is rough, but it holds true from the smallest unicellular animal to the large multicellular ones, up to about ten feet in length. Whales and elephants do not have great speed. Furthermore, it holds for animals that swim, run, and fly. A tuna fish will rush along at about 47 mph, while a motile bacterium will only move at the rate of 0.000034 mph. Co-ordination is in many ways the most intriguing, the most complex, and the least well understood of all the aspects of physiology. This is the essence of the problem of control. The sensitivity of the sense organs to the external world, the gathering of this information in the central nervous system and brain, and the issuing of orders to move certain muscles as appropriate responses are among the marvels of nature. Add to that, memory and conscious thought, and the nervous system appears supreme in its complexity and in its accomplishments.

Another point should be emphasized in discussing the adult stage. Even though the maximum size of each animal or plant is the adult size, all organisms spend only part of their existence as adults. In fact, an organism is more properly its entire life cycle, from egg (or some form of asexual spore) to an adult stage, which, in turn, produces similar reproductive bodies. And this reproductive stage is roughly the same for all organisms, namely the size of a cell. This means, looking back to our original correlations, that the larger the adult stage, the more growth is necessary and the more complex the development. Furthermore, the normal functioning of the large adult presents complicated engineering problems. Hence large forms have a greater division of labor, a greater internal coordination, and a more elaborate organization. Size has significance for all aspects of life.

X. Evolution

With the concept of organic evolution firmly established, biologists have turned to the investigation of other aspects of this fascinating topic. Two essays in this unit are concerned with the basic questions of how and where life began.

Professor Cyril Ponnamperuma, research scientist, Ames Research Center, NASA, outlines the search by modern science for the secrets of life's beginnings in chemical evolution. Although a chemical origin of life has almost universal support among the scientific community, less agreement is found concerning the nature and existence of extraterrestrial life. Miss Eugenia Keller, managing editor of *Chemistry* magazine, summarizes evidence in support of the existence of extraterrestrial life, and suggests that life beyond this solar system could be similar to life as we know it.

As man is now belatedly aware, his dominance of the biosphere has created monumental problems, perhaps his own survival. It becomes vitally important to comprehend our evolutionary background for therein may be clues to insure future survival. Two essays relevant to this important consideration are by Professor D.R. Pilbeam, anthropologist, University of Cambridge, and Gabriel W. Lasker, professor of anatomy, Wayne State University.

The final essay of the unit, "The Exploitation of Animals by Man," by M.L. Ryder of the Animal Breeding Research Organization, Edinburgh, discusses another dimension of human evolution, the domestication of animals and plants. Many specialists consider plant and animal exploitation a point of evolutionary divergence between mankind and the lower animals. Our present environmental crisis may necessitate a further intensification of this exploitation.

23.

The Chemical Origin of Life

Cyril Ponnamperuma,

There is no reason to doubt that we shall rediscover, one by one, the physical and chemical conditions which once determined and directed the course of chemical evolution. AUTHOR

The problem of the origin of the universe is the greatest challenge the human intellect has ever faced. Indeed the origin of the universe, along with the origin of life and of intelligence, may be regarded as the fundamental question of our time. It is the purpose of this article to outline how modern science is searching for the secrets of life's beginnings.

While the very concept of the birth of the universe is staggering to the human mind, the unravelling of the riddle may come from a surprisingly simple observation or experiment. Abbé Lemaître's theory of an evolutionary universe arising from a primeval atom will stand or fall when an astronomer's penetrating gaze has compared the spatial density of galaxies 50 million light years away with those of 10,000 million light years away. The rival cosmology of continuous creation demands the appearance of hydrogen at several billion trillion tons per second in the observable universe. This idea will be satisfactorily proved if nuclear physicists demonstrate that energy can be converted into hydrogen at the rate of one atom per year in a volume about the size of a New York skyscraper.

In our attempt to understand the origin of life, the enormity of the enigma is paralleled only by the complexity of the possible hypotheses. It is this difficulty which prompted the physicist J. D. Bernal to write in 1949: "It is probable that even a formulation of this problem is beyond the reach of any one scientist, for such a scientist would have to be at the same time a competent mathematician, physicist, and experienced organic chemist, he would have a very extensive knowledge of geology, geophysics and geochemistry and, besides all this, be absolutely at home in all biological disciplines. Sooner or later this task will have to be given to groups representing all these faculties, working closely together theoretically as well as experimentally".

Our ancestors thought that they had found a ready solution to the problem in the belief that life arose continually from the non-living. One had only to accept the evidence of the senses: worms from mud, maggots from decaying meat, mice from old linen and fireflies from the morning dew. Aristotle had propounded the doctrine of spontaneous generation in his "Metaphysics", and this teaching was accepted by the long line of European thinkers who had turned to him as the final authority in matters metaphysical and physical. The ancient Hindu scriptures described life as having evolved from non-living matter. The Rig Veda, for example, pointed to the beginning of life from the primary elements while the Atharva Veda

Reprinted from the May 1965 issue of *Science Journal*. By courtesy of *Science Journal*, London, England.

postulated the oceans of the Earth as the cradle of all life.

The world's literature is full of allusions to the popular belief in spontaneous generation. Virgil, in his "Georgics", tells us how a swarm of bees arose from the carcass of a calf. Lucretius in "De Natura Rerum" refers to the Earth as the mother of all living things.

The Belgian chemist and physician, Van Helmont, even had a recipe for the making of mice. "If a dirty undergarment is squeezed into the mouth of a vessel containing wheat, within a few days, (say 21), a ferment drained from the garments and transformed by the smell of the grain, encrust the wheat itself with its own skin and turns it into mice . . . And, what is more remarkable, the mice from corn and undergarments are neither weanlings nor sucklings nor premature but they jump out fully formed."

Beliefs such as these could not long withstand the rigours of advancing scientific method. Francisco Reddi, a celebrated member of the Accademia del Cimento, demonstrated that the worms in putrefying flesh were larvae from the eggs of flies. His proofs were as simple as they were decisive. He showed that meat placed under a screen so that flies cannot lay their eggs on it never develops maggots.

But soon, with the invention of the microscope by Antony van Leeuwenhoek in the second part of the 17th century, the doctrine of spontaneous generation began to reappear. Some investigators were unable to explain the origin of the varied organisms which the microscope showed in infusions from animal and vegetable matter. Among these organisms they discerned nothing that resembled the process of sexual generation. Thus they were led to assume that matter, once alive, keeps after death a special vitality.

About this time the English Jesuit, John Turberville Needham, published a work in which he described experiments to support the theory of spontaneous generation. The Italian Abbé Lazaro Spallanzani counteracted by showing that nutritive broths, sealed off from the air while boiling, never develop micro-organisms and hence never rot. Needham objected that by too much boiling Spallanzani had rendered the broth and the air above it incompatible with life.

In the middle of the last century Louis Pasteur demonstrated by a series of brilliant experiments that living organisms could not arise out of non-living material. Pasteur dealt the death blow to the theory of spontaneous generation which was based on incompetent observation and the willingness to accept the evidence of the senses. Unfortunately, his work also gave rise to the misconception that the problem of the origin of life could not be approached by scientific methods. The question of life's beginning was therefore considered to be unworthy of the attention of any serious scientific investigator. But it is transparently clear that what Pasteur proved was that micro-organisms could not be grown from sterile starting material; his experiments had no bearing on the gradual formation of organic compounds leading to the emergence of biologically significant polymers.

With the growth of organic chemistry a new approach to the problem appeared to be possible. J. J. Berzelius thought that organic coumpounds were produced from their elements by laws different from those governing the formation of inorganic compounds. This led him to believe that organic compounds were produced under the influence of a vital force and that they could not be prepared artificially. In 1828 Friedrich Wöhler converted ammonium cyanate into urea, a substance hitherto obtained only from animal sources. This synthesis weakened the distinction between organic and inorganic compounds. The distinction was completely ended with the synthesis of acetic acid from its elements by A. W. H. Kolbe in 1845, and the synthesis of methane from a mixture of carbon disulphide and hydrogen sulphide by M. P. E. Berthelot in 1856.

The vitalists objected that Wöhler had obtained his ammonium cyanate from ammonia and cyanic acid, both of which were of animal origin. But already in 1781 Joseph Priestley had prepared ammonia by the reduction of nitric acid which was synthesized from its elements by Henry Cavendish in 1785. Potassium cyanide had also been obtained by K. W. Scheele by heating a mixture of potassium carbonate and carbon in the presence of nitrogen. Since potassium cyanide is readily converted into potassium cyanate, Wöhler's synthesis was truly one which started from inorganic materials. The experiments of these early synthetic chemists gradually removed the boundary between the inorganic and the organic world.

The basis of our approach to the problem is an evolutionary philosophy. The process may be traced forward from the primeval cloud of hydrogen gas or backward from the primordial germ which Charles Darwin placed at the beginning of life. The Darwinian theory of evolution has postulated the unity of the Earth's entire biosphere. According to Darwin, the higher forms of life evolved from the lower over a very extended period in the life of this planet. Fossil analysis has shown that the oldest known forms of living systems may be about 2000 million years old. Life, indeed, had a beginning on this planet. The consideration of biological evolution thus leads logically to another form of evolution, namely chemical evolution.

The evidence available from practically every field of science suggests that nature is a unity which has been divided into categories for human convenience. The division of matter into living and nonliving is, perhaps, an artificial one which is convenient for distinguishing such extreme cases as a man and a rock, but would be quite inappropriate when describing a virus particle. Indeed, the crystallization of a virus by Wendell Stanley almost 30 years ago precipitated the need for revising our definition of the terms 'life' and 'living'. These sentiments were powerfully expressed by N. W. Pirie in an essay entitled "The Meaninglessness of the Terms 'Life' and 'Living'." He compared the use of the terms living and non-living to the words acid and base as used in chemistry. While sodium hydroxide is distinctly alkaline, sulphuric acid is a powerful acid. However, in between there is a whole variation in strength. The chemist has overcome the confusion arising from the use of these two terms by inventing the nomenclature of hydrogen ion concentration. He is thus able to describe all the observed phenomena in terms of one quantity; thus a fluid may have a pH of 4 or a pH of 8. According to Pirie, we may have to invent a similar quantity in order to avoid any vagueness that might arise in applying the term life to borderline cases such as the virus.

One of the first to speculate on the conditions necessary for the origin of life was Erasmus Darwin, the grandfather of Charles Darwin. In his "Temple of Nature" he had written: "All vegetables and animals now existing were originally derived from the smallest microscopic ones formed by spontaneous vitality." Perhaps this idea had influenced Charles Darwin in his own thinking. Several years later he wrote to a friend about "Some warm little pond with all sorts of ammonia and phosphoric salts, light, heat, electricity, etc., present" in which he postulated that a protein compound was chemically formed ready to undergo still more complex changes.

About the same time, the physicist John Tyndall argued that every portion of a living organism can be reduced to inorganic matter. In his essay on vitality, published in 1866, he suggested that one could conceive of the reverse change from the inorganic to the organic, and that the special arrangement of elements in living bodies led to the phenomenon of life. In 1868 Thomas Huxley delivered a lecture in Edinburgh in which he pointed out that protoplasm was substantially the same over the whole range of living things. To him, the existence of life depended on certain molecules such as carbonic acid, water and nitrogenous compounds. These compounds were lifeless but, when brought together, gave rise to protoplasm.

To the Russian biochemist A. I. Oparin, more than to anyone else today, do we owe our present ideas on the scientific approach to the question of the origin of life. In clear and scientifically defensible terms, he pointed out in 1924 that there was no fundamental difference between a living organism and brute matter. The complex combination of manifestation and properties so characteristic of life must have arisen in the process of the evolution of matter. The first English edition of his book "The Origin of Life" was published in 1938. A fourth edition of the book has recently appeared under the title of "The Chemical Origin of Life".

In 1928 the British biologist, J. B. S. Haldane, expressed his own ideas on the origin of life. He attributed the synthesis of organic compounds to the action of ultraviolet light on the Earth's primitive atmosphere. He suggested that organic compounds accumulated till the primitive oceans had the consistency of a primordial 'soup'. Twenty years after the appearance of Haldane's paper in *The Rationalist Annual,* Bernal theorized before the British Physical Society in a lecture entitled *The Physical Basis of Life.* "Condensations and dehydrogenations are bound to lead to increasingly unsaturated substances, and ultimately to simple and possibly even to condensed ring structures almost certainly containing nitrogen, such as the pyrimidines and purines. The appearance of such molecules makes possible still further synthesis. The primary difficulty, however, of imagining processes going thus far is the extreme dilution of the system if it is supposed to take place in the free ocean. The concentration of products is an absolute necessity for any further evolution."

The growth of biochemical and geochemical knowledge in the 20th century has stimulated a fresh interest in this age-old problem. The possibility of discovering extra-terrestrial life has given a further impetus to its experimental study. The first international symposium on the origin of life was held in Moscow in 1957. Ideas presented were grouped under headings which described five stages in the evolution of life from non-life: primary formation of primitive organic compounds on the Earth; the transformation of primary organic compounds on the Earth; the origin of proteins, nucleoproteins and enzymes; the origin of structure and metabolism; and the evolution of metabolism. So much experimental work has since been done in laboratories all over the world that a second international symposium was convened in Wakulla Springs, Florida, in the autumn of 1963. The highlight of this meeting was the presence of both Oparin and Haldane, pioneer thinkers in the field.

Almost half a century ago, Jacques Loeb had studied the effect of electric discharges on mixtures of hydrocarbons and had suggested that a large number of organic compounds could be formed. However, the first experiment designed with the set purpose of testing some of the hypotheses on the origin of life was that of Melvin Calvin and his associates who, in 1951, treated water and carbon dioxide in the Berkeley cyclotron and obtained significant yields of formaldehyde and formic acid. In 1953, Stanley Miller, then a graduate student in Harold Urey's laboratory, assembled a sample of the assumed primeval terrestrial atmosphere consisting of methane, ammonia, water vapour and hydrogen, and exposed it to an electric discharge simulating lightning. Amino acids and other organic compounds found in living systems were formed.

Since this classic experiment, several investigators have entered the field. Notable among them are Sydney Fox of Florida State University and John Oro of the University of Houston. The majority of publications have dealt with the formation of amino acids and components of nucleic acid from a wide variety of conditions which may be considered pre-biological. Fox's work centred on the origin of proteins. A plausible answer seems to have begun to take shape. Proteinoids have been obtained by the thermal polymerization of 18 amino acids. These proteinoids have a distinct tendency to form microspheres having diameters in a bacterial range. Starting with ammonium cyanide, Oro has synthesized adenine and a number of biochemical intermediates of purines.

A starting point for any such experimental work must take into account cosmic abundances. Astronomical spectroscopy reveals that the most abundant elements in our galaxy are, in the order of rank: hydrogen, helium, oxygen, nitrogen and carbon. With the exception of helium these are, indeed, the basic elementary constituents of all living organisms. We know from chemical equilibria that – in the presence of hydrogen – carbon, nitrogen and oxygen must exist in their reduced forms as methane, ammonia and water. The equilibrium constants for these reactions at 25°C are all of considerable magnitude. It is this atmosphere of methane, ammonia, water vapour and small amounts of hydrogen which we shall consider as the primitive atmosphere of the Earth.

The energies available for the synthesis of organic compounds under primitive Earth conditions were ultraviolet light from the Sun, electric discharges, ionizing radiation and heat. While it is evident that sunlight is the principal source of energy, only a small fraction of this is in the wavelength below 2000Å which could have been absorbed by the methane, ammonia and water (one Ångstrom is 10^{-8} cm). However, the photodissociation products of these molecules could absorb energy of higher wavelengths. Next in importance as a source of energy are electric discharges, such as lightning and corona discharges from pointed objects. They occur close to the Earth's surface and, hence, would transfer the reaction products to the primitive oceans more efficiently. A certain amount of energy was also available from the disintegration of uranium, thorium, and potassium-40. While some of this energy may have been expended on the solid material, such as rocks, a certain proportion of it was available in the oceans and the atmosphere. Heat from volcanoes was another form of energy that may have been effective but, in comparison to the energy from the Sun, this was only a small portion and perhaps not widely distributed.

In our experiments with ionizing radiation, we have found that the electron beam from the linear accelerator at the Lawrence Radiation Laboratory of the University of California, Berkeley, provided us with a convenient source of electrons simulating potassium-40 on the primitive Earth. When a mixture of methane, ammonia and water was irradiated with electrons for about one hour, resulting in a total dose of approximately 10^{11} ergs, the largest single non-volatile compound formed was adenine. The production of adenine in this experiment was significant in the light of the multiple role played by this chemical in biological systems. Not only is it a constituent of both deoxyribonucleic acid (DNA) and ribonucleic acid (RNA), but it is also a unit of many important cofactors in living organisms.

In our experiments with electric discharges, tesla coils and luminous tube-transformers were used to simulate lightning on the primitive Earth. In a typical experiment, lasting 48 hours, most of the methane was converted into more complex organic compounds. In this experiment some of the constituents of the nucleic acid molecule have been identified.

The action of heat on the Earth's primitive atmosphere was duplicated by passing a mixture of methane and ammonia, in the presence of water vapour, through a heated Vycor tube at about 1000°C. The effluent gases were absorbed in water. Analysis of the water-soluble material revealed the presence of several amino acids. This result has recently been reported by Fox, who identified 14 of the amino acids commonly present in protein. Analysis of the gas fraction has shown that a great portion of the methane was converted into higher hydrocarbons, including ring compounds such as benzene, toluene and anthracene.

Chemosynthesis by meteorite impact on planetary atmospheres has been suggested as a possible pathway for primordial organic synthesis. The reaction is probably a result of the intense heat generated momentarily in the wake of the shock wave following the impact. In a very preliminary experiment simulating these conditions, by firing a ballistic missile into a mixture of methane, ammonia and water vapour, we have been able to identify some amino acids and a few ultraviolet absorbing compounds which may be of biological significance.

In all the experiments just described, whether using ionizing radiation, electric discharges or heat, one of the primary products appears to be hydrogen cyanide. The second product is formaldehyde. S. L. Miller and Urey identified both these compounds in their early work. C. Palm and Calvin recorded similar results. Our own experiments point to the same conclusion. In subsequent experiments we have, therefore, used hydrogen cyanide and formaldehyde as our starting materials.

When a dilute aqueous solution of hydrogen cyanide is exposed to ultraviolet light, a wide variety of organic compounds can be formed. Among these have been identified adenine, guanine and urea. Adenine and guanine are the two purines in the nucleic acids. Urea is an important biochemical intermediate. The reaction with hydrogen cyanide appears to proceed even without an external source of energy. When a solution of hydrogen cyanide is left standing at −10°C, it is spontaneously converted into more complex organic compounds. In experiments with very dilute solutions of formaldehyde as starting material, the two sugars, ribose and deoxyribose, have been identified.

The same forms of energy that have been instrumental in the production of the sugars and of the purines have now been demonstrated to be responsible for the synthesis of nucleosides, nucleotides and peptides. When a solution of adenine and deoxyribose, in the presence of the cyanide ion, is exposed to heat or ultraviolet light, the nucleoside deoxyadenosine is rapidly formed. This result confers a unique role on hydrogen cyanide. Not only is it a pathway for the purines but also, in the synthesis of nucleosides, it performs the function of a catalyst.

Since the Earth's primitive atmosphere was reducing in nature, some ultraviolet light in the region between 2400Å and 2900Å must have reached the surface of the early oceans. The activation of dissolved purines, pyrimidines and their nucleosides which absorb ultraviolet light in this range may have been a step in the synthesis of nucleoside phosphates.

To test this hypothesis, several experiments were performed. The conversion of adenine to adenosine, adenosine to adenosine monophosphate, adenosine monophosphate to adenosine diphosphate, and the diphosphate to the triphosphate has been experimentally established. In these experiments, ethyl metaphosphate was used as a source of phosphorus. It can be expected that this was not the most abundant source of phosphorus on the primitive Earth but it is likely that other, more abundant, phosphate salts can substitute for ethyl metaphosphate.

The synthesis of peptides under abiological aqueous conditions has been considered to be a matter of great difficulty. In our laboratory, however, we have recently found that when an aqueous solution of glycine and leucine was exposed to ultraviolet light in the presence of cyanamide, the dipeptides glycyl-glycine, glycyl-leucine, leucyl-glycine, leucyl-leucine were formed. The action of cyanamide may be analogous to that of dicyclohexylcarbodiimide which has been extensively used by H. G. Khorana and his co-workers for the synthesis of nucleotides.

These results show that, under simulated primitive Earth conditions, molecules of biological significance can be synthesized. They lend support to the hypothesis of chemical evolution because the starting materials are the simple constituents of the primordial atmosphere; the sources of energy used are those that are most likely to have existed under primitive Earth conditions; the concentrations of materials used are very small; and the conditions are aqueous.

Recent studies in quantum biochemistry by B. Pullman and A. Pullman of Paris have thrown new light on some very significant aspects of chemical evolution. It is a striking fact that many of the molecules which are essential to living systems are conjugated systems exhibiting the phenomenon of electronic delocalization. In the nucleic acids, for example, the purines and pyrimidines are conjugated systems. Although the proteins do not at first sight, appear to enjoy this property, a closer look shows that the matrix of hydrogen bonding which exists in a protein molecule provides a certain measure of electronic delocalization. In the high energy phosphates, there is interaction between the mobile electrons of one phosphoryl group with those of another. The porphyrins, such as chlorophyll and haem, which are of paramount importance in living systems, are highly conjugated molecules. This leads to the following conclusions:

(a) Evolutionary selection used the most stable compounds.
(b) On account of electron delocalization these compounds were best adapted for biological purposes.
(c) The possibility of life as we know it was made more probable by the appearance of the compounds.

There is no reason to doubt that we shall rediscover, one by one, the physical and chemical conditions which once determined and directed the course of chemical evolution. We may even reproduce the intermediate steps in the laboratory. Looking back upon the biochemical understanding gained during the span of one human generation, we have the right to be quite optimistic. In contrast to unconscious nature which had to spend billions of years for the creation of life, conscious nature has a purpose and knows the outcome. Thus the time needed to solve our problem may not be long.

FURTHER READING

The Chemical Origin of Life, by A.I. Oparin (Charles C. Thomas, Springfield, 1964)

The Origins of Prebiological Systems, edited by S.W. Fox (Academic Press, New York, 1965)

La Genese de la Vie, by A. Dauvillier and E. Desguin (Hermann et Cie, Paris, 1942)

"Organic Compound Synthesis on the Primitive Earth" by S. L. Miller and H. C. Urey (in *Science,* **130,** 245, 1959)

"The Origins of Life" by G. Wald (in *Proceedings of the National Academy of Sciences,* **52,** 595, U.S.,1964)

24.

Origin of Life on Other Planets?

Eugenia Keller

Life beyond the earth might not be far different from that on earth. With surprising uniformity, the most abundant elements found so far in extraterrestrial environments are in decreasing order of abundance, hydrogen, helium, oxygen, nitrogen, and carbon. Four of these elements (helium excepted) comprise 99% of living matter as we know it. AUTHOR

The number of galaxies in the universe is unknown, but some 100 million are within range of terrestrial telescopes. Our own galaxy, the Milky Way, contains an estimated 100 to 200 billion stars, many of which are now thought to contain a planetary system, although stellar planets have never been observed. Either they are too small to be seen or they are obscured by the brilliance of the central star. However, when the enormous number of stars is considered, the conclusion that life exists in other parts of the universe is inescapable.

For life to exist on any planet, several conditions must be met: First its sun must be in the second stage of evolution, known as the main sequence. In the first stage, cosmic dust aggregates through gravitational forces. In the main sequence or second stage, pressures thus created have generated sufficient heat to start thermonuclear reactions in the interior and consequently outward pressures that balance inward gravitational pressures. In these reactions, hydrogen is burned to helium and the larger the star, the faster hydrogen is consumed. Finally, when the hydrogen consumed equals about 12% of the total mass of the star, nuclear reactions spread to the surface. Then the star explodes and becomes a red giant. Such an event would kill all life in any planetary system. Finally, by a mechanism unknown, the red giant becomes a white dwarf—a hot, faint, dense object incapable of supplying sufficient energy for the maintenance of life.

For the most massive stars, the main sequence lasts for only 1 to 10 million years, an insufficient time for life to evolve. In contrast, much smaller stars can remain on the main sequence for as long as 100 billion years. But if the star is too small, the surrounding area for planetary orbits where conditions are favorable to life is only a narrow band. The history of a star 1.2 times the mass of our sun is postulated thus: The condensation period or first stage requires some 10 million years; the main sequence about 8 billion years; the red giant stage, some 100 million years; and finally after several thousand years, a white dwarf.

Any planet supporting life is probably the size of earth—if smaller, it would be unable to hold an atmosphere because gravitational forces would be insufficient. If too much larger, its atmosphere would probably be too dense for penetration of sunlight, considered, necessary to supply

Reprinted by permission from the April 1969 issue of *Chemistry*.

an external source of energy. If the temperature is too low, chemical processes necessary for emergence of life would be too slow; and if too high, biochemical polymers would be unstable. Further, water would probably have to be liquid, rather than ice or steam, and the radiation pattern of the sun would have to be similar to ours—that is, wavelengths of 3000 to 11,000 millimicrons. Shorter wavelengths destroy macromolecules, denature proteins, and depolymerize nucleic acids. Longer wavelengths do not activate photochemical reactions.

Another condition for origin of life is a fairly stable environment, which means that the planet must follow a stable orbit. Many stars are clustered in groups of two, three, or even more. Where such grouping occurs, planets at some point in their orbits are likely to be either too near or too far away from a star for life to exist.

The most efficient means of perceiving distant objects is with electromagnetic radiation, and such radiation should be in the range absorbed least by the atmosphere. Electron transfer causes ultraviolet light to be absorbed, and infrared is absorbed through molecular vibration. Thus light used by life on distant planets should be in the visible spectrum. Although wavelengths in the radio-frequencies beyond 3 cm. probably would be available, perception with these frequencies is difficult. For an eye to have the resolving power equal to that possible at 5000 angstroms, an eyeball about ½ mile in diameter would be needed. To conceive of such an eye requires an active imagination.

Life beyond the earth might not be far different from that on earth. With surprising uniformity, the most abundant elements found so far in extraterrestrial environments are, in decreasing order of abundance, hydrogen, helium, oxygen, nitrogen, and carbon. Four of these elements (helium excepted) comprise 99% of living matter as we know it. The most common condensation products of these elements are hydrogen, water, ammonia, and methane. These materials are said to have been present in the earth's primitive atmosphere and are believed to be the materials from which life arose *(5)*.

Suggestions have been made that a form of life could exist based on a chemistry entirely different from terrestrial carbon chemistry. This possibility, when considered along with the adaptability of life on earth, can lead to some fantastic concepts, a situation which science fiction writers have taken full advantage of. But sober-minded scientists are more conservative. Where temperatures are less than about 400°K., they say, a solvent seems necessary to form complex organic molecules. By far the most desirable substances for this purpose are water, liquid ammonia, and liquid hydrofluoric acid, in order of their desirability. However, because of its low cosmic abundance and other reasons, hydrofluoric acid is less probable than either water or liquid ammonia. On the other hand, ammonia is an excellent solvent and, on low temperature planets where it is abundant, systems of organic chemistry could develop with NH_2 replacing the OH group in water. Where temperatures exceed 400°K. and atmospheric pressures are moderate, no appropriate solvent seems available and thermostability of organic molecules would not be sufficient for an organic chemistry to evolve.

George Wald *(12)* argues against the possibility of an ammonia-based life thus: An obvious disadvantage of ammonia, he says, is that it remains liquid over a much narrower range than water. Retaining liquidity demands more uniform temperatures than planets are likely to have. Also with temperatures never rising above −44°C., which is the boiling point of ammonia (its melting point is −77°C.), chemical processes would be much slower than on earth. In such an environment, processes that led to the origin of life on this planet within perhaps a billion years would require some 64 billion years. This is more than the estimated age of our galaxy and much longer than a star such as our sun could remain on the main sequence.

Another factor is important. If ice did not float, the waters of the earth would long ago have frozen solid, except perhaps for a thin layer at the top. Even long periods of warm weather could not have thawed them. Solid ammonia, however, very definitely sinks in its liquid counterpart. This is a distinct disadvantage for origin of life.

Furthermore, in photosynthesis, plants split the water molecule and give off oxygen which we breathe. If ammonia replaced water, the comparable by-product would probably be nitrogen which we can breath, but which, because of its inertness, could not support respiration which is essentially combustion.

Another suggestion is that perhaps somewhere a life based on silicon chemistry exists. Like carbon, silicon can form long chains and complex molecules. Also, the earth's surface contains about 135 times more silicon than carbon, and diatomaceous earth, found in fairly large deposits, is composed of the skeletal remains of microorganisms which used silicon in their cell walls.

But, according to George Wald, silicon cannot replace carbon in living organisms. Its chains are susceptible to attack in the presence of oxygen, ammonia, and water. Silicon also has a fatal disability to form multiple bonds, a property that can be understood by comparing carbon dioxide with silicon dioxide. In carbon dioxide, double bonds between carbon and oxygen completely saturate their combining ability. The molecule is a gas; it dissolves in and combines with water. These are the forms in which living organisms obtain it. In silicon dioxide, however, the silicon remains singly bonded to oxygen and forms silicon–oxygen chains. The result is a huge polymer such as a crystal of quartz, so hard because it can be broken only by breaking covalent bonds.

Phosphorus and sulfur used by life on earth form unstable compounds, but in their case this property is an advantage. Because they are susceptible to attack, they can participate in energy-forming and group-transfer reactions. This constitutes their principal contribution to cellular metabolism.

The Panspermia Theory

In the scientific climate established by Darwin and, especially by Louis Pasteur, who discredited the spontaneous generation theory regarding origin of life, Svante Arrhenius in 1907 proposed the panspermia theory which holds that life is ubiquitous in the universe. He imagined that simple living things propelled by radiation pressure (light when bounced from a wall creates pressure, although infinitesimal) drift through space. Balloon studies have detected microorganisms at great altitudes from the earth.

Propelled by radiation pressure from the sun, an organism 0.2 to 0.6 micron in diameter would in space attain high speeds, reaching the Martian orbit in weeks, Jupiter's in months, and the nearest star in thousands of years. With no collisions, it could cross the galaxy in a few hundred million years. However, rather than travel in a straight line it would wander at random as it collided with interstellar particles. Arrhenius thought that the earth might have been seeded with life in this manner.

Of course, one argument against this theory is that radiation would destroy the spores. However, the Russian scientist, Imshenetsky, proved that microorganisms can be protected from radiation. In air, he suspended spores on tiny mineral particles and subjected them to ultraviolet radiation. In a control experiment, he omitted the mineral particles and showed that the mineral particles provided protection.

Also, in a more recent experiment, certain pathogenic spores have been shown to thrive on ionizing radiation *(1)*.

How to Identify Extraterrestrial Life

The search for life on other planets should begin first with heterotropic types which are those capable of metabolizing organic molecules; second, with chemoautotropic types, those which can metabolize simple molecules such as hydrogen, methane, ammonia, sulfur, or iron; third, with organisms capable of photosynthesis; fourth, with optically active materials (those that rotate a plane of polarized light either to the right or left). On earth, only (+)-glucose, the type that rotates a plane of light to the right, is found and not its mirror image, (−)-glucose. Likewise, all proteins, except some found in cell walls of bacteria, have a similar characteristic—the same nonsymmetrical arrangement of atoms denoted as L in the molecule $RCH(NH_2\ CO_2H)$. Enzymes contain only L amino acids, a fact that supports the lock and key model used to explain the high specificity of enzymes. Because L and D molecules are shaped differently, enzymes can distinguish between them through their ability to attack only molecules of a certain configuration.

Any biochemistry which does not distinguish D and L type configurations is probably very primitive. Therefore if a racemic mixture is found (composed of equal quantities of D and L types and therefore not affected by a plane of polarized light), perhaps on Mars, the form of life present would probably approximate the first forms on primitive earth. An exciting thought is the possibility of finding *l*-glucose or D amino acids.

Recently, an interesting explanation for the origin of L amino acids was proposed *(1)*. When beta particles from strontium-90 slow down, they emit gamma rays which are circularly polarized to the left. D and L-tyrosine in alkaline solution decompose at the same rate. But in the presence of strontium-90, D-tyrosine appears to decompose faster. On primitive earth perhaps strontium-90 or another radioactive isotope was responsible for the selection of L amino acids.

Conceivably, optical activity could have arisen also through natural selection. As biochemical reactions became more complex, enzymes by using the lock and key fitting process could have distinguished between D and L structures. Then those organisms synthesizing L amino acids might have had the advantage of better ability to adjust to their environment. After generations this type would survive and the D type would become extinct. The L type would continue not because of an advantage but rather because the pattern had been woven into the fabric of life early. This situation, if true, could be interpreted as an historical accident. But if L amino acids are found to predominate in extraterrestrial life also, then this conclusion would not be valid.

Nature of Extraterrestrial Life

Carl Sagan, an astronomer at Harvard University, has speculated about life on other planets. The types present, he says, can vary widely, but certain restrictions must be assumed. The organisms must be large enough for at least simple metabolic functions. The smallest organism known on earth capable of independent reproduction of itself is 10^{-5} microns across. On the other hand, if the organism is too large, it would be unable to support its own weight. But, buoyancy of the atmosphere can influence size. For example, some marine animals such as whales are larger than land animals. Also, the organism cannot be too long because of the time required for neural impulses to be conveyed. Size is limited, too, by the forms of respiration available — insects on earth are small because oxygen in their bodies is distributed by diffusion, a method much less efficient than circulation.

Most earth animals have two, four, or six legs, and some such as snakes have none. On other planets, different methods of propulsion might have been developed, such as the type of ramjet propulsion used by certain protozoa. More than two eyes might be present, perhaps one in the back of the head. Some animals on earth during the Mesozoic Age seem to have had three eyes, and some physiologists believe that the pineal gland in human beings is the vestige of an eye in the middle of the forehead. Some representations of Buddha show a third eye, although this might be only symbolic of an inner eye.

For life on other planets, a mechanism for chemical analysis would be useful,

CALCULATING NUMBER OF ADVANCED SOCIETIES IN THE UNIVERSE

I. S. Shklovskii and Carl Sagan have proposed an equation for calculating the number of intelligent societies in our galaxy which have devised a technology far in advance of our own. The equation is:

$$\text{No. societies} = R f_P \, n_e \, f_t \, f_i \, f_c \, L$$

where R is the mean rate of star formation during the lifetime of the galaxy. This is assumed to be 10 because its age is estimated at 10 billion years and the number of stars it contains is taken as 100 billion stars. f_P is the fraction of stars containing planets, a situation thought to be fairly common and therefore the fraction is 1. n_e is the mean number of planets around each star having environments favorable to life. Around our sun one is known for sure and therefore n is taken as 1. f_t is the fraction of planets where intelligent life has arisen. In our solar system this has occurred at least once and therefore f_t is assumed to be 1. f_i pertains to the number of planets containing intelligent beings with manipulative ability. Here the probability decreases considerably, and therefore f_i is assumed to be only 0.1.

When assessing f_c the matter becomes more complicated. This factor applies to the extent of communication carried on among intelligent inhabitants of the planet. For example, would Western technology have been developed without contact with the East? f_c therefore is assigned 0.1. L is the lifetime of a civilization after it has become capable of interstellar communication. Not a single example is known, but two possible assumptions can be used: First a technical civilization destroys itself soon after reaching the stellar communication stage; in other words L is less than 100. The other assumption is that the civilization learns to live with itself soon after reaching the interstellar communication stage, and therefore its lifetime can be measured in terms of stellar-evolutionary time scale. Therefore L can be greater than 10^8 years.

The first assumed value of L gives the number of technical civilization more advanced than ours as fewer than 10. The second gives 10^7. When these two values of L are averaged, the number of societies then becomes 10^6. This means that about 0.001% of the stars in our galaxy possess planets having civilizations more advanced than ours. But it should be remembered that a one-way trip to our nearest neighbor would require several hundred light years.

probably similar to our sense of smell and taste. Pressure sensors such as our sense of touch are probably present, but whether or not hearing is used depends on composition and temperature of the atmosphere, factors which determine the speed of sound.

In thin atmospheres where sound travels slowly, a system of communication based on emitted light probably would have evolved. On earth this method is used among fireflies and certain marine animals. Or perhaps a method based on other wavelengths of electromagnetic radiation would have evolved.

Devices for intake and processing of food and discharge of waste products could vary from planet to planet, depending on nature of the food chain and the interrelationship among organisms. There is no reason to expect the same combinations found on earth where organs for vocalizing, breathing, and eating are combined, to a certain degree, as are organs of excretion and reproduction.

Intelligent Life

The type of environment that gives rise to intelligent life is difficult to speculate about. That leading to development of human beings is unknown. Some believe that during the Pleistocene ice ages, the severe climates, recession of forest, and scarcity of food imposed new habits of dress, eating, and habitat. In other words, without the ice ages, human society might not have evolved.

The assumption seems reasonable, Sagan says, that civilizations in advance of our own exist. Our sun is midway in its main sequence, and man-like creatures have inhabited earth for only 0.1% of its history, have been civilized for only one millionth, and have had ability to receive radio signals from outer space for only one billionth of its history. Therefore older civilizations further along in the main sequence of their sun should exist.

Based on our own experience, other assumptions can be made about advanced extraterrestrial civilizations: They strive for control of the universe. Toward this end, we have learned how to place a ring around our planet—the Air Force has proved this ability by putting needles in orbit. We have made this small planet the second most powerful radio source in the solar system and some day it might be as powerful as the sun.

We have put satellites in orbit and perhaps other civilizations have too. In this connection, Phobos, one of the two moons orbiting Mars, is much smaller than some scientists think it should be, and according to estimates it will crash on Mars in some 10 to 20 million years.

An explanation for this was proposed in 1959, by a Russian scientist, I. S. Shklovskii, working at the Sternberg Astronomical Institute of the Soviet Academy of Sciences in Moscow. He suggested that Phobos is hollow and therefore responds to small amounts of drag. The suggested reason for the hollowness is that, some millions of years ago, it was placed in orbit by intelligent beings, possibly for spying on celestial neighbors, including us. Perhaps the sands of Mars have drifted over evidences of a mighty civilization.

The laws of physics are universal and discoveries on earth probably have been made also on other planets. For example, the radio-frequency line of hydrogen, 1420 megacycles per second, can be considered the fundamental frequency of nature. However, for a technical civilization more advanced than ours, this might not be the preferred means of interstellar communications. For communication across distances greater than 3000 light years, this frequency would be strongly absorbed by interstellar hydrogen. Perhaps other frequencies exist that are multiples of 1420 such as 2840 or 4260.

If a signal were received, two-way interstellar radio contact could be established, followed by exchange of information. But radio signals can travel no faster than the speed of light, and even a brief exchange of salutations would require tens or hundreds of years, depending on the distance.

Artificial signals can be detected also by systematic observations. That a sending planet revolves about its sun must be assumed. Therefore, because of the Doppler effect, a periodic variation in frequency must occur unless compensated for by the transmitting civilization. Orbital velocities of the planet must be tens of kilometers per second, a situation which would produce frequencies with periods that could vary several months to several years, depending on the planet's period of revolution. Then by analyzing uncompensated signals, information about the length of the year for that civilization could be obtained.

Transmission frequency should vary also with rotation of the sending planet about its axis. Because rotating velocity is likely to be less than the orbital velocity, frequency variations would be smaller, but nevertheless detectable. In this way, length of the day could be determined.

Still other things could be learned. Through spectral analysis, composition of the star which is probably on the main sequence could be determined. By knowing the orbiting time, distance of the planet from the star could be determined and, based on luminosity of the star, a rough estimate of planet surface temperature would be possible. Further, by knowing the velocity of rotation and length of day, the planet's radius could be determined.

Such observations would be equivalent to spying on other planets. Perhaps other planets are spying on us. On earth, large scale radio communication has been possible for some 40 years and perhaps early transmissions such as an aria sung by Enrico Caruso are traveling forever through space at the speed of light. If an advanced civilization is within 20 light years of our sun, it might have received these signals 20 years ago. If this is true and they chose to reply, their signals should be received soon. If they are farther away, we will have to wait longer.

Another thought is pertinent. Our ionosphere transmits two general channels in commercial use – the television band and the high frequency of AM, about 1000 and 1400 kilocycles per second respectively. Thus, the first aspect of us that other civilizations might detect is our prattle of commercials on TV.

The possibility of interstellar communication using laser beams should not be overlooked. Earth's atmosphere absorbs completely ultraviolet at wavelengths less than 2900 angstroms, making communication impossible below this region. But if a laser operating at perhaps 1500 angstroms could be sent aloft in a satellite, light intensity some tens of thousand of times greater than that of the sun could be achieved. However, such a laser would be difficult to construct because reflectivity of mirrors necessary for pumping laser beams abruptly declines at ultraviolet wavelengths.

Manned Interstellar Travel

At a speed one third that of light, some 3000 years would be required to travel 1000 light years, and a round trip to the galactic center would require some 60,000 years. To make such trips within human lifetimes, two approaches are possible: slowing down body metabolism or taking advantage of the relativistic retardation of time.

Slowing down metabolism is possible on earth – for example, as in hibernation of certain animals, seedlings, bacterial spores, insects in egg and pupa stage. Storage by freezing has been accomplished in the laboratory for human blood and sperm, but not for a whole human body. Density of ice is lower than that of water, and its expansion during freezing causes severe damage to cells. However, one solution might be possible. Another kind of ice is formed at 3000 atmospheres of pressure and $-40°$ C. Under these conditions, ordinary ice becomes ice II having a density nearly that of water.

According to the theory of relativity, passage of time on a spacecraft traveling close to the speed of light would be very slow. The crew would become only slightly older and move more slowly; the effect in fact would be metabolic inhibition. In terms of relativistic time, only a few years would be required to reach the nearest star; 21 years to reach the galactic center; and 28 years to reach the spiral galaxy beyond the Milky Way. Time on the home planet would of course go on as before, and before the travelers reached their destination, their own civilization might have long since vanished.

Suggested Reading

(1) *Chemistry*, "How It Is on Venus," **37**, 3 (June, 1964); "Life on Mars Theory Revived," **40**, 32 (May 1967); "Life on Jupiter," **40**, 24 (June 1967); "Bacteria That Thrive on Gamma Radiation," **41**, 44 (July-August 1968): "Origin of Optically Active Substances," **41**, 29 (December 1968).

(2) Gamow, G., "Birth and Death of the Sun," Viking, New York, 1945.

(3) Huang, S. S., "Life Outside the Solar System," *Sci. Am.* **202**, 55 (April 1960).

(4) Hurlburt, E. H., "Signals from Outer Space," *Science Teacher* **31**, 9 (November 1961).

(5) Keller, E., "The Origin of Life," *Chemistry* **41**, 6 (December 1968); **42**, 12 (January 1969).

(6) Lovelock, J. E., Hitchcock, D. R., "Detecting Planetary Life from Earth," *Sci. J.* **3**, 56 (April 1967).

(7) Mellor, D. P., "Origin and Evolution of Chemical Elements, '*Chemistry* **37**, 6 (February 1964); **37**, 12 (April 1964).

(8) National Academy of Sciences, "Extraterrestrial Life: An Anthology and Bibliography," National Academy of Sciences-National Research Council, Washington, D. C., 1966.

(9) Shklovskii, I. S., Sagan, Carl, "Intelligent Life in the Universe," Holden Day, Inc., San Francisco, 1966.

(10) *The Sciences*, "Is Anyone Out There," **6**, 2 (March 1967).

(11) *Time*, "Mystery from Wayout," **84**, 38 (September 3, 1964)

(12) Wald, G., "The Origins of Life," in *Scientific Endeavor*, Rockefeller Press, New York, 1966.

25.

Human Origins

D. R. Pilbeam

Tool-making in fact has had far more than a passing importance in human evolution. Most authorities now accept the idea that tools have, so to speak, made man. AUTHOR

Perhaps the most famous British Association meeting ever was held at Oxford just over 100 years ago, in 1860. The meeting followed the publication in the preceding year of Charles Darwin's *The Origin of Species*. This book not only divided biological opinion on the one hand from popular and religious opinion on the other, but also divided biologists themselves. The debate at Oxford between Huxley, Hooker, and other Darwinian protagonists and Samuel Wilberforce, the Bishop of Oxford, is well known.

During the debate, Wilberforce facetiously asked Huxley whether he was descended fron an ape on his grandfather's or grandmother's side. Huxley is said to have replied:

"If I am asked whether I would choose to be descended from the poor animal of low intelligence and stooping gait, who grins and chatters as we pass, or from a man, endowed with great ability and a splendid position, who should use these gifts to discredit and crush humble seekers after truth, I hesitate what answer to make."

Reprinted from the March 1968 issue of *Advancement of Science* by permission of the British Association for the Advancement of Science and the publisher, Academic Press. This article formed the text of the Evening Discourse delivered 4 September 1967 at the Leeds meeting of the British Association.

Although Darwin's book covered general evolutionary theory, it seems to have been the human implications which caused most anxiety. It bothered not only biologists and bishops but others too, for in 1860 the wife of a canon of Worcester Cathedral is reputed to have said to her husband, "Descended from the apes! My dear, we will hope that it is not true. But if it is, let us pray that it may not become generally known." How could man, the fallen angel, have such brutish ancestors!

In 1871, Darwin wrote another book, *The Descent of Man*, this time devoted specifically to human evolution. In his characteristically cautious manner, he pointed out that man's ancestors, although not identical to, nor even closely resembling, the living apes, might nevertheless, when discovered, be described as "apes". He continued.

"We are naturally led to enquire, where was the birthplace of man . . . ? In each great region of the world the living mammals are closely related to the extinct species of the same region. It is therefore probable that Africa was formerly inhabited by extinct apes closely allied to the gorilla and chimpanzee; and as these two species are now man's nearest allies, it is somewhat more probable that our early progenitors lived on the African continent than elsewhere."[6]

So, almost 100 years ago, Charles Darwin had concluded first, that man's earliest ancestors were closely related to the apes, and second, that these early human ancestors most probably originated in Africa.

I want to tell you here about some of the exciting discoveries which have been made in the past few decades, discoveries which are beginning to throw light on those Darwinian speculations of a century or more ago. However, before talking about human origins, I think that I should try to explain some of the words that I shall be using.

Zoologists divide the mammals into a number of basic groups known as "orders". The order to which man belongs, together with his closest relatives, the apes, monkeys, and prosimians, and their ancestors, is called Primates. Orders are further subdivided. The living great apes – chimpanzees, gorillas and orang-utans–are grouped into a "family" known as the Pongidae. Their extinct ancestors are also included in this family, and it therefore has a time dimension. Living and extinct men are classified in another family, the Hominidae. We can describe the contents of these families by the more convenient colloquial terms "pongids" and "hominids"; these are broadly equivalent to the words "apes" and "men", although they are used – we hope – in more precise ways.

As Darwin noted, the Hominidae and Pongidae are more closely related to each other than either is to any other primate family. The exact time at which the early primates first split into human and ape stocks has been debated for more than 100 years. Estimates have fluctuated, some authorities favouring an ancient split, others a very recent one. Although we still do not know the exact time of the bifurcation, we do know that it had occurred in all probability at least 20 million years ago; but more of that later.

I have discussed the terms "pongid" and "hominid". One other remains to be discussed. Few, I think, will have any doubts about the meaning of the word "human", human is what we are, and we are members of the species *Homo sapiens*. But when did we become human? If we accept the fact that evolution has occurred, and that evolution applies as much to us as to any other species, then clearly, somewhere in our past, there are ancestors, that differ from us in their anatomy and behaviour. There will be no sharp jumps in this evolutionary record, and as we discover more and more fossils, as we fill in the gaps, our record of human evolution will become ever more continuous. With continuity, of course, definitions and boundary lines become harder to draw, and so it is with "human". However, it is generally agreed that the term "human" should be applied to the first hominids to be recognizable as members of the genus *Homo*.

The hominid fossil record is most complete during the last two million years or so. Before that it is exceedingly sparse. This two-million year period falls within the epoch called by geologists the Pleistocene (Fig. 1). The Pleistocene is now known to have begun about two million years ago. (Only recently, one million years has been regarded as a liberal estimate.) Before the Pleistocene we have the Pliocene, which began 11 or 12 million years ago, and before that the Miocene, beginning more than 25 million years ago. However, it is to the Pleistocene hominids that we turn first.

Men physically indistinguishable from ourselves first appeared about 40,000 years ago. Before that our ancestors did not look exactly like us. Although there is a danger in labouring this point, if we accept the fact that evolution has occurred in the human line, it is surely not too illogical to expect that at some stage our ancestors will *not* resemble us. I am emphasizing this because a few anthropologists seem to forget this point when constructing human evolutionary trees.

One of the distinctive characteristics of living *Homo sapiens* is the ability to walk

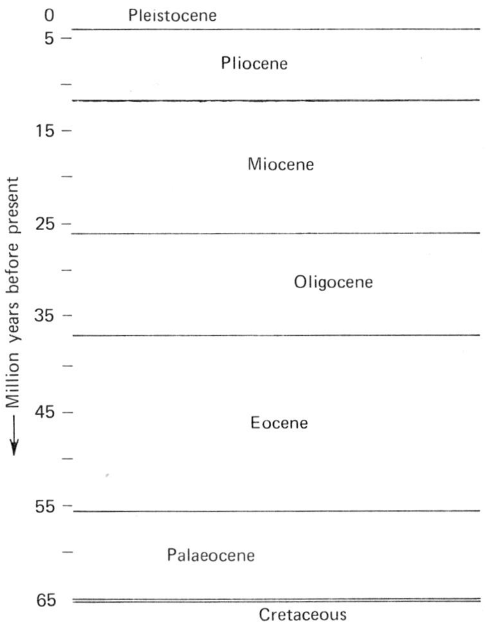

Figure 1 Geological epochs of the Tertiary and Quaternary periods, following the Cretaceous period. The earliest primates appeared in the Cretaceous, the first monkeys and apes in the Oligocene, and the first hominids in the Miocene. Men of the genus *Homo* are not known before the Pleistocene.

and run bipedally – that is, on two legs. Bipedal running and walking probably evolved quite early on in hominid evolution, perhaps as an aid to more efficient hunting. Certainly, by the time of Neandertal man, hominid hunting had become very efficient.

Modern man is also characterized by his enormous brain. The average skull volume, [2] the normal estimate of brain size, for both sexes of all races of *Homo sapiens* is 1320 cm^3. This is three times as great as the average volume for apes. The average for gorillas is around 500 cm^3, and for chimpanzees around 400 cm^3. Only very rarely does one find capacities about 700 cm^3 in gorillas or above 500 cm^3 in chimpanzees. Most normal human beings have capacities between 900 and 1800 cm^3. Of course there are microcephalics with smaller volumes, but these are almost always mentally subnormal. At least 900 cm^3 of brain would seem to be an absolute minimum for the type of intelligence found in *Homo sapiens*; this is well outside the ape range.

This relatively enormous brain of ours has clearly been very important to us. If we want to find just one character which distinguishes us from all other animals it is surely the magnitude and quality of our intelligence. (I do not wish to get involved here in a discussion of the meaning of "intelligence". However it is defined, we certainly are more intelligent than any other animal.) It is our great brain size which enables human social behaviour to be so complex. Anthropologists term this specifically human complexity "culture". Broadly, culture includes those parts of human behaviour which are transmitted by learning from one generation to another. These include the legal, political, and religious institutions of societies, the technological traditions of tool-making, as well as many other features. Some other mammals have complex social behaviour depending on individual learning, but human learned behaviour is infinitely richer. Humans communicate by means of vocal language while other animals do not; human language is a wonderfully efficient way of conveying information, information not only about the here and now, but about yesterday and tomorrow, and over-there-out-of-sight.

For much of our existence, we have been stone-age hunters. At a conservative estimate, this phase of our history lasted for a quarter of a million years. Little more than 10,000 years ago, some groups of men discovered a new way of living. They took to domesticating plants and animals, and turned from hunting to farming, from a shifting to a settled existence. This change is called by archaeologists, the *Neolithic*, following the Paleolithic, or stone age.

Populations which are settled and which can rely on regular supplies of vegetable and animal food can increase greatly in size, and they can support far more non-productive, specialized politicians, priests, soldiers, inventors, artists, and craftsmen than before. The Neolithic heralded a great burst in human social evolution, changes which are still going on. From the first farmers to the hydrogen bomb took just 10,000 years, and this "progress" was due not to organic evolution, not to a change in brain structure, but to new social inventions, to cultural evolution; two very different types of evolution. There is little reason to doubt that our brains are structured in a way essentially similar to those of our ancestors 10,000 years, or even 50,000 years ago. It is as though once the brain had reached a certain evolutionary threshold, no more physical changes followed.

Stone-age *Homo sapiens* was a superb big-game hunter, making up for his comparative lack of size, strength, and speed with his intelligence, planning, and co-operation. His stone, wood and bone tool kits were of great complexity; he buried his dead, and has left behind carvings and paintings of great beauty. This much we know from the archaeological record. Doubtless his social, political and religious institutions were as well-developed as his technology. We are dealing then with men very similar to ourselves, or at least to the non-literate peoples of the world today.

The best known fossil hominids are those caricatures of cave-men, the Neandertals.[3] It is rather improbable that the Neandertalers from Western Europe were ancestral to any modern men. However, the ancestors of modern man were obviously alive at that time, living probably in Eastern Europe, Western Asia, and elsewhere; it is unlikely that they would have looked very different from the Western European Neandertals, but for convenience we shall use the term "Neandertaloids" to describe those which were living at this time (50,000 and more years ago) in Eastern Europe, Asia and Africa, retaining "Neandertal" for Western European men.

It can be inferred from their skeletons that these Neandertaloids were bipeds as efficient as ourselves. Although their jaws, teeth and faces were larger than ours and their skulls shaped differently, their brains were as large as those of modern man. They too buried their dead, and apparently believed in an after-life. They understood the uses of fire. Their tools were varied and well-made, though not as beautifully made as those of later man.

Although they differ a little from us anatomically, most anthropologists would now classify the Neandertaloids as *Homo sapiens* too, in this way conveying the opinion that the similarities between modern men and the Neandertaloids are more important than the difference. Viewed in the perspective of 20 million years or more of hominid evolution, these earliest *Homo sapiens* are close to us not only in time but also in anatomy and in what we can infer of their behaviour.

Before about 150,000 years ago the picture is more indistinct, and there are many gaps in the hominid fossil record. It becomes less hazy around half a million years ago during the period known as the Middle Pleistocene. Hominid fossils of this age are known from a number of localities in the Old World – from Java, China, Europe and Africa. Doubtless men were living in other places too, although they are yet to be uncovered.

Like *Homo sapiens*, these men were apparently perfectly efficient upright walkers. Their brains however were smaller than ours; the average cranial capacity at this time was less than 1000 cm^3. They were almost certainly less "intelligent" than us, and had less highly developed cultures. This is reflected in the relative crudeness of their stone tools, compared with those of *Homo sapiens*. However, these men were efficient hunters who also knew how to use fire. They have been classified as members of the genus *Homo*, although they are usually put into another species, *Homo erectus* separate from *Homo sapiens*.

Now, we should be very careful when we say that *Homo erectus* evolved into *Homo sapiens*. What we are really doing is taking the last one million years or so of the hominid line and chopping it into two pieces with a dividing line somewhere between one-quarter and one-half million years ago. We have divided an evolutionary continuum in a quite arbitary way (see Fig. 2). The latest *Homo erectus* populations will obviously be virtually identical in structure to the earliest *Homo sapiens*. The important point to remember is that we are dealing at any one time-level with just one species of hominid, distributed through most of the Old World. This species has evolved through time, and has changed so much between the middle and late Pleistocene that we need to use different names to describe the different segments of the lineage. Unfortunately, giving different names to different segments of a lineage immediately creates a sharp difference where none, of course, exists.

In fact, the *erectus/sapiens* boundary has been decided by the historical accidents of discovery. The hominid line has been fairly well sampled in fossil terms for the period back to about 150,000 years, and for the period around one-half million years ago, and there has been little in between. So it has been easiest to draw the boundary in that gap. Had the gap in the record occurred elsewhere, we would have had different boundaries and consequently different limits for our species.

You can no doubt imagine the arguments that will follow when we find intermediate fossils. Are these the last of *Homo erectus* or the first *Homo sapiens*, or even a new species? As long as we can be sure that the new fossils are part of the single lineage which is evolving from the *erectus* to the *sapiens* level, it really matters very little what they are called. Of course, these arguments also apply to hominid fossils before *Homo erectus*.

Until 1925 no hominids were known from deposits of early Pleistocene age, that is the geological period between approximately one and two million years ago. It was claimed that the discoveries which had been made at Piltdown in Sussex in 1911 and 1912 represented an early Pleistocene type having a brain-case very close to that of modern man, although with a jaw like that of an ape. In the 1950's the Piltdown discoveries were shown to be fakes.[20] The evidence for the existence of modern-looking *Homo sapiens* before 40,000 years ago is now practically non-existent.

In 1925, a South African anatomist, Raymond Dart, described a new type of early Pleistocene hominid, called by him *Australopithecus africanus*.[4] The first specimen came from Taung in South Africa. Since 1925, a large number of specimens of this species has been recovered from other sites in South Africa, and today we know quite a lot about its anatomy.

When Dart first published his description of *Australopithecus*, only the skull was known. He noted that the brain size was small, only 500 cm^3. You will remember that volumes of this order are found in the living apes, and that they are never found in modern man, nor in the more primitive *Homo erectus*. However, Dart also observed that the teeth were man-like rather than ape-like. He believed that certain other features of the skull were also more typical of hominids than of pongids. But at this time Dart, along with many other anthropologists, believed that a large brain was the most distinctive hominid trait, and that since *Australopithecus* had an ape-sized brain, it could not be a hominid. However, he did place it in a new family, intermediate between the pongids and the hominids.

The reaction of the European anthropological establishment to Dart's discovery

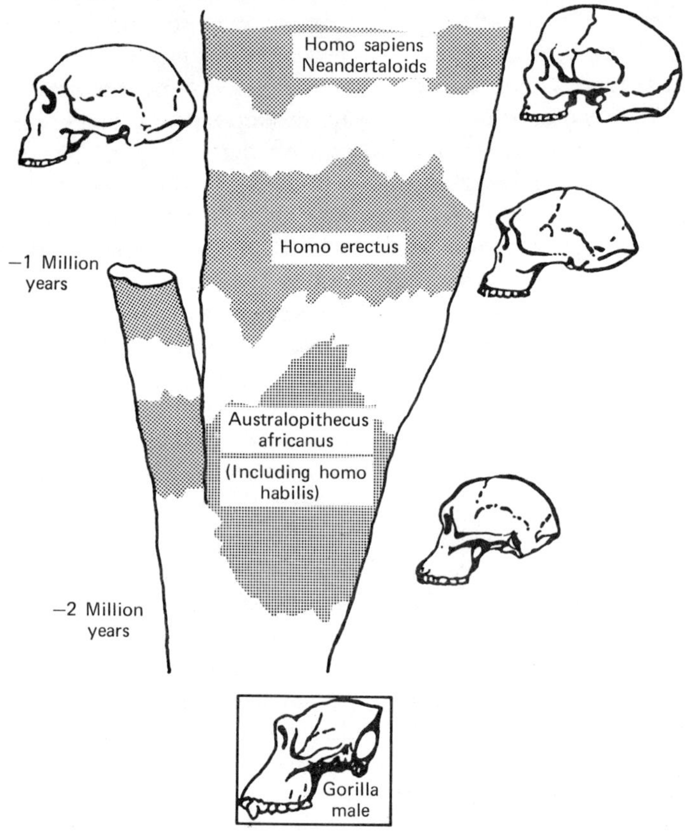

Figure 2 An idealized representation of Pleistocene hominid evolution, the stippling representing known fossils. The pictures on the right are those of listed species; top left is a Neandertal from Western Europe. The distinct hominid lineage represented by the column on the left is not mentioned in the text. It became extinct half a million years ago.

was sceptical in the extreme, and it was written off as just another ape.[11] Remember that at this time most anthropologists believed that early Pleistocene hominids, like Piltdown, resembled *Homo sapiens*. It was generally accepted that early hominids had large brains and ape-like jaws and teeth. In fact this is the complete opposite of the correct picture. Our early Pleistocene ancestors had small, ape-sized brains, and dentitions like later hominids.

Australopithecus africanus was a small, lightly-built hominid standing perhaps no more than 4 ft 6 in tall with an average brain volume of 500 cm^3. However, the smallness of his brain does not make him an ape, any more than a horse with a cow-sized brain becomes a cow.[15] Unlike the apes, *Australopithecus* had a dentition which is essentially human. The incisors were small and vertically placed in the jaws, the canines small and incisor-like, and the premolars and molars, although large, were similar to those of *Homo erectus* and *Homo sapiens*.

In apes, particularly the males, the canines are large and projecting teeth, important in display behaviour and in fighting. The incisors, particularly in the fruit-eating chimpanzee and orang-utan, are relatively very large indeed. This impressive battery of front teeth is used for cutting and chewing the tough fruits and shoots which cannot be prepared with the hands. Of course, it is still true that apes do use their hands in a wide range of activities, including the collection and preparation of a variety of food material.

The man-like structure of the dentition of *Australopithecus* implies that this early hominid did not use its teeth in the way in which pongids do. This point was in fact noted by some early workers, who inferred that *Australopithecus* probably used tools, both as weapons for offence and defence and as cutting and digging implements for obtaining food.[1] In more recent times, further discoveries have strengthened this view.

Archaeological evidence suggests that *Australopithecus* made and used stone and bone tools. To what end? Dart collected very carefully from caves the animal bones associated with these hominids and was able to show that, in all probability, these were remains of animals hunted, killed and eaten by *Australopithecus*.[5] It is difficult to visualize how *Australopithecus*, with his unimpressive front teeth, could have hunted and killed on this scale without using stone and bone tools.

A number of bones of the pelvis and leg show that *Australopithecus* almost certainly did not move in a quadrupedal – four-footed – or modified quadrupedal fashion like the monkeys and apes. In a number of basic ways, the pelvis and the leg resemble those of later hominids. *Australopithecus* was probably a bipedal animal, although not as efficient an upright walker as ourselves. Nonetheless, he was well on the way to being an efficient biped, and he was therefore probably a very mobile animal, covering a good deal of territory.

Finally, the geological setting of the fossils shows that these early hominids were essentially open-country, savannah dwellers, unlike the basically forest-living apes.

The original South African *Australopithecus* specimens came from deposits which, unfortunately, cannot be dated with absolute accuracy. It is possible, though, that the oldest remains may be as much as two and a half million years old, or more. Recently, more early Pleistocene hominids have been recovered by Dr. and Mrs. Louis Leakey from the magnificent East African deposits at Olduvai Gorge in Tanzania.[13] The earliest of these finds are almost two million years old. Other fossils from the Gorge provide a series which runs right through to the middle Pleistocene with specimens of *Homo erectus* dated at around one-half million years.

The early Pleistocene hominids from Olduvai have been described by Leakey, Tobias and Napier as a new species of the genus *Homo, Homo habilis*.[14] Like *Australopithecus africanus* this species contains hominids with small, ape-sized brains and man-like teeth and skeletons. One particularly fine specimen is that of a set of fossil foot bones, which show quite clearly that *Homo habilis* was an upright biped and not a quadruped.

He was also a tool-maker, producing a surprising variety of stone implements. The Leakeys are fortunate enough to have excavated complete campsites with tools and animal remains preserved, as well as evidence that *habilis* built shelters at his campsites. This last is a discovery of great significance, suggesting that, by two million years ago, hominid bands were spending periods of time in one place, hunting from and living at a single camp.

Now, there has been a great deal of argument about the exact position of *Homo habilis* in human evolution. My own view is briefly as follows. Middle Pleistocene and later hominids are found throughout much of the Old World, and it is not unreasonable to assume that the immediate ancestors of these hominids would also have been widespread. The South African early Pleistocene *Australopithecus africanus* fossils are sampled from this ancestral species. So too are the *Homo habilis* specimens from East Africa. I believe that both *Australopithecus africanus* and *Homo habilis* are simply geographically separate populations sampled from one species. No doubt early Pleistocene populations of this species from other parts of the Old World are still waiting to be uncovered.[18]

In summary, the early Pleistocene hominids, living one and a half to two and a half million years ago, were quite efficient bipedal runners. This method of locomotion seems to have enabled them to hunt with some success, to cover a large amount of ground in pursuit of prey. Generally, it seems they inhabited open country in the tropics. In spite of their small "ape-sized" brains, they must have been capable of quite a high level of co-operative behaviour, and they were able to make quite a variety of simple tools.

It was quite a struggle for Dart and others to gain acceptance for the view that these early forms were in fact hominids. We have noted already one of the objections to this view; that hominids anatomically more like modern man were living at the same time as *Australopithecus*. This objection is no longer tenable. The exposure of the Piltdown forgery has exploded what we can call the "early *sapiens*" theory.

The geological picture too is being modified. When Sir Arthur Keith (President of the Association at the Leeds Meeting in 1927) was writing in 1931 he believed *Australopithecus* to be about 400,000 years old, not as we now know more than two million years; he thought that *Homo erectus* was 300,000 years old, not one half or three quarter million years.[12] This "stretching out" of the Pleistocene gives us more time into which to fit our fossils, and more time for leisurely evolution from primitive to more advanced men.

There was another reason for the general reluctance to accept *Australopithecus* as an ancestor – brain size. We have noted already the evidence indicating that a number of rather human anatomical and behavioural features seem already to have been well established two million years ago. Yet these behavioural characters are found in animals with only ape-sized brains. Let me quote here from the 1947 edition of a work first published in 1931.[10] This book was a major, if not the major, text for American and English anthropologists for many years.

"[*Australopithecus*] lacked the brain overgrowth that is specifically human and perhaps should be the ultimate criterion of a direct ancestral relationship to man . . . Because they lacked brains, they remained apes, in spite of their humanoid teeth."

Presumably, we must once have had ancestors with brains smaller than our unusually large ones. It is becoming increasingly clear that our ancestors were hominids below the neck (and indeed above if teeth be included) long before our heads were graced with well-filled and lofty brows. It is surely illogical to call any small-brained hominid an ape, if at the same time one inevitably rules out of human ancestry any such "ape-like" creature! We now realize, of course, that the early hominids were not "apes" in any meaningful sense of the word.

Finally, we have what might be called, after the canon's wife, the Worcester syndrome – a horror at the idea of having any subhuman ancestors at all.

The apes, I think, have had rather a bad deal. The Victorians seem to have thought of them as brutish creatures. This view is certainly not held by those who have actually studied primate behaviour. Since the 1950's, there has been a great rebirth of interest in living primates. Of special interest to us are the studies of chimpanzees and gorillas in their natural habitats in Africa. (We should never forget that the behaviour of animals in zoos is not necessarily typical of their normal behaviour.) Of the field studies, the most important for us is the work of Jane Goodall (now

Baroness von Lawick-Goodall). Miss Goodall has been studying a group of chimpanzees in Tanzania, on and off, since 1960.[8]

The social behaviour of these apes is extremely complex. Individually, they are highly intelligent animals. They live in small, rather loosely-constituted bands of varying composition. Thus one finds groupings of females with infants and juveniles, of males and females, of juveniles, and of males. The composition of these small groups is fluid; an individual may leave one and join another at any time, without interference.

A wide range of relationships are clearly recognized; mothers and offspring greet one another affectionately when reunited, and interactions within and between these small bands are generally very friendly. As in human groups, facial expressions and body postures are important in social intercourse.

In Tanzania, where the animals live in open woodland, 50 or so chimpanzees may range over 10 or more square miles searching for food. Some students believe that it is this larger grouping of animals which makes up the chimpanzee "troop", rather than the small bands. Within the "troop", the social structure is constantly in flux, yet its overall coherence is maintained because individuals are able to recognize relationships with a large number of other animals, with whom they are not necessarily always in contact.

Chimpanzees sleep in the trees in nests which they make up each night. A new feeding and a new sleeping area are chosen each day as the troop moves on; unlike human hunters, chimpanzees are not tied to one home base for long periods of time. The all-male bands are the most adventurous, travelling widely in search of food. When a new source is found, males will often drum on trees or on the ground, their noise forming a focus for other chimpanzees in the area.

Yet this amazingly complex social structure, and the multitude of very flexible inter-individual relationships, are all possible with an ape-sized brain. So, at least as far as this very general point is concerned, the study of chimpanzees in the wild can help us in palaeoanthropology.

Australopithecus, the early Pleistocene hominid, had a brain somewhat larger than that of the chimpanzee, particularly if body size is considered; there is no reason to believe that the social behaviour of this early hominid was any less complicated than that of the chimpanzee. This is not to say that chimpanzees behave like *Australopithecus*! Of course, the difference between ape and human brains is not just a matter of size. Brain cells in man are larger than those of apes, and are more complex. New tracts of cells have been added and the internal organization of the human brain remodelled. So although early hominids may have had ape-sized brains, these brains probably functioned in rather different ways from those of apes and produced different behavioural patterns.

As I have already noted, the apes are mainly forest-dwelling vegetarians, while *Australopithecus*, like later hominids, was a hunter living in open country. He was probably not as efficient a hunter as *Homo erectus*, nor would his social behaviour have been as highly developed. However, it seems probable that these early hominids were already occupying campsites for a few days at a time, and there might well have been some division of labour between the more mobile hunting males and the females who remained in camp caring for the young and gathering and preparing vegetable foods.

I have mentioned briefly, almost in passing, the phenomenon of tool-making in hominid social life. Tool-making in fact has had far more than a passing importance in human evolution. Most authorities now accept the idea that tools have, so to speak, made man. There has been, to use fashionable terms, a constant feedback between an increasingly complex cultural, tool-making environment and, via an enlarging brain, increasingly complex behaviour.

Now, man is one of the few social mammals in which the social group is organized to obtain food. The food which is obtained, particularly animal food, can only be utilized if natural objects like stones are used to prepare it. With a dentition which is not designed for slashing and cutting, tools become vital for dismembering carcasses, and, for instance, for smashing bones to obtain the edible marrow.

Tool-using and tool-making have therefore been important catalysts in human evolution. Object manipulation is one component of the play of young humans not found in the play of other primates. As juveniles and adults we generally exhibit a high degree of manual dexterity, and this dexterity has become built into our brains over several millions of years of evolution – tool-making evolution.

One of the most startling discoveries made by Jane Goodall during her field studies has been the fact that chimpanzees prepare tools to aid themselves in obtaining food. They will, for example, pick a grass stalk and poke this into a hole in a termite mound. Withdrawing the stalk covered with termites, they will then pass it between their lips, stripping off the termites. Small twigs are also used, in which case leaves and small twiglets are removed beforehand, leaving a single trimmed tool. These tools may be carried for some distance and time before being used. Chimpanzees have also been observed using a previously chewed leaf as a sponge to mop up water from a tree hollow. Tool-use, however, is not a vital aspect of chimpanzee survival. For hominids though, tools *must* be used for individuals and groups to survive.

Meat-eating has also been observed by Miss Goodall. She has seen a number of male chimpanzees occasionally band together to trap, kill and eat Red Colobus monkeys. The meat is also shared out to other chimpanzees who have "begged" for a share. Although some other primates do catch and eat animal food, this is nevertheless very rare; the sharing out of animal food has not been reported in any other primates except chimpanzees. Food-sharing does of course occur in social carnivores like wolves or Cape Hunting Dogs. We, too, are social carnivores.

It should be noted that these behavioural patterns have been observed only in chimpanzees living in open woodland, and have not so far been reported for animals living in forests. This may be because of insufficient work on forest-living chimpanzees, or because vegetable food is not so plentiful in open woodland.

The late Professor Hall has recently suggested that the most important uses of objects by primates are as weapons rather than as tools in purely economic circumstances (for obtaining and preparing food).[9] Let us consider again the teeth of *Australopithecus, Homo erectus* and *Homo sapiens*. The canines are small, just like the incisors. Compare these to the dagger-like canines of male apes and monkeys. Why do hominid males *not* possess these dental weapons? Most anthropologists would argue that other objects were available for personal and group defence. These "other objects" are tools, used as weapons.

The apes do occasionally use objects in their display behaviour, when they are aggressive or excited. Males may often pick up and hurl stones, clods of earth, sticks, or handfuls of vegetation. If these sorts of behavioural patterns were present in the very earliest hominids, it is not difficult to understand how regular club-using or spear-throwing might have developed. Unlike those primates that are vegetarians, a social hunting primate would soon learn to integrate the economic and the aggressive use of tools.

It seems that tools must have been used for some time before the early Pleistocene because the front teeth of *Australopithecus* are already reduced in size and are human-like in shape. Inanimate objects must have played an important part in hominid life long before the Pleistocene.

Now, let us summarize some of the points I have made. I said earlier that we should not worry too much about what names we are going to give to the various stages of hominid evolution. Instead, we should concentrate on the various trends in that evolutionary sequence.

If we look at the Pleistocene hominid fossils from oldest to youngest, from *Australopithecus* to present day *Homo sapiens*, we can see that bipedal walking and running became steadily more efficient. Hominids also became taller – from an average of perhaps 4 ft 6 in to one of over 5 ft 6 in; their brain size increased – from an average of about 500 cm^3 to over 1300 cm^3. Teeth became smaller, and with this change the face also shrank, the skull becoming rounded rather than elongated fore and aft.

Technological advance is also marked. Tool-making becomes more refined, and the tool-kits become more complex. In the middle Pleistocene the use of fire appears in northern latitudes. The size of the animals hunted increases; from middle Pleistocene time on, the remains of large mammals in large numbers are found at campsites.

This increasing technological competence is one reflection of a general improvement in behavioural, social, or culture efficiency. Other features of society, such as religious and political institutions and language, would have been developing too. Unfortunately, these are not preserved as archaeological fossils. Nevertheless, we have noted that the Neandertaloids buried their dead with ceremony 50,000 years ago and more, which implies belief in an after-life. Paintings and sculpture appear in the latest Pleistocene.

Many anthropologists would like to use the first appearance of manufactured tools as a sign of the first appearance of the genus *Homo* – the first sign that hominids had become human.[16] There are a number of objections to this. Unfortunately, stone tool-making is only one restricted facet of human behaviour. Also, as we go further and further back in time, manufactured tools become more difficult to recognize, and adequate sites more difficult to find. Finally, whichever hominid features we study, whether they are anatomical or behavioural (inferred from anatomy and archaeology), we see that these features change gradually through time. To draw a meaningful line across one or several of these trends is impossible. If arbitrary boundaries must be drawn, I believe that they should be rigid, time boundaries. For example, we might agree to draw the boundary between *Australopithecus africanus* and *Homo erectus* at exactly one million years ago.

I said before that we should not make the mistake of thinking that chimpanzee social behaviour can be used as a replica of early hominid behaviour. However, Vernon Reynolds has recently suggested that the social structure of the living apes may not be so very different from that of the very earliest hominids.[19] He argues that the behaviour of the three species of living ape is similar in a number of basic features and that these features are likely to be ancient, possibly shared with the very earliest hominids when these branched off from the line leading to the apes. His ideas are as follows.

Groups of exploratory males might have become converted into hunting parties, and the general peacefulness of social interactions would facilitate the development of co-operative behaviour between these male bands. Hominids would begin to live less in forests and more in open country where hunting is easier. Gradually, bipedalism would have become the most important mode of locomotion. During this time, too, the use of tools as weapons must have developed. As it concentrated on hunting in one particular area, the hominid group would tend less and less to move its campsite every night.

I want now to describe very briefly some of the exciting work which has been done during the past eight years with pre-Pleistocene hominids and pongids.[17] The majority of these fossils are between 10 and 20 million years old and probably cover at least part of the period during which these important behavioural changes mentioned above occurred.

There is definite evidence for the existence of a hominid in the late Miocene and early Pliocene, between 10 and 15 million years ago. This hominid is *Ramapithecus*; the evidence consists of just a handful of jaws and teeth from India and Kenya (see Fig. 3). Unfortunately no skeleton has been recovered. What is preserved shows that the canines and incisors were already reduced in size 14 million years ago. Like those of the early Pleistocene, these much older hominids did not possess large canines. In fact, in the parts preserved, *Ramapithecus* and *Australopithecus* are surprisingly similar. We have no record of any stone, bone, or wooden objects associated with *Ramapithecus*, although it is unlikely that he could have lived successfully without tools.

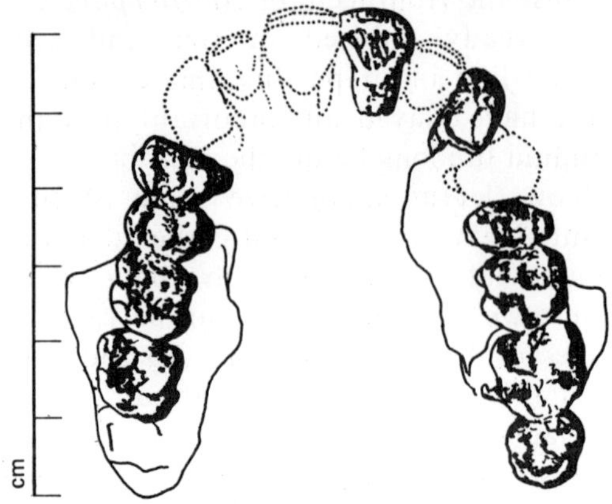

Figure 3 A composite reconstruction of *Ramapithecus punjabicus* using specimens from India and Kenya. The dental arcade is rounded as in later hominids, the canine small, and the incisor human-like in form. Age: approximately 14 million years.

So we have now pushed back our knowledge of the earliest hominids from two or three million years ago to more than 10 million years. Clearly, long before the appearance of *Homo sapiens*, or even of the genus *Homo*, the hominids had been in existence as a lineage separate from the apes. This discovery is, as I have said, quite a new one. Although a number of earlier workers have proposed similar hypotheses, for the first time we have fossil evidence to support our theories.

Living alongside *Ramapithecus* in India, 14 million years back, were animals ancestral to the living Asian great ape, the orang-utan. Both hominids and pongids moved from Africa to Eurasia, along with many other animals, when a connection between the two landmasses was formed in the Middle Miocene, say 16 million years ago.

African deposits a little older than this – of middle and early Miocene age, some 20 million years old – have yielded some very interesting fossil apes, species which are ancestral to the gorilla and to the chimpanzee. Dr. Leakey has recently suggested that certain fossils from the same deposits are hominids rather than pongids. I am unable to agree with him in this and I prefer to regard these particular specimens as ancestral orang-utans before that lineage left Africa.

However, we have noted that at this time – 20 million years ago – three separate species of a genus called *Dryopithecus* were living which were most probably ancestral to the chimpanzee, the gorilla, and the orang-utan. I think that these ancient pongid species were already too specialized, already too committed to their ape-dom, to have produced the hominids.

Some day therefore we must expect to find hominids even earlier than *Ramapithecus*; it is possible that the hominid and

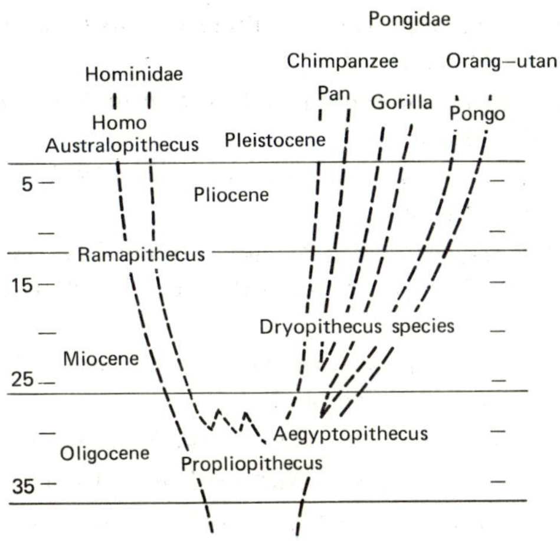

Figure 4 A tentative scheme of hominid and pongid evolution during the past 30 million years.

pongid lines have been separate for 30 million years or more (see Fig. 4). One might ask why so few of these early hominids have been recovered. Possibly because there were so few of them about. Why so few? Perhaps because they were creatures with a very low population density, perhaps because they were already hunters.

Of course, this is the sheerest of sheer speculation. However, I think it more likely than not that in the years to come we shall learn that our hominid peculiarities of behaviour and anatomy have very ancient origins indeed.

In this paper I have tried to cover, very briefly, a little of what is known of human evolution. When I began I quoted from a lady who could not bear to think about her animal origins. Let me finish by quoting in reply some of Darwin's own eloquent words:

"Man may be excused for feeling some pride at having risen, though not through his own exertions, to the very summit of the organic scale; and the fact of his having thus risen, instead of having been aboriginally placed there, may give him hope for a still higher destiny in the distant future . . . (However) . . . with all these exalted powers – Man still bears in his bodily frame the indelible stamp of his lowly origin."

Like Darwin, I prefer to regard men as risen apes rather than as fallen angels.

References

1. Alsberg, P. (1933): "The Taung puzzle. A biological essay". *Man,* **34**, 154-159.
2. Ashton, E. H. and Spence, T. F. (1958): "Age changes in the cranial capacity and foramen magnum of hominoids", *Proc. Zool. Soc. Lond.* **130**, 169-181.
3. Coon, C. S. (1963): *The Origin of Races.* London: Jonathan Cape.

4. Dart, R. A. (1925): "*Australopithecus africanus*: the man-ape of South Africa", *Nature, Lond.* **115**, 195-199.
5. Dart, R. A. (1957): "The osteodontokeratic culture of *Australopithecus prometheus*", *Transv. Mus. Mem.* **10**, 105 pp.
6. Darwin, C. (1879): *The Descent of Man, and Selection in Relation to Sex,* p. 155. London: John Murray.
7. Darwin, C. (1879): *The Descent of Man, and Selection in Relation to Sex,* p. 619. London: John Murray.
8. Goodall, J. (1965): "Chimpanzees of the Gombe Stream Reserve". in *Primate Behaviour,* ed, by I. Devore. New York: Holt, Rinehart and Winston.
9. Hall, K. R. L. (1963): "Tool-using performances as indicators of behavioral adaptability", *Curr. Anthrop.* **4**, 479-487.
10. Hooton, E. A. (1947): *Up from the Ape.* New York: Macmillan.
11. Keith, Sir A. (1925): "The Taung skull," *Nature, Lond.* **116**, 11.
12. Keith, Sir A. (1931): *New Discoveries relating to the Antiquity of Man.* London: Williams and Norgate.
13. Leakey, L.S.B. (1965):*Olduvai Gorge,* vol. 1. Cambridge University Press.
14. Leakey, L. S. B., Tobias, P. V. and Napier, J. R. (1964): "A new species of the genus *Homo* from Olduvai Gorge", *Nature, Lond.* **202**, 5-7.
15. Le Gros Clark, Sir W. E. (1967): *Man-Apes or Ape-Men?* New York: Holt, Rinehart and Winston.
16. Oakley, K. P. (1961): *Man the Tool-maker.* London: British Museum (Natural History).
17. Pilbeam, D. R. (1967): "Man's earliest ancestors," *Science J.* **3**, 47-53.
18. Pilbeam, D. R. and Simons, E. L. (1965): "Some problems of hominid classification," *Am. Scient.* **53**, 237-259.
19. Reynolds, V. (1966): "Open groups in hominid evolution," *Man,* new ser. **1**, 441-452.
20. Weiner, J. S., Oakley, K. P. and Le Gros Clark, W. E. (1953): "The solution of the Piltdown problem," *Bull. Br. Mus. Nat. Hist.* Geol. Ser. 2, no. 3.

26.

Human Biological Adaptability

Gabriel W. Lasker

Adaptation can . . . be involved in (i) major evolutionary events, (ii) growth of the individual, and (iii) behavioral and physiological changes lasting only hours or minutes. AUTHOR

The theme of the U.S. effort in the International Biological Program is listed as "Man's survival in a changing world," and the whole of the International Biological Program has been described as focusing on ecology, especially huma[illegible] ecology.

The Human Adaptability Project is one of the principal aspects of the International Biological Program. Inte[illegible]ationally about 50 nations are participating in the Human Adaptability Project [illegible]tudies of adaptation of many different peoples to a wide variety of environments. Of the integrated research programs constituting the U.S. contribution to the Intern[illegible]onal Biological Program, five studies [illegible]orm the Human Adaptability group while eight constitute the environmental management group (*1*). The former consis[illegible] of (i) the "International Study of Circu[illegible]olar People Including Eskimos" involving adaptations to cold; (ii) the "Popula[illegible] Genetics of the American Indian" emp[illegible]izing adaptations to life under p[illegible] [illegible]onditions on the tributaries o[illegible] [illegible] and Orinoco rivers; (iii) th[illegible] [illegible] [illegible]uman Populations at Hi[illegible] [illegible] [illegible]he Andes of Peru and i[illegible] [illegible] [illegible]t coordinated with studi[illegible] [illegible] [illegible]n highlands, and in the Himalayan and Tien Shan mountains; (iv) "Nutritional Adaptation to the Environment", and (v) the "Ecology of Migrant Peoples." A sixth program in chronobiology is being prepared.

Because adaptation has a variety of connotations that are different in different disciplines, this surge of activity and the participation of so many physical anthropologists in it warrant an attempt to explain a physical anthropologist's view of human adaptation. I shall not attempt to review here man's adaptation to infectious diseases, man's cultural adaptations, the story of the evolution of man, or the abundant physiological literature on various homeostatic adaptations.

Adaptation is the change by which organisms surmount the challenges to life. In the broadest sense biological adaptation encompasses every necessary biological process: biochemical, physiological, and genetic. Adaptation can therefore be involved in (i) major evolutionary events, (ii) growth of the individual, and (iii) behavioral and physiological changes lasting only hours or minutes. Human adaptation covers both functional processes and the structures on which they depend. It differs from human biology as a whole chiefly by its limitation to the concern with how the organism relates to the circumstances it must meet to live.

Adaptation implies its antithesis: if one way of functioning is adaptive, another is

Reprinted from *Science*, V[illegible] [illegible]80-1486, December 19, 196[illegible] [illegible] 1969 by the American Association fo[illegible] [illegible]dvancement of Science.

less adaptive or disadaptive under comparable circumstances. From this springs the idea of adaptive selection, the central theme of the Darwinian theory of evolution – the natural selection of better-adapted organisms and the extinction of the less well adapted through reduced fertility or earlier death. In this sense adaptation is a modification in structure or function that enables an organism to survive and reproduce. The term can apply to a particular organ or the whole individual and to entire populations or whole species. The more different the individuals or species are, the more able we are to identify the relation of the anatomical differences to different behavior and different adaptation to the environment. Conversely, the peoples of the world today are so similar that it is often difficult to relate specific structural differences to the specific environmental differences man encounters throughout the world. Adaptation occurs at three levels: (i) selection of genotype, (ii) ontogenetic modification, and (iii) physiological and behavioral response. As one goes from interspecific differences to individual differences within the human species, the chief emphasis shifts from the first level to the second and third. This can be exemplified by anthropological studies of human adaptation to altitude, cold, heat, migration, and other circumstances.

Altitude Adaptation

Adaptation of man to high altitudes involves numerically a relatively small problem. Only about 25 million people (that is, less than 1 in 100) of the world's people live in high mountains. But high altitudes, with their low atmospheric oxygen tension, present an environmental problem that could not be modified by human inventions until the present century when bottled oxygen and other such therapies were available for treating mountain sickness. Men living in the mountains use drugs such as alcohol and coca (the plant that yields the narcotic cocaine) to lessen their psychological burden, and this may alter the nature of their response and hence the impact of the conditions. However, the extent to which these drugs ameliorate the physiological burden of the altitude seems to be slight although consumption of alcohol can raise the foot temperature of the highland Indian and increase his comfort during the cold of night *(2)*.

When individuals climb from sea level to an elevation of 4000 meters or more, there are large differences in the extent of the response and some individuals may even die of pulmonary edema. However, a usual response is an increased rate of breathing and an increased pulse rate under comparable work loads. After a few days at that altitude there is some short-run "adaptation," including increase in hemoglobin concentrations, but there still are difficulties in working. Families who continue to live there incur increased risks of miscarriages, birth defects, and infant deaths. Individuals reared at these elevations achieve more adequate adaptation and the risks are lower. Those born into populations genetically adapted to the altitude apparently do better still. Newman and Collazos and, more recently, Baker *(3)* report that in the Peruvian Andes growth and skeletal maturation is retarded; the consequent relative stunting is possibly an advantage. Chest measurements do not follow this trend toward small dimensions, however; Indian boys in the high mountains of Peru, while developing more slowly than coastal dwellers in other respects, develop a larger thorax and greater lung capacity *(4)*.

Mountain dwellers thus show the three chief modalities of adaptation: (i) short-run physiological changes; (ii) modifications during growth and development; and (iii) modification of the gene pool of the population. It is probable that the well-adapted mountain dweller suffers some relative shortcomings when at sea level, but since the Indians who migrate from the Andes to the cities on the coast suffer some of the same kinds of social disabilities as Appalachian Mountain folk do in the core cities of the United States, analysis of purely biological status is complicated by the concomitants of social status, and the results of studies of such people are difficult to interpret. One example of the fact that good genetic adaptation at one elevation may be bad at another is the case of sickle-cell and thalassemia traits. Heterozygosis confers a degree of immunity to endemic malaria in low-lying areas; but at very high elevations even the heterozygotes may have hemolytic crises.

Cold Adaptation

Man in the arctic provides another example of adaptations. In the arctic, however, people build houses, wear clothes, and light fires. These cultural traits constitute the predominant adaptations, and they are available for anyone to borrow. Thus the Eskimos have designed fine arctic clothing, and European and American explorers have copied their parkas and mukluks. Furthermore, Eskimos have developed behavioral patterns to meet crises. William Laughlin, F. A. Milan, and others are contributing much to our knowledge of Eskimo adaptations. If one Eskimo falls into the water, a companion will immediately share half his dry clothing – enough to get both men home, cold but alive.

Despite cultural adaptations there are times when biological differences count. Baker *(5)* gives an example of the Yahgan at the cold, southern tip of South America: we should not envision one native dying of exposure in a snowstorm while a better-adapted companion survives; instead, cultural modes ordinarily modify the biological conditions; the family of the ill-adapted individual dies of starvation huddled at their campfire while the well-adapted counterparts comfortably collect shellfish in the frigid water. To study cold-adaptation in man one must take account of indoor as well as outdoor temperatures, and also of activities, clothing, and shelter.

For life, man must maintain a core temperature close to 37°C. In the cold he does this by reducing circulation to the extremities (which therefore drop to lower temperatures than the trunk). But the vessels of the extremities periodically dilate; this cold-induced "cycling" is a widespread phenomenon among mammals and must be adaptive in some way (perhaps through decreased chance of frostbite) although it costs loss of stored heat. Body heat is generated by metabolizing food, burning it up as fuel. Shivering is an involuntary activity that increases the production of heat. All peoples of all places respond to cold in much the same ways. However, some people from cold climates (notably the Central Australian aborigines) have been reported to meet cold sleeping conditions by having the extremities cool off more relative to the trunk, and also by lowering core temperatures; but the subjects of these studies chewed tobacco and the leaves of *Duboisia* which contains an alkaloid posion. This cultural practice rather than genetic constitution may account for the difference in response *(6)*. The short-run adaptations differ somewhat in those inured to the cold and in those new to it, but it is not definitely known how much genetic capacity for acclimatization, if any, differs among peoples.

In other species of animals, arctic forms tend to differ in predictable ways from those found in more southerly areas. One of the differences, fur, has no counterpart in man. The Eskimos, for instance, are relatively devoid of body hair. Arctic forms have small body-surface area relative to body mass. Some heat is lost in breathing, but most heat loss is through the skin; therefore the surface area of the skin (and hence the size of the individual, which largely determines surface area) relates to dissipation of heat. Body mass consists of metabolizing tissues (which produce heat) and fat (much of it just beneath the skin where it may help to insulate); hence increases in weight cause increased heat production and retention. Newman applies to man the two rules (or laws) which express these relations between body size and form with temperature *(7)*. Bergmann's rule *(7)* states that, in bodies of the same shape, the larger one has relatively smaller surface area; cold-adapted animals therefore tend to be large. Allen's rule *(7)* states that short extremities further increase the ratio of mass to surface area and the cold-adapted forms have relatively short limbs. In man, general body size as measured by weight or stature is, on the average, positively correlated with climate – especially the temperature in the coldest month. In continuous populations of large land areas of the Northern Hemisphere, including China, Europe, and the contiguous states of the United States, there is a gradient from larger average size in the north to smaller in the south *(8)*. Nevertheless these dimensions vary considerably in any one place, and there are numerous exceptions to Bergmann's rule. For instance, Eskimos are small but squat and have relatively short limbs. Some people of East Africa provide an instructive example; they are very tall, but slim, and their limbs are exceptionally long *(9)*. In any case, Roberts showed that the relation of weight to climate applies on a world scale, Schreider extended this to the ratio of surface area to weight, and Newman concluded that temperature accounts for almost 80 percent of the variance in average body mass of different populations throughout the world, when an appropriate allowance for stature is included *(10)*. In mankind in general, tall stature is achieved primarily by growth of the limbs (trunk length being much less variable). The real test of the applicability of the rules of Bergmann and Allen to man should, therefore, come in genetically determined tendencies to depart from this nutritionally determined pattern, and to find large size with short limbs in the arctic and short stature with long limbs in the tropics.

Other details of morphology may also relate to heat balance. Nose form seems to be adapted to the degree of need to moisten the air one inhales. Noses are narrower in colder zones. Although correlations of nasal dimensions with degree of prognathism and shape of the dental arch complicate interpretation, the average ratio of the width to the length of the nose (nasal index) of populations throughout the world is highly correlated with climate – especially with vapor pressure, the amount of moisture in a given amount of air *(11)*. Of the various Mongoloid peoples, Eskimos have the narrowest noses. But it is unwise to assume that every morphological feature found in Eskimos is directly a protection against cold. For instance, the large, broad Mongoloid face of the Eskimo is more exposed to cold than the smaller face of the European, but frostbite of the face is rarely serious in either people *(12)*.

High altitudes are also cold, and altitude studies must deal with the influence of both altitude and temperature. Some individuals adapted to altitude and cold maintain warmer hands and feet than nonadapted controls. This may serve to maintain more oxygen in these tissues as well as meet the challenge of low temperature *(13)*.

Heat Adaptation

The regulation of human responses to excessive heat involves at least two distinct types of environment – dry heat and humid heat. In a hot place when a person works hard, the temperature is extreme, the humidity is high, or the sunlight is excessive, it puts a strain on the temperature-regulation system of the body. After about a week of acclimatization to repeated heat stress a subject will improve his tolerance for these conditions through increased sweating and decreased cardiovascular strain *(14)*.

Most species of primates live in tropical forest environments. In the view of many, man's ancestors at one stage also lived in a hot, humid zone with little movement of air and with little direct sunlight. Under those circumstances the heat is well tolerated during rest but hard work produces heat stress.

In drier and more open country at low latitudes, where it is possible that, at a later stage, man's progenitors evolved their upright posture and a hunting-gathering economy, sweat evaporates more readily but sunlight on man and on the objects about him add a severe radiant heat load to the problems of head adaptation. Sweating remains important. The upright posture reduces the surface area exposed to direct sunlight compared to that of a quadruped. Whether hair, at least long, straight hair could be an added protection against heat is problematical. In any case nakedness, far from being an advantage, as some anthropologists have mistakenly claimed, would have prevented man from adapting to life under desert-like conditions until he achieved ready access to water through use of vessels or learned to wear loose clothing of some sort *(15)*. Man cannot drink very much water at one time but can sweat more per hour than any other mammal so far tested. Before man learned how to carry water with him, human occupation of open plains and savanna therefore required behavior that would make it easy to reach drinking water frequently.

Although man is unique in his heat-adaptive mechanisms, there is little innate difference between human groups in their ability to respond to heat stress. Short-term acclimatization aside, there is little evidence of population differences. Since heat absorption and dissipation are surface phenomena, the search for possible population differences is logically concentrated on the area of the skin and the nature of its structures; pigment granules, hair, and sweat glands. Groups inhabiting the tropics are, as already noted, generally composed of small individuals. This increases the surface area relative to mass so that heat produced by activity can be more readily dissipated. Although even moderate activity in a hot environment normally results in some heat storage, excess heat must be dissipated sooner or later and this is more significant than the somewhat greater capacity for heat storage of larger individuals. In hot deserts the people are generally lean, another way of increasing the relative surface area and facilitating heat dissipation.

Pigment is not a simple question. Inhabitants of equatorial zones are dark. This is true of Melanesia, Australia, South India, and Africa, although in other respects the peoples of these areas are very different. Common adaptive modification rather than close common origins therefore accounts for the similarity in skin color. Numerous theories have been advanced to explain why dark skin color is adaptive in hot climates, and there is still no general agreement. Although light skin reflects more radiant heat, dark skin must protect the body better. Among other things dark skin inhibits sunburn, and sunburn interferes with the sweating response. Dark skin is also less susceptible to skin cancer, and it prevents the synthesis of too much vitamin D. In zones with much sun in summer, and little in the winter, the ability to tan in summer would therefore be an advantage. Marjorie Lee and I *(16)* subjected individuals of various groups to measured amounts of untraviolet light and we measured changes in the amount of light of different wavelengths reflected from the skin. We found that the capacity to tan varies considerably between individuals – even those with similar initial pigmentation.

The chief pigment of skin is melanin, and control of its production is evidently polygenic but based on few alleles at few loci. Livingston *(17)* shows by computer simulation that on these assumptions one can easily account for the sort of skin color distributions one encounters in going from north to south in Europe, the Near East, and Africa, if there is a small differential advantage of presence of dark color in tropical Africa and vice versa in Europe. His calculations include a variety of assumptions, including migrations between populations. They do not explain its basis, but they do show that adaptative natural selection for skin color reasonably accounts for the known skin color distribution. Low rates of selection and recent migration are consistent with the facts that after some hundreds of years Europeans in the tropics are still light and Negroes in the United States are still dark. After several thousand years, however, American Indians in the tropics are slightly darker than those of North America to about the extent Livingstone's calculations would predict. Furthermore, the selective advantages postulated in the model are of a relatively low order so it is hardly surprising that we still lack direct evidence of the nature and amount of selection that takes place.

The rôle of human differences in hair with respect to heat tolerance is hard to calculate. Sheep with thick wool can thereby stand very heavy exposure to direct sunlight. Perhaps the retention of head hair in human beings is related to the crown of the head being the most exposed part in the noonday sun, but the influence of hair of different color and form remains to be established experimentally. Essentially nothing is known concerning when man's progenitors began to have such bald bodies or whether this antedated the invention of cloaks and hats.

It was once believed that Negroes had more or more efficient sweat glands than whites. Careful counts of areas of skin show no such thing. The only known difference is that obese individuals have fewer sweat glands per given area of skin than thinner individuals. Persons of different body build seem to have approximately the same total number of sweat glands, however *(18)*. After all, ability to maintain body temperature at a constant level is as important to an Eskimo trotting beside his sled dogs as to a caravaneer astride his camel.

Nutritional Adaptation

Nutritional adaptation not only depends on the resources available but also on the mode and degree of utilization. The utilization of food resources is a culturally-determined matter. Hunters, fishermen, and food collectors select only certain of the available foodstuffs. Agriculture and animal husbandry in any one place is limited to a very small number of edible species by the knowledge and culturally determined technology of the group. We still know relatively little about differences in nutritional requirements between populations. There do seem to be some inherent differences of this kind. Size itself is the most significant variable. Furthermore basal metabolic rate seems to be inherently high in populations which have always had a diet of good quality and low in some poorly nourished populations. Thus in one region in South China where a meager diet, 90 percent of which was from grain (mostly polished rice), has been reported, basal metabolic rates are also low. Neither the immediate diet nor that usual in the population explain all the variability in metabolic rates, however *(19)*. Newman, in considering human bodily adjustments to nutrition, reports several instances where ingestion of necessary nutrients is a very small fraction of the supposed minimum requirements *(7)*. He concludes that tolerances must vary over the world and that some human populations have adapted to levels of intake that would be fatal in others. On the other hand, ability to get by on less food may be more general. Throughout history most populations of the world have been subjected to repeated famines, and mankind must therefore share adaptive mechanisms to meet food storages.

In addition, time apparently has been adequate for some special adaptations in subpopulations. For instance, difficulty in digesting milk sugar (lactose) has been reported as occurring more frequently in some populations than in others, and the Chinese, whose cultural practices do not include use of milk from domestic animals, have a high rate of lactose intolerance; symptoms from ingesting monosodium glutamate, on the other hand, were reported not in those who have used this cooking powder for many generations but in Americans who call it "the Chinese restaurant syndrome" *(20)*. One takes it for granted, on theoretical grounds at least, that where specific nutrients are scarce, populations with lower natural requirements for them would thrive, but this must be a minor advantage compared with the possession of adaptive cultural practices for acquiring such nutrients.

Lack of much food (hypocaloric diet) often occurs in the very environments, the tropics, where soils are leached of minerals and foods of animal origin are scarce. Furthermore, the food ingested must often feed parasites as well as the person. Thus the increase in body weight, from tropics to frigid zones, noted earlier under consideration of Bergmann's rule, may be a response to nutritional stress as well as to cold and heat stresses.

Plasticity

The adaptations discussed so far are mostly of two kinds; those genetically entrenched in the population by repeated natural selection and those dependent on a capacity to acclimatize in the short run (whether the capacity is equally shared by all subpopulations of the species or not). There is also a third intermediate type of alteration, modification of an individual during his growth and development. It may be thought of as a special case of acclimatization, but since the process is essentially irreversible after adulthood, it deserves separate consideration and it may be separately designated as plasticity.

Early in the present century some anthropologists first asked whether the traits that were being used to characterize races could be influenced directly by environmental factors. They did not mean Lamarck's theory of inheritance of acquired characteristics, which had already been discredited; nor did they mean Darwin's description of the selectivity of the natural environment. The question was: Would people grow up to be physically different if they lived differently? In 1905, Walcher, a German obstetrician, showed that, when babies are regularly placed on the backs of their heads, they become broader-headed than do babies who are customarily placed on their sides *(21)*. In the same year, Fishberg compared measurements of the cephalic index and stature of Jews in various parts of Europe and in the United States and found that the two groups differed in these respects from each other *(22)*. If cephalic index and stature could change in immigrants of at least one European group, how could these measurements be used as major criteria of race?

At this time many Americans were concerned with the assimilation of immigrants. Franz Boas exploited this concern in getting support from the United States Immigration Commission for a survey of the physical measurements of immigrants *(23)*. Whatever may have been the expectations of the commission, Boas characteristically set himself a concrete problem and defined it operationally. Are the American-born children of European immigrants significantly different from their parents in such characteristics as cephalic index and stature? If so, do they also differ in the same respects from their immigrant brothers and sisters? Of the various groups Boas studied, his largest samples were of Central European Jews and Italians from Sicily. In both groups the offspring born in America tended to be taller than their parents; but in the Jews the cephalic index decreased, and in the Sicilians it increased. The measurements of the immigrant brothers and sisters, correlation with the length of the time the parents had been in the United States, and smaller studies of other nationalities led to the conclusion that the changes were the result of some aspect of the American environment that tends to bring about an American type with tall stature and medium cephalic index.

Some anthropologists attempted to explain Boas' findings on the basis of some selection, perhaps a self-selection of immigrants. To examine the possibility of such selection of immigrants, as well as the possibility of changes in the offspring of immigrants, Shapiro, in collaboration with Hulse and Lessa, undertook studies of two national groups in Hawaii, the Chinese and Japanese. Besides migrants and persons born in Hawaii, his study included nonmigrants in the Orient, for whom Shapiro coined the word "sedentes." Of the Chinese study only a brief preliminary report ever appeared; the Japanese study, however, has been fully presented *(24)*. The Japanese born in Hawaii were taller and broader-headed, and they differed significantly from Japanese immigrants in this and numerous other respects. However the immigrants were also different from the sedentes in many dimensions – including many in which those born in Hawaii differed from the immigrants. Shapiro explained the difference between sedentes and migrants on the basis of selection, and that between migrants and Hawaiian-born persons on the basis of factors in the environment during the growth period. The general findings of Boas and Shapiro have been confirmed by a number of studies *(25)*.

Nevertheless, a number of problems remained. In the various studies, the immigrants measured were older than the subjects born in America, and to some extent this fact might explain the differences. There is a tendency for individuals to decrease somewhat in height after the age of 30 or so, and, in addition, people have been getting taller from generation to generation all over the world; both in Japan and the United States, for example, adult sons are taller than their fathers (and also taller than their fathers were at the same age). Such increases have been going on for 100 years and possibly for 200 or more years. Bowles has shown that Harvard sons of Harvard fathers were taller than their fathers at the same age. The sons are larger in most other bodily measurements also, markedly so in the length of their thighs and forearms. Damon has followed some of these same families for four generations and finds no further increase in stature in the last generations, however, and he suggests that it is a straw in the wind that the secular increase in height has ended for economically favored American men *(26)*.

To take account of secular trends as well as the effects of migration, Goldstein undertook a study in Mexico of Mexican parents and their adult children and a parallel study of Mexican immigrants to the United States and their adult children born in America *(27)*. He found that the immigrants were larger in the usual respects than the sedentes, that the younger adults in Mexico were larger than their like-sexed parents, and that those born in the United States were larger than their parents. The last difference was the most pronounced, however, and seemed to indicate a growth factor that is especially strong in the United States.

Subsequent studies of Mexicans have confirmed that Mexicans who grow up in the United States are taller and larger in other ways than those brought up in Mexico and that this is not the result of initial differences between those who migrate and those who do not *(28)*.

On the other hand, studies of Italian Swiss in Italy and California demonstrate three factors leading to increased dimensions in the United States: (i) A selective migration of larger individuals; (ii) a tendency for those whose parents come from different communities to be larger than offspring of endogamous marriages, a finding confirmed in some but not in all other studies of the question; and (iii) plasticity in response to environmental conditions that are different in the two locations. *(29)*.

What is there in the environment of the United States that accounts for the greater growth here? The pattern of plastic changes is constant in all the studies of migrants, but such studies do not isolate specific causes *(30)*. Nevertheless there is little evidence of significant influences of variation in temperature, altitude, or hygiene. The biggest single factor is diet, and the most significant aspect of diet is not the quality nor the vitamin content but simply the quantity of food. In the United States hypercaloric diets (too much food) are apparently more of a problem for health than any shortage or dietary deficiency. Coronary heart disease and other circulatory disorders are one of the chief causes of death, and the life insurance companies have found that, on the average, persons of normal weight for height live longer than the overweight and that underweight individuals live still longer.

Large body size would seem to be an advantage in the face of hypercaloric diets; the extra size achieved during growth would provide a large frame and more metabolizing tissue to consume high caloric diets in adulthood without creating an excess that can only turn to fat. This may be as important as a small size achieved through the growth period is to an adult who must survive on a diet of very few calories per day. But there are other factors in the equation. Individuals adjust their food intake, and even in times of scarcity individuals of different size may have different abilities to acquire food. Cultural factors are important, too, through the enforcement of various modes of food distribution and norms of rationing during shortages.

Growth

The adaptation of organisms is not only to adult form – important as viability of adults is to the survival of the population. Each developmental stage must also be viable, of course. In man growth itself requires considerable nutrients and is therefore part of the nutrient-energy economy. Tanner has surveyed the influence of environments on the growth process itself *(31)*. During a period of starvation, a child slows down his growth, thereby conserving food energy. But he also tends to delay his maturation, thereby preserving the possibility of making up most of the loss if the period of deprivation is not too long and adequate food again becomes available. These temporary interruptions of growth in size with age may have little effect on body form. In virtually every country studied, growth of children was retarded as a result of the wartime food shortages of 1944-1945, but the characteristic population norms in proportion and eventual adult size were much less, if at all, affected.

Adverse conditions tend to retard growth, but not equally in all respects. Weight is, of course, immediately affected and height somewhat more slowly. As already noted, there is a general stunting and retardation of bony development in children in the Andes, but early and marked chest development occurs. Greulich found that in Japan children of all ages are, on the average, smaller than American white children in standing height, sitting height, and weight. The Japanese children are also slower in the ossification and maturation of specific bones. In these respects, however, North American-born children and also Brazilian-born Japanese children approach American standards *(32)*.

Different criteria of maturation show different degrees of influence by nutrition and disease. Eruption of the teeth seems more resistant to modification than skeletal development (usually assessed on x-rays of the wrist). Sexual development, especially menarche (the age at which girls have their first menstruation) is retarded about 2 years in populations of lower social status or who for other reasons have poor nutrition; menarche occurs earlier as conditions ameliorate *(33)*. There is always a positive correlation between different signs of growth and development, however. Growth is thus channelized by genetic factors – no doubt slightly different in different populations – and adaptation is achieved by retardation within the channel in the face of acute stresses on the growth process. When the stress is chronic there may also be some change in the channels.

Some individuals, viewing the way racial differences have been used in the past as an excuse to enslave or even to annihilate whole populations, try to deny the existence in man of any adaptive differences. But differences exist both between individuals within the population and, on the average, between populations. Adaptive variability is a great asset to the species, hence, at least potentially, to future generations of every population. Throughout evolutionary history, however, adaptive traits which were inherent in certain genotypes and differently distributed in various subpopulations have probably given way to species-wide plasticity – an adaptive capacity available to any member of the species if he is subjected to certain conditions during development. Plasticity in turn must tend to be supplanted by the capacity to ever more rapid adaptation and acclimatization. The ready reversibility of acclimatization in the individual and of plastic traits in the subpopulation permit adaptation of genetically similar individuals in diverse environments. This must have been a necessity for man as he became one of the most widely distributed species and as he frequently changed his environments with ever-increasing rapidity.

The three modes of adaptation (selection, plasticity, and acclimatization) overlap and intergrade in populations and in individuals. The increasing importance of ready response does not eliminate selective pressures. All three modes are still operating and, in this day of rapidly changing conditions, provide a safeguard for the species. At a time when human genetic engineering is discussed and may be implemented, it is well to understand the significance of human variability and human adaptability. Social and political policies that provide opportunities for all will best accommodate varying individuals according to their biological capacities and needs.

Summary

Adaptation is an aspect of virtually all questions of human biology. Besides their interest in evolution through adaptive selection of the primates, including man, physical anthropologists are concerned with biological adaptability as a human attribute. In this sense adapation has been examined at three overlapping levels: (i) those represented by differences in the extent of inherent capacities in subpopulations long exposed to different conditions, such as differences in the inherited determinants of body form and skin pigment in peoples in different climatic zones; (ii) adaptations acquired during the growth period of the individual such as residual stunting and reduced caloric needs in individuals receiving low caloric diets throughout childhood; and (iii) reversible acclimatization to the immediate conditions such as the changes which make it easier to work at high altitudes after the first few days there. Greater resilience to change is achieved if adaptations are reversible in each generation or within a lifetime. This implies an evolutionary tendency to shift human adaptability from genetic selection to ontogenetic plasticity to reversible adaptability.

References and Notes

1. Research Studies Constituting the U.S. Contribution to the International Biological Program (National Academy of Sciences, Washington, D.C., 1968), part 2.
2. R. B. Mazess, E. Picon-Reategui, R. B. Thomas, M. A. Little, *Aerosp. Med.* **39**, 403 (1968); J. M. Hanna, "The effects of coca chewing on exercise in the Quechua of Peru," *Hum. Biol.,* in press; M. A. Little, "Effects of alcohol and coca on foot temperature responses of highland Peruvians during a localized cold exposure," *Amer. J. Phys. Anthropol.,* in press.
3. M. T. Newman and C. Collazos, *Amer. J. Phys. Anthropol.* **15**, 431 (1957); P. Baker, *Science* **163**, 1149 (1969).
4. A. R. Frisancho, "Human growth and pulmonary function of a high altitude Peruvian Quechua population," *Hum. Biol.,* in press.
5. P. T. Baker, *ibid.* **32**, 3 (1960).
6. P. F. Scholander, H. T. Hammel, J. S. Hart, D. H. LeMessuria, J. Stem, *J. Appl. Physiol.* **13**, 211 (1958); C. S. Hicks, *Perspect. Biol. Med.* **7**, 39 (1963); R. W. Newman, personal communication; *J. Appl. Physiol.* **25**, 277 (1968).
7. M. T. Newman, *Ann. N.Y. Acad. Sci.* **91**, 617 (1961).
8. G. W. Lasker, *Anthropol. J. Inst. His. Phil. (Chungking)* **2**, parts I and II, 58 (1941); C. S. Coon, *Races of Europe* (Macmillan, New York, 1939); R. W. Newman and Ella Munro, *Amer. J. Phys. Anthropol.* **13**, 1 (1955).
9. The great size of the Watusi is a notable exception to Bergmann's rule which Baker *(5)* thinks may be due to their having been a ruling caste – hence perhaps less subject to the stress of work in the heat. In any case, these and other data from Africans show the complexity of the interaction between different ecological principles and quite unknown factors which must be integrated in the determination of physique, and which leads to many exceptions in the application of any one explanation. D. F. Roberts and D. R. Bainbridge, *Amer. J. Phys. Anthropol.* **21**, 341 (1963); J. J. Hiernaux, *La Diversité Humaine en Afrique Subsaharienne* (Inst. Sociol. Univ. Libre, Bruxelles, 1968).
10. D. F. Roberts, *Amer. J. Phys. Anthropol.* **11**, 533 (1953); E. Schreider, *Nature* **165**, 286 (1950); E. Schreider, *ibid.* **167**, 823 (1951); M. T. Newman, *Ann. N.Y. Acad. Sci.* **91**, 617 (1961).
11. J. S. Weiner, *Amer. J. Phys. Anthropol.* **12**, 615 (1954); M. H. Wolpoff, *ibid.* **29**, 405 (1968); E. V. Glanville, *ibid.* **30**, 29 (1969).
12. A. T. Steegmann, Jr., *Hum. Biol.* **39**, 131 (1967); "Thermal responses to cold in Hawaii," *Amer. J. Phys. Anthropol.,* in press.
13. P. Baker, *Science* **163**, 1149 (1969).

14. J. S. Weiner, *J. Roy. Anthropol. Inst. Gt. Brit. Ireland* **94**, 230 (1964).
15. R. W. Newman, "Why man is such a sweaty and thirsty naked animal: A speculative review," *Hum. Biol.,* in press.
16. M. M. C. Lee and G. W. Lasker, *ibid.* **31**, 252 (1959).
17. F. B. Livingstone, "Polygenic models for the evolution of human skin color differences," *ibid.,* in press.
18. A. S. Kip, "Measurement and regional distribution of functioning eccrine sweat glands in male and female Caucasians," *ibid.,* in press.
19. Reviews of the anthropological significance of basal metabolic rates by E. A. Wilson [*Amer. J. Phys. Anthropol.* **3**, 1 (1945)], N. A. Barnicot [*Sci. Progr. London* **153**, 124 (1951)], and D. F. Roberts [*J. Roy. Anthropol. Inst. Gt. Brit. Ireland* **82**, 169 (1952)] indicate that the world distribution of basal metabolic rates is inversely related to mean annual temperature even when due allowance is made for differences in stature and weight; there remains some doubt concerning what fraction of this adaptation is inherent, because rates change in migrants.
20. Among asymptomatic subjects, A. E. Davis and T. Bolin [*Nature* **216**, 1244 (1967)] report that all 15 Chinese, 4 of 5 individuals from the Indian subcontinent, and 2 of 12 Caucasians had abdominal pain and diarrhea after ingesting 80 grams of lactose. The Chinese restaurant syndrome was described by H. H. Schaumburg, R. Byck, R. Gerstl, and J. H. Mashman [*Science* **163**, 826 (1969)].
21. G. Walcher, *Zentralb. Gynaekol.* **29**, 193 (1905).
22. M. Fishberg, *Ann. N.Y. Acad. Sci.* **16**, 155 (1905).
23. F. Boas, *Changes in Bodily Form of Descendants of Immigrants. Senate document 208, 61st Congress, 2nd Session. Washington, D.C. (1910).*
24. H. L. Shapiro, "The Chinese population in Hawaii." Preliminary paper prepared for the 4th general session of the Institute of Pacific Relations, New York (1931), pp. 3-29; *Migration and Environment.* (Oxford Univ. Press, New York, 1939).
25. P. K. Ito, *Hum. Biol.* **14**, 279 (1942); G. W. Lasker, *Amer. J. Phys. Anthropol.* **4**, 273 (1946); F. P. Thieme, *Pap. Mich. Acad. Sci. Arts Lett.* **42**, 249 (1957).
26. G. T. Bowles, *New Types of Old Americans at Harvard and at Eastern Women's Colleges* (Harvard Univ. Press, Cambridge, 1932); A. Damon, *Amer. J. Phys. Anthropol.* **29**, 45 (1968).
27. M. S. Goldstein, *Demographic and Bodily Changes in Descendants of Mexican Immigrants* (Univ. of Texas, Inst. of Latin Amer. Studies, Austin, 1943).
28. G. W. Lasker, *Hum. Biol.* **24**, 262 (1952); *ibid.* **26**, 52 (1954); ———and F. G. Evans, *Amer. J. Phys. Anthropol.* **19**, 203 (1961).
29. F. S. Hulse, *Arch. Suisses Anthropol. Gén.* **22**, 103 (1957).
30. B. A. Kaplan, *Amer. Anthropol.* **56**, 780 (1954).
31. J. M. Tanner, in *The Biology of Human Adaptation,* P. T. Baker and J. S. Weiner, Eds. (Clarendon Press, Oxford, 1966), pp. 45-66.
32. W. W. Greulich, *Amer. J. Phys. Anthropol.* **15**, 489 (1957); P. B. Eveleth and J. A. de Souza, *Hum. Biol.* **41**, 176 (1969).
33. R. J. W. Burrell, M. J. R. Healy, J. M. Tanner, *Hum. Biol.* **33**, 250 (1961); K. P. Sabharwal, S. Morales, J. Méndez, *ibid.* **38**, 131 (1966); V. G. Vlastovsky, *ibid.* **38**, 218 (1966); K. Bojlén and M. W. Bentzon,

ibid. **40**, 69 (1968); L. A. Malcolm, "The growth and development of the Bundi child of the New Guinea highlands," *ibid.,* in press.

27.

The Exploitation of Animals by Man

M. L. Ryder

The discovery . . . that food could be produced by domesticating animals and cultivating plants has been regarded . . . as a far more important advance than the discovery of fire or the splitting of the atom, for it enables mankind to establish large settled communities which provide the basis for civilization. AUTHOR

The Quest for Food

The exploitation of animals by man had its origin at the stage in his evolution at which man could be regarded merely as an animal preying on other animals to satisfy his hunger. This crude exploitation during the Palaeolithic age, although it led to progressively better methods of food gathering, culminating in the hunting and fishing cultures of the Mesolithic period, yielded little more than flesh for food, skins for clothing and bones from which weapons and tools could be made.

Although other animal products were available, man required the leisure of the more settled life following domestication to enable him to exploit other products and develop complicated uses. We must not, however, belittle the achievements of peoples such as the eskimos and the Australian aborigines who never progressed beyond the Mesolithic stage.

The culture of the aborigines of Australia has mainly been studied from its modern expression, and Butzer (1965) warns that as such surviving food-collecting societies are in poor environments they cannot illustrate the optimal exploitation of earlier times. Only recently has archaeology begun to yield dated evidence of the Australian past (Mulvaney, 1961). The men hunted animals, and caught fish with a wide variety of traps, nets, spears and boomerangs, while the women collected shell-fish, small reptiles and their eggs, and birds' eggs. Ryder (1964a) found fragments of emu egg shell preserved among remains from ancient aboriginal shelters, presumably owing to the thickness of the shell. The aborigines had bone implements and shell knives, teeth and feather ornaments, and used skin cloaks in the colder parts of Australia. They cooked food by roasting it near a fire.

The Maoris of New Zealand can be regarded as a Neolithic people because they grew some crops. Golson (1959) has shown that when the Polynesians first colonized New Zealand in the eleventh century they hunted the large flightless bird, the moa, and used its feathers to make decorative cloaks. As a result by the seventeenth century the moa had become very scarce, and a new source of food had to be found. New settlements were established near the coast, and thereafter the food remains indicate mainly shell-fish, fishes and seals. A similar shift to fishing has been noted in parts of prehistoric Europe following the decline of larger game, although, that man was entirely responsible for the Palaeolithic extinction

Reprinted from the May 1966 issue of the *Advancement of Science* by permission of the author and the British Association for the Advancement of Science.

of the large European mammals, was questioned by Butzer (1965).

Molluscs, both terrestrial and aquatic, were probably collected very early in man's evolution, and their shells provided ornaments and a useful raw material. The artifacts, e.g., knives and fish-hooks, of some cultures, notably on Pacific islands, were based almost entirely on shell (Ryder, 1963a). A few archaeological experiments have been made on the possibility of "living off the land" by the collection of animal and plant food. Reed (1962) cooked different species of local snail in Persia, and tested their edibility. He found that, whereas ancient man had apparently rejected one of the two local species, both were equally palatable.

Singer (1957) in seeking to trace the origins of biological knowledge, considered that the hunters of the Old Stone Age had what he termed as "anatomical instinct." He illustrated this point by referring to a carving of some animal heads in deer antler which showed the detailed contours of muscles beneath the skin. He also depicted Palaeolithic cave drawings of animals in which the position of the heart is indicated, to show that the hunters who made the drawings knew how vital the heart is to life.

Dart (1959) suggested the continuity of a type of bone implement for about a million years when he showed that what he considered were crude food scoops made by shattering the cannon bones of antelopes by Australopithecine man in South Africa, were similar to apple-scoops made from sheep cannon bones commonly used by toothless people for eating apples in nineteenth century Britain. Such scoops, which could be regarded as primitive spoons, were also used for coring apples, and, in earlier times for the feeding of infants. Recent scoops made from large bird bones in New Guinea are to be seen in the Queensland Museum.

Domestication

Food gathering has the disadvantages that, in general, the wild products of nature are not sufficiently abundant to support more than a very small population, and that almost the whole of one's energy is consumed in searching for food, so that little time is left for material or mental advancement. The discovery in what Gordon Childe named the Neolithic Revolution, that food could be produced by domesticating animals and cultivating plants, has been regarded by Sir Mortimer Wheeler as a far more important advance than the discovery of fire or the splitting of the atom, for it enabled mankind to establish large settled communities which provided the basis for civilization.

We are still a long way from a complete understanding of the way in which domestication came about. The older theories tended to ignore the biological background of the problem in their efforts to seek a purposeful economic or religious explanation. We now know (e.g., Flannery, 1965) that domestication must be considered in an environmental setting and from an ecological point of view.

Historical geographers suggest that domestication was stimulated by the desiccation following the last Pluvial period in western Asia. Zeuner (1963) emphasized the biological basis of domestication and pointed out that the Mesolithic peoples who initiated domestication would have found it easier to obtain food by hunting instead of making domestication experiments that would not reward their efforts for several generations. Indeed, the Red Indians of North America had wild sheep in their fauna yet never to our knowledge domesticated them.

Zeuner detailed the various levels of natural association that occur between different animal species, starting with symbiosis. A development of symbiosis is scavenging in which the wolf for instance probably scavenged the camps of Mesolithic man for food, man no doubt gaining from the association by the removal of waste. It is likely that such a voluntary association led to man's first domestic animal, the dog, and it illustrates another important basis for domestication, viz., that the animals concerned usually form social groups. And since man himself is a social animal, Zeuner considered that the initial phases of domestication can be regarded as instances of social overlap between man and the animal being domesticated. Although such associations of animals with man have usually led to their domestication by man, the primitive nomadic association between man and the reindeer was regarded by Zeuner as social parasitism.

The next animal to be domesticated, about 9000 B.C. (Reed, 1962b) was the goat, followed by the sheep, which emerged as the economically better animal making an immense contribution to the development of civilization in many countries. These herding animals went through a taming rather than a parasitic stage and complete the group domesticated in the pre-agricultural phase of man's development. The wild ancestors of goats and sheep have a relatively narrow distribution in the mountains of western Asia, but sheep, at any rate, come down on to the plains in winter so that their nomadism is seasonal.

Hilzheimer (1936) suggested that domestication of the sheep could have taken place during the seasonal migrations when flocks of sheep associated with man for protection. This may be the origin of transhumance, i.e., the movement of flocks up into the hills in summer, and back to the lowlands in winter, that is still practiced in southern Europe and parts of Asia, where Hilzheimer considered the animals, following their leader, migrate as much by instinct as at the instigation of their owners who follow behind.

Zeuner considered that the credit for domesticating sheep should go to the dog rather than to man, and he illustrates the natural ability of the dog family to round up ruminants with a remarkable air photograph showing a wolf driving a herd of caribou. Man here therefore used one animal to exploit another.

The domesticated goat and sheep provided man with a permanent supply of meat, and allowed him to develop new products such as milk and wool. At this early Neolithic stage man's crops would attract wild herbivores such as the pig and ox, and these animals would become not scavengers but robbers of the crops. But man already having had experience of domestication was quick to domesticate such trespassers and probably had enclosures available to pen them. Both the pig and ox seem to have been domesticated in the Near East about 6000 B.C. Further east other bovids such as the yak and buffalo were domesticated.

The mention of cattle leads one to think of milk, and the milk complex between man and animals. Drinking from the udder is practiced by some primitive peoples, and the women often suckle young animals. It is possible therefore that domestication of some animals was assisted in this way with young brought into the home.

Less is known about the next group of animals to be domesticated, the main use of which is transport. The first seems to have been the ass which was domesticated in Mesopotamia, next the horse and camel, which were domesticated by the nomads of the Asiatic plains and deserts, and finally the elephant which was first put to work in the tropical forests of India.

Honey was virtually the only source of sweetness available to man until the recent cultivation of sugar cane, and was no doubt collected in Palaeolithic times along with the manna of the Bible, which has been identified as the crystallized honey of a stick insect(Montgomery, 1959). Beekeeping seems to have begun in the Neolithic period, and possibly suggested itself from the swarming habit of bees, which were probably given a container such as a hollow tree trunk to live in. Australian aborigines collect ants swollen with honey. The keeping of silk worms began in China possibly about 3000 B.C. The spinning of silk followed the spinning of wool and flax.

The following classification of the stages leading to domestication is a modification of that suggested by Zeuner; although the process was no doubt more complicated, these stages provide a useful summary.

1. Loose contacts of animals with man and no control of breeding.
2. Confinement to the human environment, with breeding in captivity.
3. Selective breeding by man towards certain characteristics, and occasional crossing with wild forms.
4. The modern planned development of different breeds with desirable economic characters, e.g., meat, wool, milk, coupled with disregard for and even extermination of, wild and primitive, domestic, ancestors.

Changes Following Domestication

Domestication can be regarded as an intensification of the exploitation of animals, and the chief way in which man has exploited livestock towards his own ends is through selective breeding, yet we know hardly anything of the way in which primitive man brought about changes.

Domestication did not in itself produce any new inheritance, but the changed environment may have caused a reduction in natural selection. This could have caused increased variation so that a greater range of types could survive, and become fixed by inbreeding. This could have resulted from the small size of the first domestic flocks. Evidence from skeletal remains from early Neolithic sites in the Near East shows a high proportion of young males being eaten (Flannery, 1965), which would tend towards inbreeding by leaving fewer males for mating and could have led to the first artificial selection. But the extent to which selection was at first only incidental to killing for other purposes is not clear. Whatever the cause, evolution became more rapid after domestication, and the animals changed more, and did so in a shorter period of time than under natural selection.

Primitive man lived much closer to his animals than does the modern farmer. This closeness would enable him to observe the variations in his animals, and simple truths such as "like begets like" which provided the basis for selective breeding long before Mendel. Just as Paleolithic man appeared to have an anatomical instinct, it may be that Neolithic man acquired a breeding sense.

One of the first changes evident in the remains of Neolithic sheep, which reached Britain about 3000 B.C., is the reduction in body and horn size. This could have taken place through breeding on a lower plane of nutrition. To what extent man is responsible for other big changes summarized in the following list is not clear:

1. Wild sheep have a short tail, whereas modern domestic sheep have long tails which may have arisen during man's selection for a fat tail. Other differences in body proportions are mainly due to modern selection towards a good mutton carcase.
2. Wild sheep have an overall brown colour, whereas domestic sheep are mainly white.
3. The woolly undercoat of the wild sheep has developed at the expense of the bristly outer coat.
4. Wild sheep undergo a complete spring moult, whereas domestic sheep have almost lost this tendency to shed wool.

Given wild sheep today, man would be very hard put to it to produce such changes. And it is remarkable that when sheep illustrations first appeared in Mesopotamia soon after 3000 B C. the major changes seem already to have taken place.

Regarding the subsequent divergence to give different breed types, it is possible that previously too much emphasis has been placed on the environment alone. The continued in-breeding owing to isolation, with some selection, in different environments is likely to have contributed more to changes. Once divergence had taken place, human migrations allowed wider out-breeding between different types than occurred in the wild state, and inter-crossing in this way has certainly been an important factor in the development of modern breeds. It is probable too, that advantage has been taken of occasional mutations giving new characters.

The wild sheep has a woolly undercoat, and a bristly outercoat composed of fibres known as kemps about 6 cm long. Following domestication there has been a gradual loss of the kemp to give a range of fleece types from hairy ones in which there is often a proportion of kemps, to fine-woolled types with no kemp or hair at all. Changes in the fleece from the Bronze Age onwards have been followed by making microscopic examinations of archaeological specimens of skin and wool (Ryder, 1964b).

A unique sheepskin with a white fleece intact came from a frozen burial mound in the Altai mountains of central Asia. This was from a Scythian grave dated 400 B.C. and the fleece was a hairy type not unlike that of some British Mountain (and also leather and parchment) can throw light on no kemp fibres at all. Sections of such skin remains (and also leather and parchment) can throw light on fleece structure because the wool roots remain protected within the skin (Ryder, 1963b).

Figure 1 shows the different fibre groupings of the main fleece types. The coarse fibres tend to be arranged in groups of three, with the finer ones between, so the diagram shows only one group of each fleece type. That at the top is the wild type, and next comes the hairy domestic sheep, in which the kemps have been replaced by hairs. Below this is the medium fleece type with medium and fine fibres, and finally the fine-woolled fleece. The lower three are tracings of groups from the Dead Sea Scrolls, showing that the three main modern fleece types were in existence 2000 years ago. Cattle hairs found in leather can be similarly studied (Ryder, 1964c).

Some pieces of wool cloth with no natural colour from one of Prof. Yadin's excavations near the Dead Sea, dated second century A.D., were of particular interest because they clearly represented the an-

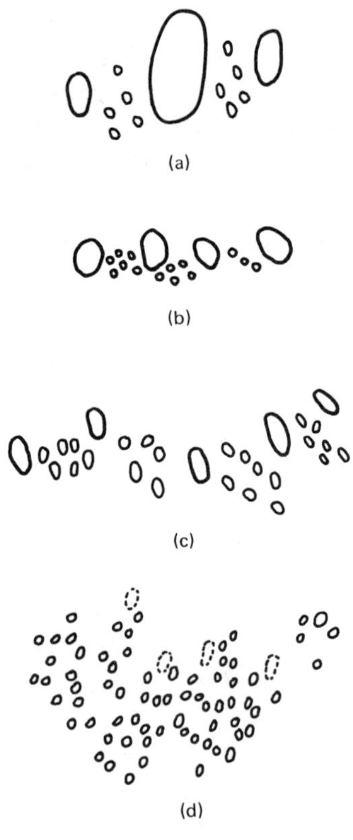

Figure 1 Wool fibre groupings (a) in wild sheep; (b), (c), (d) in parchment from Dead Sea Scrolls; (b) hairy sheep; (c) medium-woolled sheep; (d) fine-woolled sheep.

cient fine wool to which classical writers refer (Ryder, 1964b). But microscopic examination revealed that the wool contained a proportion of medium fibres (Fig. 2). This was therefore not a true fine wool, but was the medium wool whose grouping is shown in Fig. 1. These specimens showed that fleece type could also be defined from the diameter distribution of fibres in textile remains. The term, generalized medium wool, has been given to this wool because it provides an evolutionary link between ancient and modern types, as indicated in Fig. 3 (from Ryder, 1964b); this type in fact predominated in the Dead Sea specimens.

The hairy type could have given rise to the generalized medium wool by a narrowing of the hairs to give medium fibres, and the generalized medium wool could in turn have given rise to the true fine wool by a further narrowing of the medium fibres into fine fibres. On another line the modern medium wool could have arisen by a coarsening of the fine fibres of the generalized medium wool.

The small Soay sheep, which on skeletal evidence seems to represent the prehistoric sheep of Europe, has a more primitive generalized medium fleece in that it tends to have some coarse fibres, and is brown. Some brown Danish Bronze Age wool dated about 1200 B.C. was not typical of

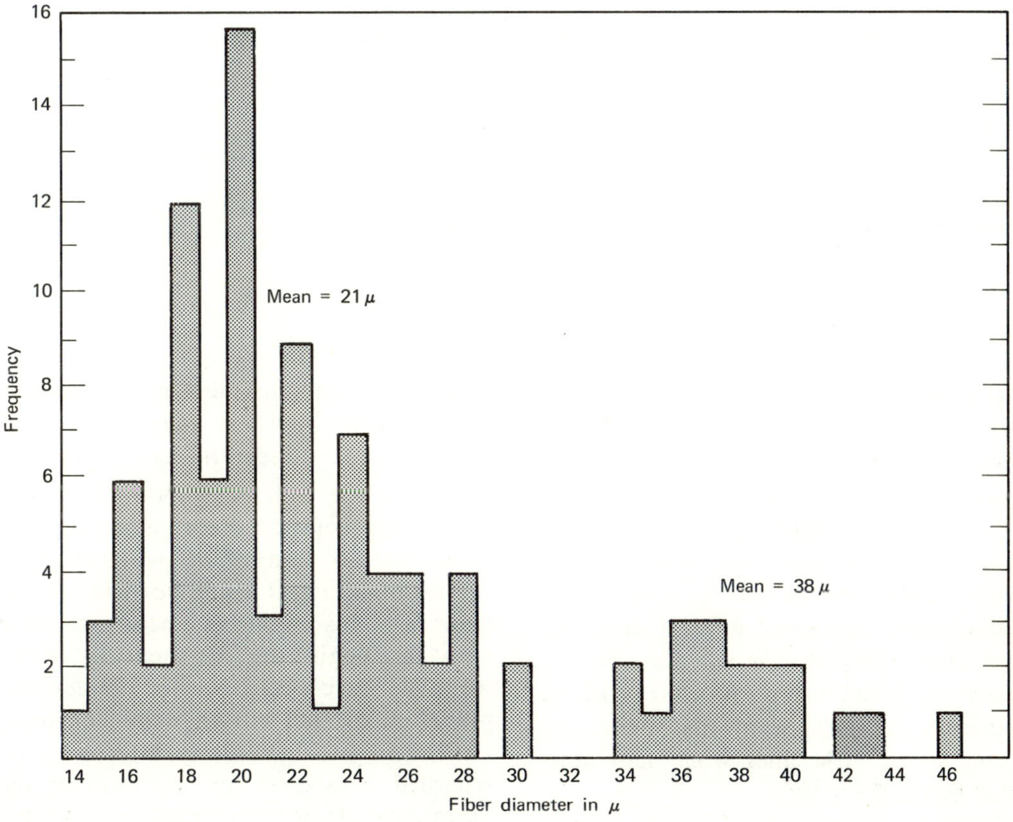

Figure 2 Distribution of fibre diameters in wool from cave near Dead sea dated 130 *A.D.* (1μ one thousandth of mm).

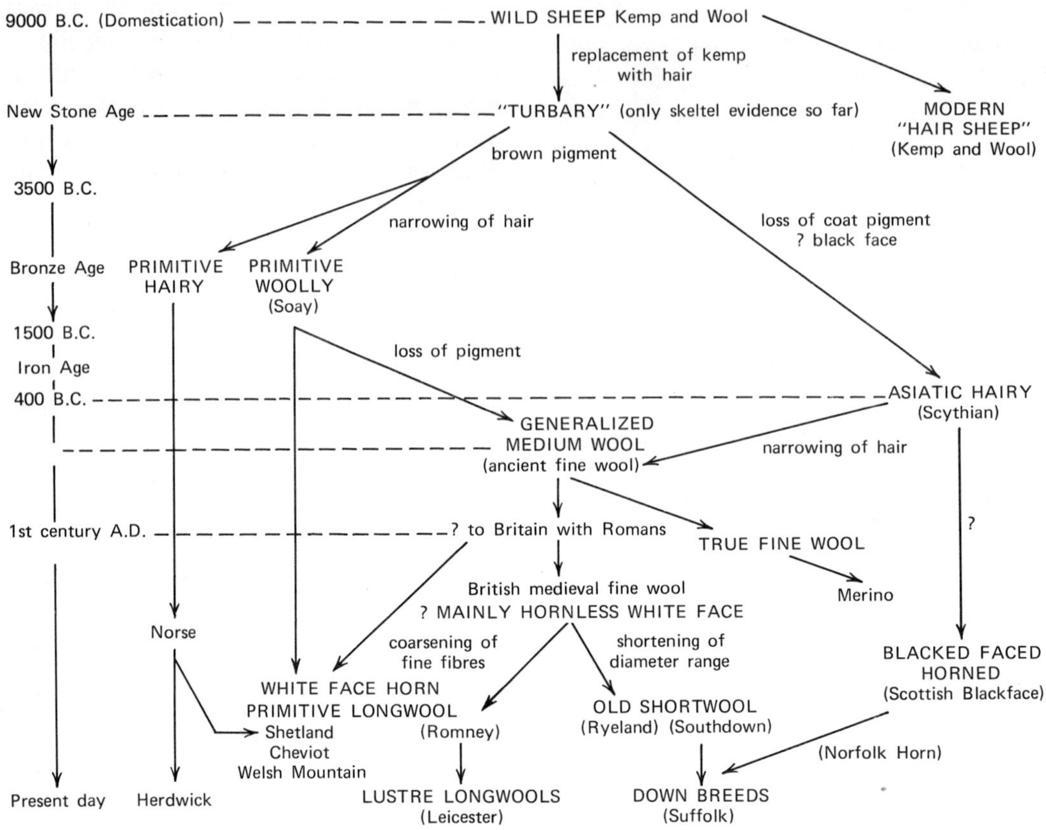

Figure 3 Suggested lines of evolution of fleece types in domestic sheep.

any known fleece type because it had a very high proportion of fine fibres. This apparent anomaly may be explained by observations (Ryder, unpublished) made at St. Kilda on moulting Soay sheep; although the fine wool and some of the coarser fibres shed in spring, most of the coarser fibres remained to shed later. As wool from sheep that moult must be obtained by plucking as is done still in Shetland, it is apparent that primitive breeds would yield finer wool than indicated by the fleece type.

There are records of sheep being plucked in Sumeria, and there is evidence that a bronze comb was used to assist the process. Shears seem to have been invented in the Iron Age about 1000 B.C. Sheep that moult can deprive one of their wool, and so man would wish to select against moulting. It would be of interest to know how this might have been associated with the beginning of shearing. Pliny seems to have been the first to describe plucking, carried out then by "barbarians", and he, like many later writers, regarded it as cruel; in fact most of the fibres have already been released from the skin before the animal is plucked.

Animal Products

There have been several recent reviews on early agriculture, e.g., Reed (1962b), Flannery (1965), and so the rest of this paper will be devoted to the wide variety of products that man has obtained from animals, particularly after domestication. When we see hard bone objects in museums it is easy to forget that many of those items of man's equipment that were made from the soft parts of animals are rarely preserved.

Emphasis will be placed on the sheep as it can supply all man's needs including transport; sheep even carry loads of 12 kg up Himalayan tracks that no larger animal could negotiate, and are very important to primitive communities living in a harsh environment.

The first source of food from animals was *meat*, but after domestication an animal would tend to be more valuable alive, and so not only would milk be developed as food, but greater efforts would be made to preserve meat, particularly for the winter, although past accounts of the seasonal killing of livestock may have exaggerated the extent of this practice. Hunters and fishers probably discovered that flesh dried in the sun would resist putrefaction, and it may be that preservation by smoking is associated with cooking. Salt would be rare, and so its use in preservation probably came much later.

Some African tribes obtain *blood* for food from living cattle, blood being collected at slaughter in this country for black puddings, and not all have thought it necessary to kill their livestock to obtain meat from them. Abyssinians in the eighteenth century used to cut flesh from a cow to eat raw, the skin being afterwards pinned over the wound, and covered with clay.

The use of *bone* (including antlers and teeth) was well developed before domestication in the making of hunting implements such as spears, and fish hooks. Indeed the use of bone may have preceded the use of stone (Dart, 1962). Jaws were used as tools by Australian aborigines, and pigs tusks have been used as currency in New Guinea. Some tools such as the antler picks used with shoulder-blade shovels and ribs used as skin scrapers have undergone little preparation, whereas others were intricately carved. Man probably soon learnt that bone can be softened by soaking it in water, making it more easily cut with a flint knife.

Early potters used the end of a bird bone, as well as shells to print a decorative pattern, and ground shells, as well as fish bones were used in mortar. Bone has been used for such a wide range of tools (e.g., drills, spinning whorls and weaving combs) that it would be tedious to attempt a complete list. Hollow long bones were used to make primitive flutes as early as the Paleolithic period. The Neolithic dwellers of Skara Brae in treeless Orkney used whale bones for roof and door supports in their stone houses, and since then perforated whale vertebrae have been used as livestock tethering rings. Perforated fish vertebrae have been used as spinning whorls in Africa. According to Herodotus the Scythians, inhabitants of another treeless area, used bones as fuel. In the Middle Ages in Britain, ox cannon bones were shaped into skate runners.

The organic collagenous element of bones can be boiled out to give gelatin and glue. If on the other hand the mineral part is removed with acid, the collagen remains as edible gristle. Bones were in fact prepared to be eaten in this way as "strjugur" not so long ago on Icelandic sheep farms, the acid coming from a complicated blood and meal preparation in a sheep's stomach (Adalsteinsson, 1956).

Skin, too, is composed of collagen, and skins were almost certainly first used untanned as in the rawhide of whips and ropes, and in later parchment. The first form of tanning to produce leather seems to have been with fish-oil, this technique, along with smoking, may have originated in Paleolithic times. Tanning with oak-bark extract began in the Neolithic period (Ryder, 1963b). The hair can be removed by allowing putrefaction, but removal is speeded with the use of alkali, an early source of which was stale urine (another animal product).

Skin, like bone, has a wide range of uses, starting with clothing, oiled skins in fact being used until recently by Shetland fishermen. Footwear, the biblical sling, bags, and liquid containers were also made from skin. Cooking may have been carried out in skin, but putting hot stones into a bag of water (Ryder, 1966). It has been suggested that the first pottery vessels followed the shape of a skin bag. Bags made from a whole sheepskin were used to store grain in Scotland, and skin riddles were used in cleaning the grain. Skins may have been used in boat construction as early as Mesolithic times, just as more recently boats have been made by stretching skins over a wooden frame. In fact Johnstone (1964) considers that skin boats were the main vessels of the Neolithic period. Assyrian soldiers used inflated sheepskin floats to help them in swimming, and by Roman times leather was being used for sails as well as in tents. Once the horse became important, a whole range of straps and harnesses were developed.

Stretched rawhide in the form of a coarse parchment was used to cover drums, and the sounding boxes of other musical instruments, as least as early as 2000 B.C. in Egypt. Skins were used there for writing even earlier, although 200 B.C. is the date associated with the "invention" of parchment at Pergamum. This very durable writing material is merely stretched and dried (usually sheep) skin (Ryder 1964d). If a similarly stretched skin is oiled, it becomes translucent, and lamb skins prepared in this way have been used as windows.

Milk was probably one of the first new products following domestication, and man possibly soon noticed that animals would let down their milk more readily in the presence of their young. In seventeenth-century Scotland a calf skin stuffed with straw was put into the byre to assist milking, and similar milking customs are known to go back to prehistoric times over a wide area (Amoroso and Jewell, 1963). Although we have been warned by Proudfoot (1965) about projection from more recent practices into the past, it is likely that many similar customs are very ancient.

It would not be long before the souring and curdling of milk was discovered, and the preparation of this curd into cheese that could be stored for winter food must have started very early. There are illustrations of cheese-making from ancient Mesopotamia, and European evidence from stone curd strainers does not date earlier than the late Bronze Age. But cloth has been used in cheese-making recently in Europe, and earlier evidence for this and similar methods would not survive.

It has been suggested that the way of speeding up curdling with the use of the enzyme, rennet, was discovered by carrying milk in a sheep's stomach. It is also possible that the way of making butter was discovered by carrying milk in a skin bag. The making of butter probably originated with the nomads of central Asia, and came later than cheese. Butter was stored in the ground in places as far apart as Pakistan, where it is now preferred rancid, and Scotland, where it turns up periodically in peat as "bog butter". Various other milk products, such as yoghurt, have originated in Asia, where milk is fermented to preserve it, just as in the West, fruit juice is fermented for storage as wine. These processes are dependent on the right micro-organisms being present at the right time, and so there must have been many accidents in the past. Our exploitation of micro-organisms to produce more complicated substances than alcohol (which can be used as a fuel) is likely to become more important as our natural resources decline. For instance, yeast protein can be produced from paraffins.

The main relationships between cheese and butter are indicated in Fig. 4. Milk illustrates the interrelationship of different farming operations. Thus cheese-makers in the past used to keep pigs in order to utilize the by-product, whey, as food. The introduction of the plough horse into Scotland in the eighteenth century released cows for milking so that ewes were no longer milked, and as a result the lambs grew better.

Wool is another important product of domestication, and was probably first used as a felt, which is frequently formed naturally during the shedding of the fleece even in wild sheep. Felt is still used by Asiatic nomads for tents, boots and mats and the use of wool in this way enabled cold regions to be occupied.

Wild and primitive sheep often rub a shedding fleece into long strands of wool, and these strands may have suggested to Neolithic man the spinning of wool (Ryder 1964e). As weaving, in basketry, was already known before domestication this could have led to the great Neolithic invention of cloth manufacture. Flax has usually been thought of as the first fibre to be spun, but this may be because only flax has been preserved (Ryder 1965). Much preparation is in fact required in order to obtain the fibres from the flax stem.

An interesting use of wool quoted in the Bible (Judges 6:23) was the collection of dew to obtain water in dry areas. A fleece was spread on the ground overnight, and the moisture wrung out. Sheep were frequently washed in a stream before shearing in the past, use being unwittingly made of the natural detergency of the suint (sweat) in the fleece to wash out the associated grease. The extraction of wool grease, during wool scouring, for use in protection against metal corrosion, and as a source of valuable chemicals seems to be quite modern. Hair has been used as a binder in plaster, and in paper, and although animal hair has been used in textiles, it has either been in short supply (camel hair) or inferior to wool (most goat hair).

The *horn* of hooves, and the horny covering of the bony core of horns, is composed of keratin, which is the same substance as that of hair and wool, and this has provided another raw material until recent times. Horn casings were probably used first as drinking cups and containers,

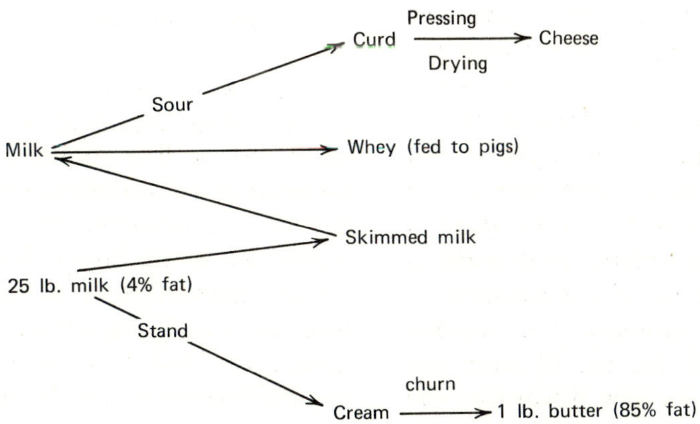

Figure 4

a use that has persisted into recent times in powder horns, strickle horns (containing sand and grease to sharpen scythes) and livestock dosing funnels. The North American Indians carried fire in tinder packed into the horn of a Bighorn sheep. Zeuner (1963) illustrated a prehistoric digging hook made from a goat horn in the Canary Isles.

The chief value of horn is that it can be softened by boiling in water, after which it can be cut, flattened or moulded into various shapes. It was thus the first plastic. One of the earliest examples of this shaping is a horn ladle found in a Bronze Age beaker in Scotland and horn was used for the handle of Bronze Age daggers. By the Middle Ages the horner's craft was well developed, with specialized tools and there was a wide range of products (Wenham, 1964). Thin sheets of horn are in fact transparent, and from the use of horn in lantern windows comes the corruption into "lanthorn".

Fat varies in amount and consistency between species, that of sheep being particularly hard. The tail of fat-tailed sheep is a delicacy in the East, but fat has uses other than food, such as in tallow candles and in the making of soap.

Gut like skin, contains much collagen. The alimentary tract of animals provides a source of containers and tubes, and also strings for musical instruments, sheep gut, and not cat gut in fact being used, at any rate in recent times. Gut-dressing was a well developed trade in the past, ox gut being prepared for sausage skins. The Scottish haggis, which is cooked in a sheep's stomach, and therefore perhaps to be regarded as a primitive sausage, is probably of great antiquity. This method of cooking allowed the use of liver and other internal organs now discarded, and although a haggis is now cooked by boiling it in a pan, the original method may have been to suspend it directly over a fire. Some eastern kebabs are wound around with gut before being roasted by the fire (Ryder, 1966). The blind caecum of the sheep was used as a contraceptive sheath in France at least as early as the sixteenth century (Huxley, 1957).

Dung, was so highly valued as manure in early medieval England that it came before meat, in importance. In India today shepherds are paid to graze stubble in order to have the land manured. Dried dung has been used as a fuel in treeless areas such as Iceland and Tibet.

Other animal products were no doubt used in the past in ways that we should never suspect today. Primitive techniques and locally-made implements have almost gone from western Europe, and are fast disappearing from other parts of the world under the influence of rapid communications and mass production, so there is little time left to put on record old customs and to preserve objects no longer in use.

It has been said that the history of mankind is the history of agriculture, the two essentials to life being food and clothing. In fact the cultures of man, as well as his evolution, are governed by biological processes, and despite modern technological advances our lives are still ordered by the biological laws that bring us into the world, feed us, purify our water, dispose of our sewage and cure our diseases. We could perhaps live without technology, but we could not do without the accumulated biological knowledge that enables us to exploit the living things that are essential for our lives. Thus if we are to hope to feed the rapidly expanding human population, attention to an intensification of this exploitation should be our aim.

References

Adalsteinsson, S. (1956): Personal communication.

Amoroso, E. C. and Jewell, P. A. (1963): "The exploitation of the milk-ejection reflex by primitive peoples," in Mourant, A. E. and Zeuner, F. E. (Eds.) *Roy. Anthrop. Inst. Occasional paper No. 18.*

Butzer, K. W. (1965): *Environment and Archaeology an Introduction to Pleistocene Geography.* Methuen.

Dart, R. A. (1959): "Osteodontokeratic ripping tools and pulp scoops for teething and edentulous Australopithecines", *J. Dent. Ass. S. Afr.* **14**, 164-78.

Dart, R. A. (1962): "Substitution of stone tools for bone tools at Makapansgat", *Nature, Lond.* **196**, 314-16.

Flannery, K. V. (1965): "The ecology of early food production in Mesopotamia", *Science, N.Y.* **147**, 1247-56.

Hilzheimer, M. (1936): "Sheep", *Antiquity* **10**, 195-206.

Huxley, J. (1957): "Material of early contraceptive sheaths", *Br. Med. J.* **1**, 581-2.

Johnstone, P. (1964): "The Bantry Boat", *Antiquity* **38**, 277-84.

Montgomery, B. E. (1959): "Anthropods and ancient man", *Bull. Entomol. Soc. Amer.* **5**, 68-70.

Mulvaney, D. J. (1961): "The Stone Age of Australia", *Proc. Prehist. Soc.* **27**, 56-107.

Proudfoot, V. B. (1965): "Bringing archaeology to life", *Adv. Sci., Lond.* **22**, 125-33.

Reed, C. A. (1962a): "Snails on a Persian hillside", *Peabody Museum of Nat. Hist. Postilla* **66**, 1-20.

Reed, C. A. (1962b): "Osteological evidence for prehistoric domestication in S.W. Asia", *Z. f. Tierzüchtung and Züchtungbiol.* **76**, 31-8.

Ryder, M. L. (1963a): "Remains of fishes and other aquatic animals", in Brothwell, D. R. and Higgs, E. S. *Science in Archaeology.* Thames and Hudson: London.

Ryder, M. L. (1963b): "Remains derived from skin", in Brothwell, D. R. and Higgs, E. S. *Science in Archaeology.* Thames and Hudson: London.

Ryder, M. L. (1964a): "The biological remains from Fromm's Landing, Trench 6", in Mulvaney, D. J. *Proc. Roy. Soc. Victoria* **77**, 479-516.

Ryder, M. L. (1964b): "Fleece evolution in domestic sheep", *Nature, Lond.* **204**, 555-9.

Ryder, M. L. (1964c): "Report on the skin", in Coles, J. M., Coutts, H. and Ryder, M. L. "A late Bronze Age find from Pyotdykes, Angus, Scotland with associated gold, cloth, leather and wood remains", *Proc. Prehist. Soc.* **30**, 193-7.

Ryder, M. L. (1964d): "Parchment, its history, manufacture and composition", *J. Soc. Archivists* **11**, 391-9.

Ryder, M. L. (1964e): "The origin of spinning", *Antiquity* **38**, 293-4.

Ryder, M. L. (1965): "Report on textiles from Catal Huyuk", *Anatalian Studies* **15**, 175-6.

Ryder, M. L. (1966): "Can one cook in the skin?" (submitted for publication).

Singer, C. (1957): *A Short History of Anatomy and Physiology,* Dover: New York.

Wenham, L. P. (1964): "Hornpot Lane and the horners of York", *Ann. Rep. Yorks. Philos. Soc.* 25-56.

Zeuner, F. E. (1963): *A History of Domesticated Animals.* Hutchinson: London.

XI. Environment

Since the dangers of environmental abuse are so well documented in current literature, we tried to select readings that emphasized ecological principles. The first three examine concepts more than abuses, based on the premise that the ultimate solution to enviromental problems rests on an understanding of these concepts. F. H. Bormann, professor of forest ecology, Yale School of Forestry, and G. E. Likens, associate professor of biological sciences, Dartmouth, in ''Nutrient Cycling'' establish a small watershed with measurable parameters as a suitable unit for the study of ecosystem interrelationships. George M. Woodwell of Brookhaven National Laboratory in ''Effects of Pollution on the Structure and Physiology of Ecosystems'' maintains that we know enough about ecosystems to predict the broad effects of man's disturbances on land, water, and air. Pollution, irradiation, and pesticides all cause similar and predictable changes. An excellent account of a backlash due to man's poor management of his waterways is the paper ''Aquatic Weeds'' by L. G. Holm, professor of horticulture, University of Wisconsin, L. W. Weldon and R. D. Blackburn, research agronomist and botanist, respectively, Crops Research Division, ARS, USDA.

Since the basic cause of most ecological problems is related to excess numbers of people and the unnatural demands of supporting technology, we included two papers on population. Harrison S. Brown, professor of geochemistry at Caltech, Pasadena, is concerned about our ineffective use of the power of science and technology to solve the basic problems of mankind. Bernard Berelson, president of the Population Council, New York, in his paper ''Beyond Family Planning'' offers a series of proposals for alleviating the population problem. The urgency of the problem and the necessity for action at high levels in our government is stressed.

The final essay, ''Infectious Diseases and Human Ecology'' by A. Macdonald, professor of bacteriology, Aberdeen Medical School, Aberdeen, Scotland, examines a facet of population growth that one seldom considers. As the density of humans increases, so also does the probability of outbreaks of infectious diseases. Bacterial and viral diseases that pose serious health menaces are systematically discussed. The list is frighteningly long.

28.

Nutrient Cycling

F. H. Bormann and G. E. Likens

The small-watershed approach to problems of nutrient cycling has . . . certain advantages. AUTHOR

Life on our planet is dependent upon the cycle of elements in the biosphere. Atmospheric carbon dioxide would be exhausted in a year or so by green plants were not the atmosphere continually recharged by CO_2 generated by respiration and fire *(1)*. Also, it is well known that life requires a constant cycling of nitrogen, oxygen, and water. These cycles include a gaseous phase and have self-regulating feedback mechanisms that make them relatively perfect *(2)*. Any increase in movement along one path is quickly compensated for by adjustments along other paths. Recently, however, concern has been expressed over the possible disruption of the carbon cycle by the burning of fossil fuel *(3)* and of the nitrogen cycle by the thoughtless introduction of pesticides and other substances into the biosphere *(4)*.

Of no less importance to life are the elements with sedimentary cycles, such as phosphorus, calcium, and magnesium. With these cycles, there is a continual loss from biological systems in response to erosion, with ultimate deposition in the sea. Replacement or return of an element with a sedimentary cycle to terrestrial biological systems is dependent upon such processes as weathering of rocks, additions from volcanic gases, or the biological movement from the sea to the land. Sedimentary cycles are less perfect and more easily disrupted by man than carbon and nitrogen cycles *(2)*. Acceleration of losses or, more specifically, the disruption of local cycling patterns by the activities of man could reduce existing "pools" of an element in local ecosystems, restrict productivity, and consequently limit human population. For example, many agriculturalists, food scientists, and ecologists believe that man is accelerating losses of phosphorus and that this element will be a critical limiting resource for the functioning of the biosphere *(1, 5)*.

Recognition of the importance of these biogeochemical processes to the welfare of mankind has generated intensive study of such cycles. Among ecologists and foresters working with natural terrestrial ecosystems, this interest has focused on those aspects of biogeochemical cycles that occur *within* particular ecosystems. Thus, information on the distribution of chemical elements and on rates of uptake, retention, and release in various ecosystems has been accumulating *(6)*. Little has been done to establish the role that weathering and erosion play in these systems.

Yet,the rate of release of nutrients from minerals by weathering, the addition of nutrients by erosion, and the loss of nutrients by erosion are three primary determinants of structure and function in terrestrial ecosystems. Further, with this infor-

Reprinted from *Science*, Vol.155, pp.424-429, January 27,1967.

mation it is possible to develop total chemical budgets for ecosystems and to relate these data to the larger biogeochemical cycles.

It is largely because of the complex natural interaction of the hydrologic cycle and nutrient cycles that it has not been possible to establish these relationships. In many ecosystems this interaction almost hopelessly complicates the measurement of weathering or erosion. Under certain conditions, however, these apparent hindrances can be turned to good advantage in an integrated study of biogeochemical cycling in small watershed ecosystems.

It is the function of this article (i) to develop the idea that small watersheds can be used to measure weathering and erosion, (ii) to describe the parameters of watersheds particularly suited for this type of study, and (iii) to discuss the types of nutrient-cycling problems that this model renders susceptible to attack. Finally (iv), the argument is developed that the watershed ecosystem provides an ideal setting for studies of ecosystem dynamics in general.

Ecosystem Defined

Communities such as fields and forests may be considered as ecological systems *(7)* in which living organisms and their physical and biological environments constitute a single interacting unit. These ecosystems occupy an arbitrarily defined volume of the biosphere at the earth-atmosphere interface.

Lateral boundaries of an ecosystem may be chosen to coincide with those of a biological community, such as the edges of a forest, or with the boundary of some pronounced characteristic of the physical environment, such as the shoreline of a small island. Most often, however, the continuous nature of vegetation and of the physical environment makes it difficult to establish exact lateral boundaries on the basis of "community" or "environmental discontinuity" *(8)*. Often the investigator arbitrarily selects an area that may be conveniently studied.

From a functional point of view it is meaningless to include within the vertical limits of an ecosystem *all* of the column of air above and of soil and rock below the laterally defined ecosystem. For a working model of an ecosystem, it seems reasonable to include *only* that part of the column where atoms and molecules may participate in the chemical cycling that occurs within the system (see the "intrasystem cycle" of Fig. 1). When the biological community is taken as a determinant, the vertical extensions of the terrestrial ecosystem will be delimited by the top of the vegetation and the depth to which roots and other organisms penetrate into the regolith *(9)*. Vertical dimensions, defined in this manner, can expand or contract depending on the growth potential of present or succeeding communities. Thus, volumetric changes with time can be considered – for example, those associated with primary and secondary succession or with cliseral changes.

The Ecosystem and Biogeochemical Cycling

The terrestrial ecosystem participates in the various larger biogeochemical cycles of the earth through a system of inputs and outputs. Biogeochemical input in forest or field ecosystems may be derived from three major sources: geologic, meteorologic, and biologic. Geologic input is here defined as dissolved or particulate matter carried into the system by moving water or colluvial action, or both. Depositions of the products of erosion or mass wasting and ions dissolved in incoming seepage water are examples of geologic input. Meteorologic input enters the ecosystem through the atmosphere and is composed of additions of gaseous materials and of dissolved or particulate matter in precipi-

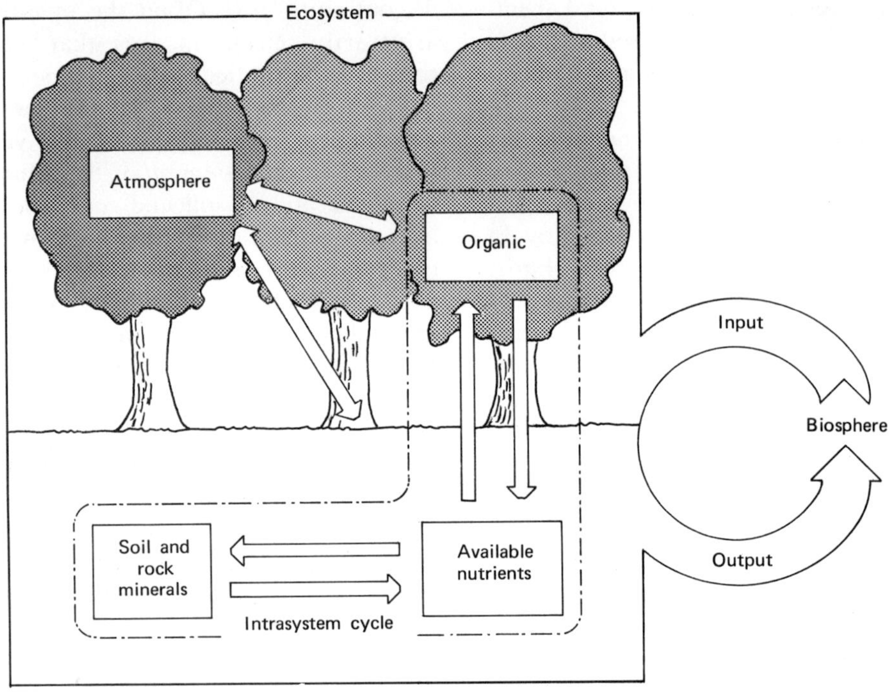

Figure 1 Nutrient relationships of a terrestrial ecosystem, showing sites of accumulation and major pathways. Input and output may be composed of geologic, meteorologic, and biologic components, as described in the text.

tation, dust, and other wind-borne materials. Chemicals in gaseous form fixed by biologic activity within the ecosystem are considered to be meteorologic input. Biological input results from animal activity and is made up of depositions of materials originally gathered elsewhere; examples are fecal material of animals whose food was gathered outside the system, or fertilizers intentionally added by man.

Chemicals may leave the ecosytem in the form of dissolved or particulate matter in moving water or colluvium, or both (geologic output); through the diffusion or transport of gases or particulate matter by wind (meteorologic output); or as a result of the activity of animals, including man (biologic output).

Nutrients are found in four compartments within the terrestrial ecosystem: in the atmosphere, in the pool of available nutrients in the soil, in organic materials (biota and organic debris), and in soil and rock minerals (Fig. 1). The atmospheric compartment includes all atoms or molecules in gaseous form in both the below-ground and the above-ground portions of the ecosystem. The pool of available nutrients in the soil consists of all ions adsorbed on the clay-humus complex or dissolved in the soil solution. The organic compartment includes all atoms incorporated in living organisms and in their dead remains. (The distinction between living and dead is sometimes hard to make, particularly in the case of woody perennial plants.) The soil-rock compartment is comprised of elements incorporated in primary and secondary minerals, including the more readily decomposable minerals that enter into equilibrium reactions with the available nutrients.

The degree to which a nutrient circulates within a terrestrial ecosystem is determined, in part, by its physical state. Gases are easily moved by random forces of diffusion and air circulation; consequently, nutrients with a prominent gaseous phase tend not to cycle within the boundaries of a particular ecosystem but, rather, to be continually lost and replaced from outside. On the other hand, elements without a prominent gaseous phase may show considerable intrasystem cycling between the available-nutrient, organic, and soil-rock compartments (Fig. 1). This internal cycle results from (i) the uptake of nutrients by plants, (ii) the release of nutrients from plants by direct leaching, (iii) the release of nutrients from organic matter by biological decomposition, and (iv) equilibrium reactions that convert insoluble chemical forms in the soil-rock compartment to soluble forms in the available-nutrient compartment, and vice versa.

Available nutrients not only enter the ecosystem from outside but are added by the action of physical, chemical, and biological weathering of rock and soil minerals already within the system. Although some ions are continually withdrawn from the available-nutrient compartment, forming secondary minerals in the soil and rocks, for most nutrient elements there is a net movement out of the soil-rock compartment. As the ecosystem is gradually lowered in place by erosion or by the downward growth of roots, new supplies of residual rock or other parent material are included; in some systems these materials may also be added as geologic input.

Hydrologic-Cycle, Nutrient-Cycle Interaction

At many points nutrient cycles may be strongly geared to the hydrologic cycle. Nutrient input and output are directly related to the amounts of water that move into and out of an ecosystem, as emphasized by the "leaching" and "flushing" concepts of Pearsall *(10)*, Dahl *(11)* , and Ratcliffe *(12)* , while temporal and absolute limits of biogeochemical activities within the system are markedly influenced by the hydrologic regime. Biologic uptake of nutrients by plants and release of nutrients by biological decomposition are closely related to the pattern of water availability. Potential levels of biomass within the system are determined in large measure by precipitation characteristics. Similarly, the nature and rate of weathering and soil formation are influenced by the hydrologic regime, since water is essential to the major chemical weathering processes [ion exchange, hydrolysis, solution, diffusion, oxidation-reduction, and adsorption and swelling *(13)*].

Biogeochemical Studies of Ecosystems

Although study of nutrient input, nutrient output, and weathering is necessary for an understanding of field and forest ecosystems *(6)*, ecologists, foresters, and pedologists have generally focused attention on the internal characteristics alone. Thus, considerable information has been accumulated on uptake, retention, and release of nutrients by the biota of ecosystems and on soil-nutrient relationships (see, for example, *6, 14*). Rarely are these internal characteristics of the ecosystem correlated with input and output data, yet all these parameters are necessary for the construction of nutrient budgets of particular ecosystems, and for establishing the relationship of the smaller system to the biosphere.

Quantitative data on input-output relationships are at best spotty. There are many data on nutrient output due to harvesting of vegetation, but for particular natural ecosystems there are only sporadic data on nutrient input in precipitation, or on nutrient output in drainage waters *(6)*. Recently, small lysimeters have been used successfully in the measurement of nutrient dynamics within the soil profile and in the measurement of nutrient losses in drainage water *(15)*. The lysimeter technique seems to be well suited for studying ecosystems characterized by coarse-textured soils with a relatively low field capacity *(16)*, high porosity, and no surface runoff. For most ecosystems, however, lysimeters are probably of limited value for measuring *total* nutrient output because (i) they are of questionable accuracy when used in rocky or markedly uneven ground, (ii) they cannot evaluate nutrient losses in surface waters, and (iii) their installation requires considerable disturbance of the soil profile.

The lack of information on the nutrient-input, nutrient-output relationships of ecosystems is apparently related to two considerations: (i) integrated studies of ecosystems tend to fall into an intellectual "no man's land" between traditional concepts of ecology, geology, and pedology; (ii) more important, the measurement of nutrient input and output requires measurement of hydrologic input and output. Unquestionably this lack of quantitative information is related to the difficulties encountered in measuring nutrients entering or leaving an ecosystem in seepage water or in sheet or rill flow, and to the high cost, in time and money, of obtaining continuous measurements of the more conventional hydrologic parameters of precipitation and stream flow. In many systems the problem is further complicated by the fact that much water may leave by way of deep seepage, eventually appearing in another drainage system; direct measurement of loss of water and nutrients by this route is virtually impossible.

Small-Watershed Approach to Biogeochemical Cycling

In some ecosystems the nutrient-cycle, hydrologic-cycle interaction can be turned to good advantage in the study of nutrient budgets, erosion, and weathering. This is particularly so if an ecosystem meets two specifications: (i) if the ecosystem is a watershed, and (ii) if the watershed is underlain by a tight bedrock or other impermeable base, such as permafrost. Given these conditions, for chemical elements without a gaseous form at biological temperatures, it is possible to construct nutrient budgets showing input, output, and net loss or gain from the system. These data provide estimates of weathering and erosion.

If the ecosystem were a small watershed, input would be limited to meteorologic and biologic origins. Geologic input, as defined above, need not be considered because there would be no transfer of alluvial or colluvial material between adjacent watersheds. Although materials might be moved within the ecosystem by alluvial or colluvial forces, these materials would originate within the ecosystem.

When the input and output of dust or windblown materials is negligible (this is certainly not the case in some systems), meteorologic input can be measured from a combination of hydrologic and precipitation-chemistry parameters. From periodic measurements of the elements contained in precipitation and from continuous measurements of precipitation entering a watershed of known area, one may calculate the temporal input of an element in terms of grams per hectare. Noncoincidence of the topographic divide of the watershed and the phreatic divide may introduce a small error *(17)*.

Losses from this watershed ecosystem would be limited to geologic and biologic output. Given an impermeable base, geologic output (losses due to erosion) would consist of dissolved and particulate matter in either stream water or seepage water moving downhill above the impermeable base. Although downhill mass movement may occur within the system, the products of this movement are delivered to the stream bed, whence they are removed by erosion and stream transportation.

Geologic output can be estimated from

hydrologic and chemical measurements. A weir, anchored to the bedrock will force all drainage water from the watershed to flow over the notch, where the volume and rate of flow can be measured. These data, in combination with periodic measures of dissolved and particulate matter in the outflowing water, provide an estimate of geologic output which may be expressed as grams of an element lost per hectare of watershed.

The nutrient budget for a single element in the watershed ecosystem may be expressed as follows: (meteorologic input + biologic input)–(geologic output + biologic output) = net loss or gain. This equation may be further simplified if the ecosystem meets a third specification – if it is part of a much larger, more or less homogeneous, vegetation unit. Biological output would tend to balance biological input if the ecosystem contained no special attraction or deterrent for animal populations moving at random through the larger vegetation system, randomly acquiring or discharging nutrients. On this assumption, the nutrient budget for a single system would become: (meteorologic input per hectare) – (geologic output per hectare) = net gain or loss per hectare. This fundamental relationship provides basic data for an integrated study of ecosystem dynamics.

Small Watersheds for Ecosystem Research

The relationship of the individual terrestrial ecosystem to biogeochemical cycles of the biosphere can be established by the small-watershed approach. Data on input and output of nutrients provide direct measurements of this relationship, while data on net loss provide, as explained below, an indirect measure of weathering rates for soil and rock minerals in relatively undisturbed ecosystems.

The small watershed may be used for experiments at the ecosystem level. This has been shown by numerous experiments concerned with hydrologic relationships (see for example, *18*). Thus, it is possible to test the effects of various experimental treatments on the relationship of the individual ecosystem to the biospheric nutrient cycles. Experiments can be designed to determine whether logging, burning, or use of pesticides or herbicides have an appreciable effect on net nutrient losses from the system. This information is not generally available at the ecosystem level.

The small watershed, with its measured parameters of hydrologic and chemical input, output, and net change, is an excellent vehicle for the study of interrelationships within a single ecosystem. Nutrient output may be related to hydrologic parameters such as seasonal and diurnal variations in stream flow, seasonal patterns of precipitation, individual rainstorms, and variations in evapotranspiration. Characteristics of the nutrient cycle may also be related to phenological events occurring within the ecosystem, such as leaf development, initiation of root growth, leaf fall, and litter turnover. In combination with current methods of biomass and nutrient analysis (see, for example, *6*), the small-watershed approach provides a comprehensive view of the status and behavior of individual elements within an individual ecosystem.

Weathering, or the rate at which an element bound in soil and rock minerals is made available, can be estimated from net losses of that element as calculated by the nutrient-budget method. Within the ecosystem (watershed), atoms of an element (one that lacks a gaseous form at ecosystem temperatures) may be located in (i) soil and rock minerals, (ii) the biota and organic debris, and (iii) the pool of available nutrients (Fig. 1). There is an intense intrasystem cycling between categories (ii) and (iii) as large quantities of ions are taken up by the vegetation each year and released by direct leaching or stepwise decomposition in the food chain. Ions are continually released to the intrasystem cycle by weathering of soil and rock material. Some of these ions, however, are reconstituted as secondary minerals. If an ecosystem is in a state of dynamic equilibrium, as the presence of climax forest would suggest (*19*), ionic levels in the intrasystem cycle must remain about the same for many years. Thus, in the climax ecosystem, net ion losses (output minus input) must be balanced by equivalent additions derived from weathering of soil and rock materials. Thus, net ionic losses from an undisturbed, relatively stable terrestrial ecosystem are a measure of weathering within the system. In a successional ecosystem (in which nutrients are accumulating in biomass and organic debris over the course of years), the rate at which an ion is released by weathering must equal its rate of net loss from the ecosystem plus its rate of net accumulation in the biota and organic debris (Fig. 1).

The watershed model allows comparison in relative importance of solution and suspended bed load in removing nutrients from an ecosystem. Nutrient matter can be removed from an ecosystem by three forms of transportation in streams: in solution in the stream water, as inorganic and organic suspended load kept in motion by turbulent flow, and as inorganic and organic bed load slid or rolled along the stream bottom *(20)*. Solution losses may be measured, as described above, from stream-flow data and periodic measurements of dissolved substances in the stream. Part of the losses of suspended matter may be estimated from stream-flow data and periodic measurements of particulate matter obtained by straining or filtering stream water as it comes over the weir. The remaining suspended matter and all of the bed load may be measured above the weir, where these materials collect in the ponding basin. These comparative measurements should be of interest not only to the ecologist concerned with ecosystem dynamics but also, since stream transportation is one of the important aspects of fluvial denudation, to geologists.

The small-watershed approach provides invaluable baseline information for the investigation of stream biology. Life-history studies of stream organisms, population studies, and shifts in community structure and diversity might be correlated with the measured physical and chemical parameters of drainage streams. Analyses of uptake, release, and transport of various nutrients by stream organisms could be made. Moreover, the vegetation of a watershed and the stream draining it are an inseparable unit functionally, and it would be of great interest to obtain information on the biological interaction between them.

Sites for Watershed Studies

Small watersheds meeting the conditions outlined above are probably common. However, even if the desired conditions are met, the investigator studying nutrient cycling is faced with the task of initiating a hydrologic study before he can attack his major problem. This is a time-consuming and expensive procedure, involving construction and maintenance of weirs, establishment of a precipitation network, and continuous collection of records, as well as land rental fees and possible road construction costs. A practical solution to this problem is inauguration of nutrient cycling studies at established hydrologic laboratories, where the required conditions exist and where hydrologic parameters are being measured and data are available.

The feasibility of this approach is demonstrated by our study at the Hubbard Brook Experimental Forest in West Thornton, New Hampshire. There, with the support of the National Science Foundation and the excellent cooperation of the Northeastern Forest Experiment Station, we are studying nutrient cycling and ecosystem dynamics on six small monitored watersheds. We have accumulated data on weathering rates, input, output, and the annual budget of several ions in this northern hardwood ecosystem. Also, studies of biomass, phenology, productivity, annual rates of nutrient turnover, and other factors are being made in one undisturbed watershed and in one in which conditions are being experimentally modified.

Preliminary data on input, output, and net change for three cations are presented in Table 1. These results allow us to add some numerical values to our ecosystem model (Fig. 2). For the calcium cycle, for example, input would be about 3 kilograms per hectare, while output (erosion) is estimated to be about 7.9 kilograms per hectare. Of this latter amount, 98 percent (7.7 kilograms per hectare) is lost in the form of dissolved substances in the stream water, while first approximations indicate that 2 percent is lost as calcium incorporated in organic matter flushed out of the ecosystem. On the basis of assumptions discussed above, it is estimated that the net amount of calcium lost, approximately 5 kilograms per hectare, is replaced by calcium released from soil and rock minerals by weathering. Hence 5 kilograms of calcium per hectare is added to the system each year by weathering.

Table 1. Budgets for Dissolved Cations in watershed No. 3 (42.4 hectares) for the period June 1963 to June 1964.

Cation	Input (kg/ hectare)	Output (kg/ hectare)	Net change (kg/ hectare)
Calcium	3.0	7.7	−4.7
Sodium	1.0	6.3	−5.3
Magnesium	0.7	2.5	−1.8

As yet we have not measured the calcium content of the soil and vegetation or the annual uptake and release of calcium by the biota. From Ovington's data *(6)*, for a beech forest in West England, which must be of about the same magnitude as values for our forest in New Hampshire, we see that 203 kilograms of calcium per hectare are localized in the trees and litter, while 365 kilograms per hectare represent exchangeable calcium in the soil. This gives a total of 568 kilograms of calcium per hectare in organic matter or as available nutrient. Assuming that our forest (Fig. 2) contains a similar amount of calcium, we estimate that a net annual loss of 5 kilograms per hectare would represent only nine-tenths of 1 percent of the total. This suggests a remarkable ability of these undisturbed systems to entrap and hold nutrients. However, if these calculations were based on actual amounts of calcium circulated each year rather than on the total, the percentage losses would be higher.

On its completion, the Hubbard Brook study will have yielded estimates, for individual elements, of many of the parameters and flux rates represented in the nutrient cycle shown in Fig. 1. These data will increase our understanding of fundamental nutrient relationships of undisturbed northern hardwood forests, and they will provide baseline information from which we can judge the effects on nutrient cycling of such practices as cutting, burning, and the application of pesticides.

Studies similar to these at Hubbard Brook could be established elsewhere in the United States. There are thousands of gaged watersheds operated by private and public interests *(17)*, and some of these must meet the proposed requirements. On selected watersheds, cooperative studies could be made by the agencies or organizations controlling the watershed and university-based investigators interested in biogeochemical cycling. Just such cooperation, between federal agencies and univer-

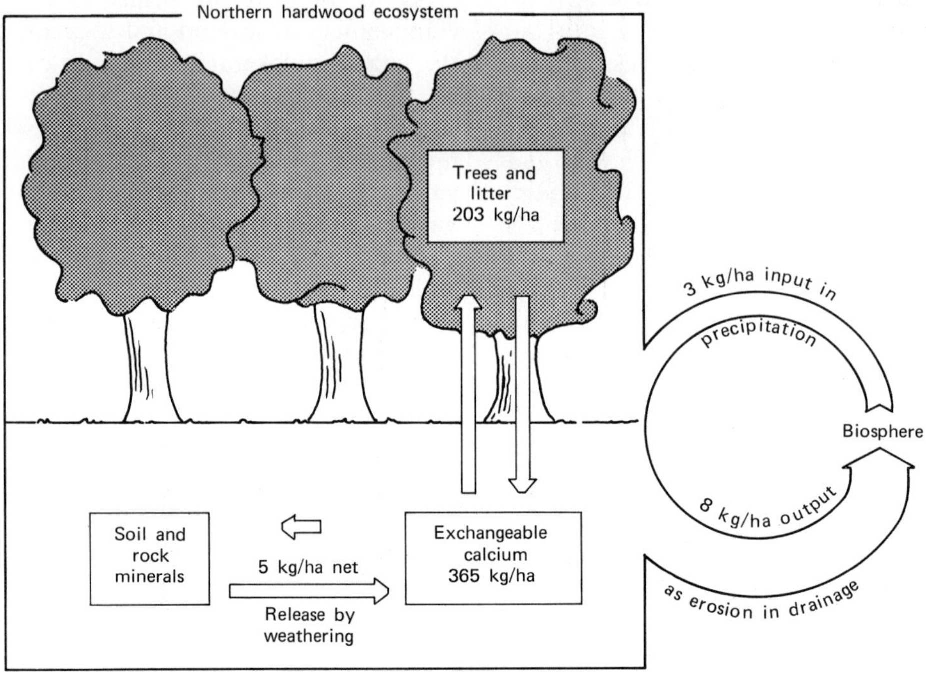

Figure 2 Estimated parameters for the calcium cycle in an undisturbed northern hardwood ecosystem in central New Hampshire [Data on trees, litter, and exchangeable calcium from Ovington(*6*)].

sities, has been urged by the Task Group on Coordinated Water Resources Research *(21)*.

Cooperative studies of this type have the advantage of providing a useful exchange of ideas between scientists in diverse fields who are working on the same ecosystem. The studies would provide a larger yield of information on a single system, the prospect of new concepts arising from the available information, and a greater scientific yield per dollar invested. Finally, cooperative studies would make available, for interpretation from the standpoint of nutrient cycling, an invaluable record of past hydrologic performance and, in some cases, of the responses of watersheds to experimental manipulation.

Conclusion

The small-watershed approach to problems of nutrient cycling has these advantages. (i) The small watershed is a natural unit of suitable size for intensive study of nutrient cycling at the ecosystem level. (ii) It provides a means of reducing to a minimum, or virtually eliminating, the effect of the difficult-to-measure variables of geologic input and nutrient losses in deep seepage. Control of these variables makes possible accurate measurement of nutrient input and output (erosion) and therefore establishes the relationship of the smaller ecosystem to the larger biospheric cycles. (iii) The small-watershed approach provides a method whereby such important parameters as nutrient release from minerals (weathering) and annual nutrient budgets may be calculated. (iv) It provides a means of studying the interrelationships between the biota and the hydrologic cycle, various nutrient cycles, and energy flow in a single system. (v) Finally, with the small-watershed system we can test the effect of various land-management practices or environmental pollutants on nutrient cycling in natural system.

References and Notes

1. L. C. Cole, *Sci. Amer.* **198,** 83 (Apr. 1958).
2. E. P. Odum, *Ecology* (Holt, Rinehart, and Winston, New York, 1963).
3. R. Revelle, W. Broecker, H. Craig, C. D. Keeling, J. Smagorinsky, in *Restoring the Quality of Our Environment* (Government Printing Office, Washington, D. C. , 1965), pp. 111-133.
4. L. C. Cole, *Saturday Rev.* (7 May 1966).
5. E. P. Odum, *Fundamentals of Ecology* (Saunders, Philadelphia, 1959).
6. J. D. Ovington, in *Advances in Ecological Research,* J. B. Cragg, Ed. (Academic Press, New York, 1962), vol. 1.
7. A. G. Tansley, *Ecology* **16,** 284 (1935).
8. L. B. Slobodkin *Growth and Regulation of Animal Populations* (Holt, Rinehart, and Winston, New York, 1961).
9. H. O. Buckman and N. C. Brady, *The Nature and Properties of Soils* (Macmillan, New York, 1960).
10. W. H. Pearsall, *Mountains and Moorlands* (Collins, London, 1950).
11. E. Dahl, "Rondane Mountain vegetation in South Norway and its relation to the environment, " *Skrifter Norske Videnskaps-Akad . Oslo* **1,** *Mat. Naturv. Kl. 1956* (1956).
12. D. A. Ratcliffe, *J. Ecol.* **47,** 371 (1959).
13. C. Bould, in *Plant Physiology, A Treatise,* F. C. Steward, Ed. (Academic Press, New York, 1963), vol. 3.
14. J. R. Bray and E. Gorham, in *Advances in Ecological Reasearch,* J. B. Cragg, Ed. (Academic Press, New York, 1964), vol. 2; K. J. Mustanoja and A. L. Leaf, *Botan. Rev.* **31,** 151 (1965).
15. D. W. Cole and S. P. Gessel, *Soil Sci. Soc. Amer. Proc.* **25,** 321 (1961); — , in *Forest-Soil Relationships in North America,* C. T. Youngberg, Ed. (Oregon State Univ. Press, Corvallis, 1965).
16. D. W. Cole, *Soil Sci.* **85,** 293 (1958).
17. C. O. Wisler and E. F. Brater *Hydrology* (Wiley, New York, 1959).
18. R. E. Dils, *A Guide to the Coweeta Hydrologic Laboratory* (Southeastern Forest Experimental Station, Asheville, N. C.,1957).
19. R. H. Whittaker, *Ecol. Monographs* **23,** 41 (1953).
20. A. N. Strahler, *The Earth Sciences* (Harper and Row, New York, 1963).
21. R. Revelle, *Science* **142,** 1027 (1963).

22. Financial support was provided by National Science Foundation grants GB 1144 and GB 4169. We thank J. Cantlon, N. M. Johnson, R. C. Reynolds, R. H. Whittaker, and G. W. Woodwell for critical comments and suggestions during preparation of the manuscript.

29.

Effects of Pollution on the Structure and Physiology of Ecosystems

G. M. Woodwell

We now know enough about the structure and function of these systems to predict the broad outline of the effects of pollution on both land and water. AUTHOR

The accumulation of various toxic substances in the biosphere is leading to complex changes in the structure and function of natural ecosystems. Although the changes are complex, they follow in aggregate patterns that are similar in many different ecosystems and are therefore broadly predictable. The patterns involve many changes but include especially simplification of the structure of both plant and animal communities, shifts in the ratio of gross production to total respiration, and loss of part or all of the inventory of nutrients. Despite the frequency with which various pollutants are causing such changes and the significance of the changes for all living systems (*1*), only a few studies show details of the pattern of change clearly. These are studies of the effects of ionizing radiation, of persistent pesticides, and of eutrophication. The effects of radiation will be used here to show the pattern of changes in terrestrial plant communities and to show similarities with the effects of fire, oxides of sulfur, and herbicides. Effects of such pollutants as pesticides on the animal community are less conspicuous but quite parallel, which shows that the ecological effects of pollution correspond very closely to the general "strategy of ecosystem development" outlined by Odum (*1*) and that they can be anticipated in considerable detail.

Reprinted from *Science*, Vol. 168, pp. 429-432, April 24, 1970.

The problems caused by pollution are of interest from two viewpoints, Practical people – toxicologists, engineers, health physicists, public health officials, intensive users of the environment consider pollution primarily as a direct hazard to man. Others, no less concerned for human welfare but with less pressing public responsibilities, recognize that toxicity to humans is but one aspect of the pollution problem, the other being a threat to the maintenance of a biosphere suitable for life as we know it. The first viewpoint leads to emphasis on human food chains; the second leads to emphasis on human welfare insofar as it depends on the integrity of the diverse ecosystems of the earth, the living systems that appear to have built and now maintain the biosphere.

The food-chain problem is by far the simpler; it is amenable at least in part to the pragmatic, narrowly compartmentalized solutions that industrialized societies are good at. The best example of the toxicological approach is in control of mutagens, particularly the radionuclides. These present a specific, direct hazard to man. They are much more important to man than to other organisms. A slightly enhanced rate of mutation is a serious danger to man, who has developed through medical science elaborate ways of preserving a high fraction of the genetic defects in the population; it is trivial to the rest of the biota, in which genetic defects may be eliminated through selection. This is an important fact about pollution hazards—toxic subtances that are principally mutagenic are usually of far greater direct hazard to man than to the rest of the earth's biota and must be considered first from the standpoint of their movement to man through food webs or other mechanisms and to a much lesser extent from that of their effects on the ecosystem through which they move. We have erred, as shown below, in assuming that all toxic substances should be treated this way.

Pollutants that affect other components of the earth's biota as well as man present a far greater problem. Their effects are chronic and may be cumulative in contrast to the effects of short-lived disturbances that are repaired by succession. We ask what effects such pollutants have on the structure of natural ecosystems and on biological diversity and what these changes mean to physiology, especially to mineral cycling and the long-term potential for sustaining life.

Although experience with pollution of various types is extensive and growing rapidly, only a limited number of detailed case history studies provide convincing control data that deal with the structure of ecosystems. One of the clearest and most detailed series of experiments in recent years has been focused on the ecological effects of radiation. These studies are especially useful because they allow cause and effect to be related quantitatively at the ecosystem level, which is difficult to do in nature. The question arises, however, whether the results from studies of ionizing radiation, a factor that is not usually considered to have played an important role in recent evolution, have any general application. The answer, somewhat surprisingly to many biologists, seems to be that they do. The ecological effects of radiation follow patterns that are known from other types of disturbances. The studies of radiation, because of their specificity, provide useful clues for examination of effects of other types of pollution for which evidence is much more fragmentary.

The effects of chronic irradiation of a late successional oak-pine forest have been studied at Brookhaven National Laboratory in New York. After 6 months' exposure to chronic irradiation from a ^{137}Cs source, five well-defined zones of modification of vegetation had been established. They have become more pronounced through 7 years of chronic irradiation. The zones were:

1) A central devastated zone, where exposures were > 200 R/day and no higher plants survived, although certain mosses and lichens survived up to exposures > 1000 R/day.

2) A sledge zone, where *Carex pensylvanica (2)* survived and ultimately formed a continuous cover (> 150 R/day).

3) A shrub zone in which two species of *Vaccinium* and one of *Gaylussacia* survived, with *Quercus ilicifolia* toward the outer limit of the circle where exposures were lowest (> 40 R/day).

4) An oak zone, the pine having been eliminated (> 16 R/day).

5) Oak-pine forest, where exposures were < 2 R/day, and there was no obvious change in the number of species, although small changes in rates of growth were measurable at exposures as low as 1 R/day.

The effect was a systematic dissection of the forest, strata being removed layer by layer. Trees were eliminated at low exposures, then the taller shrubs (*Gaylussacia baccata*), then the lower shrubs (*Vaccin-*

ium species), then the herbs, and finally the lichens and mosses. Within these groups, it was evident that under irradiation an upright form of growth was a disadvantage. The trees did vary—the pines (*Pinus rigida*) for instance were far more sensitive than the oaks without having a conspicuous tendency toward more upright growth, but all the trees were substantially more sensitive than the shrubs (*3*). Within the shrub zone, tall forms were more sensitive; even within the lichen populations, foliose and fruticose lichens proved more sensitive than crustose lichens (*4*).

The changes caused by chronic irradiation of herb communities in old fields show the same pattern—upright species are at a disadvantage. In one old field at Brookhaven, the frequency of low-growing plants increased along the gradient of increasing radiation intensity to 100 percent at > 1000 R/day (*5*). Comparison of the sensitivity of the herb field with that of the forest, by whatever criterion, clearly shows the field to be more resistant than the forest. The exposure reducing diversity to 50 percent in the first year was ~1000 R/day for the field and 160 R/day for the forest, a greater than fivefold difference in sensitivity (*3*).

The changes in these ecosystems under chronic irradiation are best summarized as changes in structure, although diversity, primary production, total respiration, and nutrient inventory are also involved. The changes are similar to the familiar ones along natural gradients of increasingly severe conditions, such as exposure on mountains, salt spray, and water availability. Along these gradients the conspicuous change is a reduction of structure from forest toward communities diminated by certain shrubs, then under more severe conditions, by certain herbs, and finally by low-growing plants, frequently mosses and lichens. Succession, insofar as it has played any role at all in the irradiated ecosystems, has simply reinforced this pattern, adding a very few hardy species and allowing expansion of the populations of more resistant indigenous species. The reasons for radiation's causing this pattern are still not clear (*3, 6*), but the pattern is a common one, not peculiar to ionizing radiation, despite the novelty of radiation exposures as high as these.

Its commonness is illustrated by the response to fire, one of the oldest and most important disruptions of nature. The oak-pine forests such as those on Long Island have, throughout their extensive range in eastern North America, been subject in recent times to repeated burning. The changes in physiognomy of the vegetation follow the above pattern very closely – the forest is replaced by communities of shrubs, especially bear oak *(Quercus ilicifolia), Gaylussacia,* and *Vaccinium* species. This change is equivalent to that caused by chronic exposure to 40 R/day or more. Buell and Cantlon (*7*), working on similar vegetation in New Jersey, showed that a further increase in the frequency of fires resulted in a differential reduction in taller shrubs first, and a substantial increase in the abundance of *Carex pensylvanica,* the same sedge now dominating the sedge zone of the irradiated forest. The parallel is detailed; radiation and repeated fires both reduce the structure of the forest in similar ways, favoring low-growing hardy species.

The similarity of response appears to extend to other vegetations as well. G. L. Miller, working with F. McCormick at the Savannah River Laboratory, has shown recently that the most radiation-resistant and fire-resistant species of 20-year-old fields are annuals and perennials characteristic of disturbed places (*8*). An interesting sidelight of his study was the observation that the grass stage of long leaf pine (*Pinus palustris*), long considered a specific adaptation to the fires that maintain the southeastern savannahs, appears more resistant to radiation damage than the mature trees. At a total acute exposure of 2.1

kR (3 R/day), 85 percent of the grass-stage populations survived but only 55 percent of larger trees survived. Seasonal variation in sensitivity to radiation damage has been abundantly demonstrated (*9*), and it would not be surprising to find that this variation is related to the ecology of the species. Again it appears that the response to radiation is not unique.

The species surviving high radiation-exposure rates in the Brookhaven experiments are the ones commonly found in disturbed places, such as roadsides, gravel banks, and areas with nutrient-deficient or unstable soil. In the forest they include *Comptonia peregrina* (the sweet fern), a decumbent spiny *Rubus,* and the lichens, especially *Cladonia cristatella.* In the old field one of the most conspicuously resistant species was *Digitaria sanguinalis* (crabgrass) among several other weedy species. Clearly these species are generalists in the sense that they survive a wide range of conditions, including exposure to high intensities of ionizing radiation – hardly a common experience in nature but apparently one that elicits a common response.

With this background one might predict that a similar pattern of devastation would result from such pollutants as oxides of sulfur released from smelting. The evidence is fragmentary, but Gorham and Gordon (*10*) found around the smelters in Sudbury, Ontario, a striking reduction in the number of species of higher plants along a gradient of 62 kilometers (39 miles). In different samples the number of species ranged from 19 to 31 at the more distant sites and dropped abruptly at 6.4 kilometers. At 1.6 kilometers, one of two randomly placed plots (20 by 2 meters) included only one species. They classified the damage in five categories, from "Not obvious" through "Moderate" to "Very severe." The tree canopy had been reduced or eliminated within 4.8 to 6.4 kilometers of the smelter, with only occasional sprouts of trees, seedlings, and successional herbs and shrubs remaining; this damage is equivalent to that produced by exposure to 40 R/day. The most resistant trees were, almost predictably to a botanist, red maple (*Acer rubrum*) and red oak (*Quercus rubra*). Other species surviving in the zones of "Severe" and "Very severe" damage included *Sambucus pubens, Polygonum cilinode, Comptonia peregrina,* and *Epilobium angustifolium* (fire weed). The most sensitive plants appeared to be *Pinus strobus* and *Vaccinium myrtilloides.* The pine was reported no closer than 25.6 kilometers (16 miles), where it was chlorotic.

This example confirms the pattern of the change – first a reduction of diversity of the forest by elimination of sensitive species; then elimination of the tree canopy and survival of resistant shrubs and herbs widely recognized as "seral" or successional species or "generalists."

The effects of herbicides despite their hoped for specificity, fall into the same pattern, and it is no surprise that the extremely diverse forest canopies of Viet Nam when sprayed repeatedly with herbicides are replaced over large areas by dense stands of species of bamboo (*11*).

The mechanisms involved in producing this series of patterns in terrestrial ecosystems are not entirely clear. One mechanism that is almost certainly important is simply the ratio of gross production to respiration in different strata of the community. The size of trees has been shown to approach a limit set by the amount of surface area of stems and branches in proportion to the amount of leaf area (*12*). The apparent reason is that, as a tree expands in size, the fraction of its total surface devoted to bark, which makes a major contribution to the respiration, expands more rapidly than does the photosynthetic area. Any chronic disturbance has a high probability of damaging the capacity for photosynthesis without reducing appreciably the total amount of respiration; therefore, large plants are more vulnerable than species requiring less total respiration. Thus chronic disturbances of widely different types favor plants that are small in stature, and any disturbance that tends to increase the amount of respiration in proportion to photosynthesis will aggravate this shift.

The shift in the structure of terrestrial plant communities toward shrubs, herbs, or mosses and lichens, involves changes in addition to those of structure and diversity. Simplification of the plant community involves also a reduction of the total standing crop of organic matter and a corresponding reduction in the total inventory of nutrient elements held within the system, a change that may have important long-term implications for the potential of the site to support life. The extent of such losses has been demonstrated recently by Bormann and his colleagues in the Hubbard Brook Forest in New Hampshire (*13*), where all of the trees in a watershed were cut, the cut material was left to decay, and the losses of nutrients were monitored in the runoff. Total nitrogen losses in the first year were equivalent to twice the amount cycled in the system during a normal year. With the rise of nitrate ion in the runoff, concentrations of calcium, magnesium, sodium, and potassium ions rose severalfold, which caused eutrophication and even pollution of the streams fed by this water-shed. The soil had little capacity to retain the nutrients that were locked in the biota once the higher plants had been killed. The total losses are not yet known, but early evidence indicates that they will be a high fraction of the nutrient inventory, which will cause a large reduction in the potential of the site for supporting living systems as complex as that destroyed – until nutrients accumulate again. Sources are limited; the principal source is erosion of primary minerals.

When the extent of the loss of nutrients that accompanies a reduction in the structure of a plant community is recognized, it is not surprising to find depauperate vegetation in places subject to chronic disturbances. Extensive sections of central Long Island, for example, support a depauperate oak-pine forest in which the bear oak, *Quercus ilicifolia,* is the principal woody species. The cation content of an extremely dense stand of this common community, which has a biomass equivalent to that of the more diverse late successional forest that was burned much less recently and less intensively, would be about 60 percent that of the richer stand, despite the equivalence of standing crop. This means that the species, especially the bear oak, contain, and presumably require, lower concentrations of cations. This is an especially good example because the bear oak community is a long-lasting one in the fire succession and marks the transition from a high shrub community to forest. It has analogies elsewhere, such as the heath balds of the Great Smoky Mountains and certain bamboo thickets in Southeast Asia.

The potential of a site for supporting life depends heavily on the pool of nutrients available through breakdown of primary minerals and through recycling in the living portion of the ecosystem. Reduction of the structure of the system drains these pools in whole or in part; it puts leaks in the system. Any chronic pollution that affects the structure of ecosystems, especially the plant community, starts leaks and reduces the potential of the site for recovery. Reduction of the structure of forests in Southeast Asia by hebicides has dumped the nutrient pools of these large statured and extremely diverse forests. The nutrients are carried to the streams, which turn green with the algae that the nutrients support. Tschirley (*11*), reporting his study of the effects of herbicides in Viet Nam, recorded "surprise" and "pleasure" that fishing had improved in treated areas. If the herbicides are not toxic to fish, there should be little surprise at improved catches of certain kinds of fish in heavily enriched water adjacent to herbicide-treated forests.

The bamboo thickets that replace the forests also reflect the drastically lowered potential of these sites to support living systems. The time it takes to reestablish a forest with the original diversity depends on the availability of nutrients, and is probably very long in most lateritic soils.

In generalizing about pollution, I have concentrated on some of the grossest changes in the plant communities of terrestrial ecosystems. The emphasis on plants is appropriate because plants dominate terrestrial ecosystems. But not all pollutants affect plants directly; some have their principal effects on heterotrophs. What changes in the structure of animal communities are caused by such broadly toxic materials as most pesticides?

The general pattern of loss of structure is quite similar, although the structure of the animal communities is more difficult to chart. The transfer of energy appears to be one good criterion of structure. Various studies suggest that 10 to 20 percent of the energy entering the plant community is transferred directly to the animal community through herbivores (*14*). Much of that energy, perhaps 50 percent or more, is used in respiration to support the herbivore population; some is transferred to the detritus food chain directly, and some, probably not more than 20 percent, is transferred to predators of the herbivores. In an evolutionarily and successionally mature community, this transfer of 10 to 20 percent per trophic level may occur two or three times to support carnivores, some highly specialized, such as certain eagles, hawks, and herons, others less specialized, such as gulls, ravens, rats, and people.

Changes in the plant community, such as its size, rate of energy fixation, and species, will affect the structure of the animal community as well. Introduction of a toxin specific for animals, such as a pesticide that is a generalized nerve toxin, will also topple the pyramid. Although the persistent pesticides are fat soluble and tend to accumulate in carnivores and reduce populations at the tops of food chains, they affect every trophic level, reducing reproductive capacity, almost certainly altering behavioral patterns, and disrupting the competitive relationships between species. Under these circumstances the highly specialized species, the obligate carnivores high in the trophic structure, are at a disadvantage because the food chain concentrates the toxin and, what is even more important, because the entire structure beneath them becomes unstable. Again the generalists or broad-niched species are favored, the gulls, rat, ravens, pigeons and, in a very narrow short-term sense, man. Thus, the pesticides favor the herbivores, the very organisms they were invented to control.

Biological evolution has divided the resources of any site among a large variety of users – species – which, taken together, confer on that site the properties of a closely integrated system capable of conserving a diversity of life. The system has structure; its populations exist with certain definable, quantitative relationships to one another; it fixes energy and releases it at a measurable rate; and it contains an inventory of nutrients that is accumulated and recirculated, not lost. The system is far from static; it is subject, on a time scale very long compared with a human lifespan, to a continuing augmentive change through evolution; on a shorter time scale, it is subject to succession toward a more stable state after any disturbance. The successional patterns are themselves a product of the evolution of life, providing for systematic recovery from any acute disturbance. Without a detailed discussion of the theory of ecology, one can say that biological evolution, following a pattern approximating that outlined above, has built the earth's ecosystems, and that these systems have been the dominant influence on the earth throughout the span of human existence. The structure of these systems is now being changed all over the world. We know enough about the structure and function of these systems to predict the broad outline of the effects of pollution on both land and water. We know that as far as our interests in the next decades are concerned, pollution operates on the time scale of succession, not of evolution, and we cannot look to evolution to cure this set of problems. The loss of structure involves a shift away from complex arrangements of specialized species toward the generalists; away from forest, toward hardy shrubs and herbs; away from those phytoplankton of the open ocean that Wurster (*15*) proved so very sensitive to DDT, toward those algae of the sewage plants that are unaffected by almost everything including DDT and most fish; away from diversity in birds, plants, and fish toward monotony; away from tight nutrient cycles toward very loose ones with terrestrial systems becoming depleted, and with aquatic systems becoming overloaded; away from stability toward instability especially with regard to sizes of populations of small, rapidly reproducing organisms such as insects and rodents that compete with man; away from a world that runs itself through a self-augmentive, slowly moving evolution, to one that requires constant tinkering to patch it up, a tinkering that is malignant in that each act of repair generates a need for further repairs to avert problems generated at compound interest

This is the pattern, predictable in broad outline, aggravated by almost any pollutant. Once we recognize the pattern, we can begin to see the meaning of some of the changes occurring now in the earth's biota. We can see the demise of carnivorous birds and predict the demise of important fisheries. We can tell why, around industrial cities, hills that were once forested now are not; why each single species is important; and how the increase in the temperature of natural water bodies used to cool new reactors will, by augmenting respiration over photosynthesis, ultimately degrade the system and contribute to degradation of other inteconnected ecosystems nearby. We can begin to speculate on where continued, exponential progress in this direction will lead: probably not to extinction – man will be around for a long time yet – but to a general degradation of the quality of life.

The solution? Fewer people, unpopular but increasing restrictions on technology (making it more and more expensive), and a concerted effort to tighten up human ecosystems to reduce their interactions with the rest of the earth on whose stabilty we all depend. This does not require foregoing nuclear energy; it requires that if we must dump heat, it should be dumped into civilization to enhance a respiration rate in a sewage plant or an agricultural ecosystem, not dumped outside of civilization to affect that fraction of the earth's biota that sustains the earth as we know it. The question of what fraction that might be remains as one of the great issues, still scarcely considered by the scientific community.

References and Notes

1. E. P. Odum, *Science* **164,** 262 (1969).
2. Plant nomenclature follows that of M. L. Fernald in *Gray's Manual of Botany* (American Book, New York, ed. **8** 1950).
3. G. M. Woodwell, *Science* **156,** 461 (1967); ——— and A. L. Rebuck, *Ecol. Monogr.* **37**, 53 (1967).
4. G. M. Woodwell and T. P. Gannutz, *Amer. J. Bot.* **54**, 1210 (1967).

5. ——— and J. K. Oosting, *Radiat. Bot.* **5,** 205 (1965).
6. ——— and R. H. Whittaker, *Quart. Rev. Biol.* **43,** 42 (1968).
7. M. F. Buell and J. E. Cantlon, *Ecology* **34**, 520 (1953).
8. G. L. Miller, thesis, Univ. of North Carolina (1968).
9. A. H. Sparrow, L. A. Schairer, R. C. Sparrow, W. F. Campbell, *Radiat. Bot.* **3,** 169 (1963); F. G. Taylor, Jr., *idid.* **6**, 307 (1965).
10. E. Gorham and A. G. Gordon, *Can. J. Bot.* **38**, 307 (1960); *ibid.,* p. 477; *ibid.* **41,** 371 (1963).
11. F. H. Tschirley, *Science* **163,** 779 (1969).
12. R. H. Whittaker and G. M. Woodwell, *Amer. J. Bot.* **54,** 931 (1967).
13. F. H. Bormann, G. E. Likens, D. W. Fisher, R. S. Pierce, *Science* **159,** 882 (1968).
14. These relationships have been summarized in detail by J. Phillipson [*Ecological Energetics* (St. Martin's Press, New York, 1966)]. See also L. B. Slobodkin, *Growth and Regulation of Animal Populations* (Holt, Rinehart and Winston, New York, 1961) and J. H. Ryther, *Science* **166,** 72 (1969).
15. C. F. Wurster, *Science* **159,** 1474 (1968).
16. G. M. Woodwell, *ibid.* **138,** 572 (1962).
17. Research carried out at Brookhaven National Laboratory under the auspices of the U.S. Atomic Energy Commission. Paper delivered at 11th International Botanical Congress, Seattle, Wash., on 26 August 1969 in the symposium "Ecological and Evolutionary Implications of Environmental Pollution."

30.

Aquatic Weeds

L. G. Holm, L. W. Weldon, and R. D. Blackburn

While we have been engaged elsewhere, the rampant growth of aquatic weeds has come to be one of our symptoms of our failure to manage our resources. AUTHORS

In the evolution of the city as a habitat, in the conversion of virgin lands to intensified farming, and in the alteration of watercourses with locks, dams, and reservoirs, man is the interloper. At his behest the natural order of things is set aside. As a result of his activities and their byproducts, new species and numbers of weeds, rodents, insects, and diseases appear where they could not, or did not, exist before. One of our priceless treasures, fresh water, is changed as civilization draws near. Its quality usually becomes poorer; it is seldom improved by man. Communities, planned and unplanned, locate on the water's edge to use navigation routes, irrigate land, and develop power. As a result the watercourses are heated, polluted, and fertilized; the levels fluctuate, and new biological pests are introduced because of man's commerce and mobility.

Several "explosions" of aquatic weeds in the great rivers and lakes of the warm regions of the world have forced us to recognize the power of such infestations. They destroy fisheries, interfere with hydroelectric and irrigation schemes, stop navigation, and bring starvation and disease problems to riverine communities. The rapid growth of weed infestations has been both spectacular and frightening, and the publicity devoted to several of these problems in the past decade has made us aware that something is wrong.

Aquatic weeds obstruct water flow, increase evaporation, cause large losses of water through transpiration, and prevent proper drainage of land. Weeds may interfere with navigation, prevent fishing and recreation, depress real estate values, and present health hazards. In the western United States, Timmons (*1*) showed that 17 states lost 1,966,000 acre-feet of irrigation water annually because of aquatic and ditchbank weeds. This water, valued conservatively at $20 per acre-foot, is worth $39,230,000 (*2*). This is enough water to irrigate 132,000 to 315,000 hectares of cropland. In the United States there never has been an evaluation of the total nonagricultural losses due to aquatic weeds. It is certain that this loss, too, would be very high.

The aquatic environment is complex and is of interest to scientists in several disciplines. The management of aquatic vegetation is not a new science, but rather an old field of botany that has been recently revitalized because of increased demand on our fresh waters and the exponential growth in problems caused by aquatic vegetation.

Reprinted from *Science*, Vol. 166, pp. 699-709, November 7, 1969.

Water Hyacinth

The scourge of some of the world's major rivers is water hyacinth, *Eichhornia crassipes* (Mart.) Solms in the family *Pontederiaceae* (Fig. 1). A native of South America, it is now widely distributed over the warm regions of the earth (Fig. 2). The plant is free-floating, has very fine roots, and produces stolons and viable seeds. The leaves are 10 to 15 cm across, bright green, shiny, and, because they are upright, serve as sails before the wind. The lovely flowers are pale lilac or mauve with a yellow patch at the center. Because of his admiration for the flowers, man has assisted the spread of this plant by cultivating it in his pools and gardens. His carelessness toward the cleanliness of his commercial and pleasure craft on land and sea has also contributed to the movement of the plant. For example, charcoal is made in the bush in Africa; and holes in the sacks used for transport are sometimes plugged with water hyacinth plants that may survive a very long journey. Large plants are used as cushions for sitting and kneeling in native canoes, to be thrown away at any place where the plant ceases to be useful. The plants catch on the sides and bottoms of river craft and thus move with the commerce of the region. Natural forces and events have also been important in the spread of water hyacinth. From the nurseries in the swamps and backwaters, great islands of the weed are flushed into the mainstream at flood time.

Water hyacinth populations increase rapidly by vegetative reproduction. In one experiment two parent plants produced 30 offspring after 23 days, and 1200 at the end of 4 months. A plant may flower at the age of 26 days and will normally produce viable seeds. Seed production varies from a few to as many as 5000 seeds per plant. The seeds sink to the bottom and may remain viable for at least 15 years. Movement of seeds between watercourses must surely take place in the mud on furbearing animals and on the legs of birds. Recent reports of the long-distance dispersal of seed by waterfowl and shore birds suggest a possibility that has never been taken seriously. These migrating birds can carry seed several thousand kilometers in one season (*3*). In the face of such vectors, once the seed of a species is on a continent, there may be little that man can do to prevent its spread.

An account of the spread of water hyacinth in the Congo and Nile rivers will illustrate the immensity of the task we face. The species had been introduced at the delta of the Nile and in Natal, South Africa, at the beginning of the century. Europeans who settled in South Rhodesia in 1937 reported the presence of water hyacinth. It is now generally distributed in Africa. In spite of the losses and human suffering caused by this plant, it was still available in 1965 for purchase as an ornamental in the street markets of Senegal. In the United States, federal law prohibits interstate shipment of water hyacinth, but it is offered for sale in the catalogs of many distributors of water garden plants.

Water hyacinth was first reported in the Congo River in 1952 and in less than 3 years it had spread 1600 kilometers from Leopoldville up to Stanleyville. In 1954 it had already begun to block transportation. Buoys were submerged and navigation channels were hidden. Fish spawning areas were blocked. Many fishing grounds were destroyed by darkness and lack of oxygen as the weed cover became more dense, and, as a result, the riverine communities were denied their principle source of protein. Using herbicides applied from ships, planes, and helicopters, by 1957 Belgian scientists had directed the cleanup of more than 1600 kilomenters of the river at a cost of $1,000,000. In spite of this massive effort, Le Brun reports that in the same year water hyacinth was still floating past Leopoldville on the way to the sea at the rate of 136 metric tons per

Figure 1 The flowers of water hyacinth *Eichhornia crassipes* (Mart.) Solms.

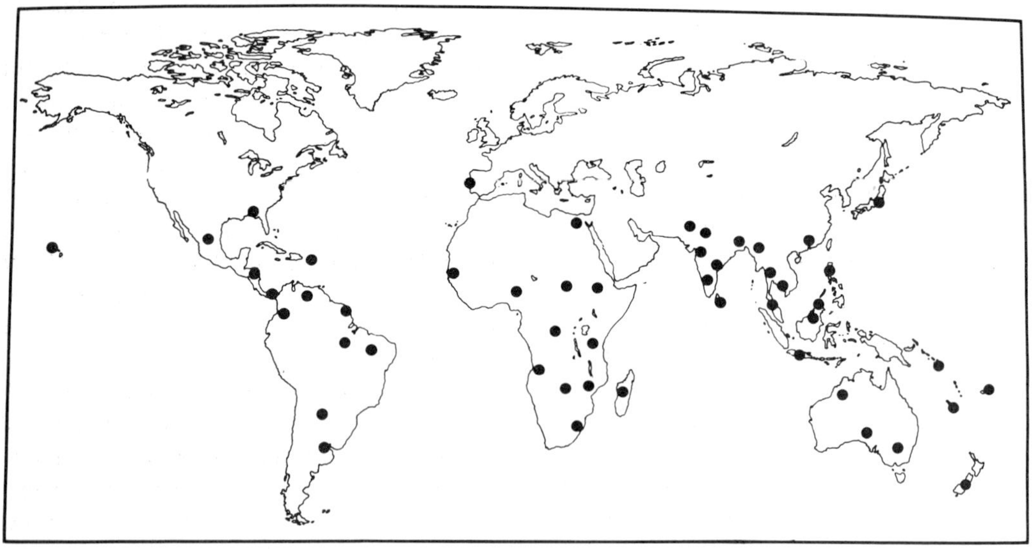

Figure 2 The distribution of water hyacinth *Eichornia crassipes*

hour (*4*). During the turbulent years following independence, the Congo government services could not maintain the weed control program, and the Congo River is again badly infested.

The history of the infestation of the

upper White Nile is equally tragic. The first report of the weed in the White Nile was in 1958. Again, the multiplication and spread took place so quickly that even the best efforts of the Sudan government could not organize a campaign in time to contain the weed. A staff of 200 workers equipped with ships, planes, and land vehicles was organized to keep the river and the harbors open. During the early 1960's, the water hyacinth team of Sudan was able to keep the weed under control in the most critical areas with herbicides. At one time the cost of this operation was $1.5 million (*5*).

In addition to all of the usual hardships caused by the weed, men now began to see the impact of its uncontrolled growth on clogged irrigation pumps and hydroelectric schemes. Some river villages that were regularly supplied with food by boat began to starve because they could not be reached. Insect vectors of human and animal diseases seek harbor in the mats of water hyacinth; so do dangerous snakes and crocodiles. In recent years the political difficulties in the south of the country have prevented workers from spraying many of the nurseries and sources of infestation in the upper White Nile. The region is again in serious difficulty. Figures 3 and 4 depict the change that took place between 1958 and 1965 above the dam at Jebel Aulia on the White Nile near Khartoum.

The people of Sudan deserve much credit for preventing the spread of water hyacinth from the White to the Blue Nile. The confluence of the two rivers is at Khartoum, and above this point a quarantine system, including many vehicle checkpoints, has been organized and is efficiently administered. By preventing the spread of the weed from the White to the Blue Nile, the Sudanese have protected the great irrigated Gezira cotton scheme.

Water hyacinth is present in the sloughs and backwaters of the Amazon River. It seldom appears to be a problem in the Amazon, perhaps due to lesser demands on utilization of this waterway, or to the presence of natural factors such as insects or plant diseases that limit growth of the weed. It does interfere with man's activities in many lakes, streams, and reservoirs of Central and South America. Many large reservoirs such as Brokopondo in Surinam, Lake Apanas in Nicaragua, and Lake Rio Lempa in El Salvador are threatened with economic disaster if dense stands of water hyacinth are allowed to develop.

The benefit of these reservoirs is dependent upon the amount of available water. However, water hyacinth consumes and wastes tremendous quantities of water through the leaves. The loss of water through evapotranspiration from the leaves has been measured as 3.2 to 3.7 times greater than free evaporation from a surface. This accounted for a loss of more than 6 acre-feet of water in a 6-month period due to a water hyacinth cover. In the dry atmosphere of India, the loss of water through water hyacinth was 7.8 times that of open water (*6*). Only a partial coverage of water hyacinth on a reservoir can result in the entire river inflow being wasted back into the atmosphere. Thus, the water is not available for hydroelectric power and irrigation.

Water hyacinth is distributed generally in Asia where it may be found in India, throughout the countries of Southeast Asia, and in the Philippine Islands. In East Pakistan some of the farming areas are covered with massive deposits of water hyacinth as the floods come down from the hills in the rainy season. The weed is in Australia and New Zealand as well.

Water hyacinth was reportedly introduced into the United States in 1884 as part of a horticultural exhibit. It spread rapidly and by 1897 was creating enough havoc to navigation to prompt an investigation. Funds for control of the weed became available in 1899. A number of mechanical removal procedures were used

Figure 3 The dam on the White Nile at Jebel Aulia near Khartoum, Sudan. The area was clean when photographed in October 1958. (Courtesy E. Buyckx)

Figure 4 The same area seen in Figure 3, showing the accumulation of water hyacinth above the dam in October 1965.

initially to control the water hyacinth, but these have largely been replaced with herbicide programs because of the great saving in time and money. Intensive control programs are being conducted by federal, state, and local agencies, drainage districts, and private interests. In spite of this large combined effort water hyacinth caused an annual loss of almost \$43,000,000 in Florida, Albama, Mississippi, and Louisiana in 1956. The types of losses are similar throughout the world (*7*).

We have indicated some of the navigational problems in Africa. The navigation of rivers and waterways in the southern United States would be just as difficult without proper control of the water hyacinth. Many rivers have to be closed to boat traffic periodically if effective control procedures have not been followed. Fishing camps and marinas are often closed, and many are forced into bankruptcy by inaccessibility.

A cover of water hyacinth on canals with a cross section of 36 to 72 centares may reduce the flow of water by half. The flow in smaller laterals and farm ditches may be almost completely stopped. Many canal systems have to be enlarged 50 percent or more to compensate for retarded flow. An additional problem is the constant accumulation of debris from decaying water hyacinth on the canal bottom. During periods of active growth, the lower leaves and roots of water hyacinth are constantly decaying and being replaced. This debris that falls to the canal bottom may amount to more than 30 centimeters per year. If the extensive water hyacinth cover is not prevented, mechanical removal of the debris becomes necessary – at a cost of $400 to $640 per kilometer (*6, 8*).

Salvinia

Some of the man-made lakes constructed in this century are so large that they are included on world maps. Lake Kariba, on the Zambesi River in Southern Africa, is an interesting example of the frustration which comes with massive efforts to arrange great alterations in the environment. The relocation of 50,000 people to make way for this lake caused considerable human suffering. The biological consequence of impounding such a large stream is in part illustrated by the story of the weed problem that later developed.

The Kariba Dam was closed late in 1958. The maximum water level was reached in July 1963, 4½ years later. Lake Kariba now covers more than 4200 square kilometers, has a maximum depth of 115 meters, and extends up to the foot of beautiful Victoria Falls.

In May 1959, floating mats of the water fern, *Salvinia auriculata* Aublet, were reported in the center of the lake. The mats grew in size and moved about with the wind. Before the end of the year a significant portion of the water near the shore was covered with *Salvinia* plants, as they lodged in branches of partially submerged trees and continued to multiply. By this time, *Pistia stratiotes* L. was associating with *Salvinia* in some of the mats. During 1960, less than 2 years after the dam was closed, it was estimated that the rafts of *Salvinia* had covered 420 square kilometers or 10 percent of the surface area (*9*). By 1965 the infestation had subsided slightly but was still estimated at 8 percent of the lake's surface (Figs. 5 and 6).

The water fern has floating leaves 3 cm long and submerged leaves so finely divided that they have a feathery appearance. In some areas the plant produces spores when it is crowded, but until 1965 only vegetative reproduction had been observed on Lake Kariba. Before formation of the lake, large infestations of *Salvinia* were present in the Upper Zambesi where they sometines interfered with the fishing of the local people. The weed was also far up the Chobe River, a tributary which feeds into the Zambesi above Victoria Falls. In 1949, Dr. O. West collected *Salvinia* just a short way above Victoria Falls. Boughey (*9*) assumed that the weed was already present in that portion of the Zambesi River which was flooded by the lake. The records show that warnings were given about the potential weed problem at the time the lake was being planned. Not until 1960 and 1961 were serious research efforts made toward control measures that would be feasible and safe for the lake (*10*). As a result of political turmoil in the area, no important measures for control or eradication have ever been taken.

Salvinia auriculata, first described as

Figure 5 A harbor on Lake Kariba completely choked with *Salvinia*. Only very powerful boats can penetrate the mat of weeds (photo taken October 1965).

from Guiana, has spread over a wide area in Central and South America, from Cuba to Argentina. It is known to be in the Cape area of South Africa, in the Congo River and the Cameroons on the West Coast, and in some of the countries of Southeast Asia. It first "exploded" and became a serious problem in Ceylon in the period before 1955 (*11*). At that time *Salvinia* had covered an estimated 8800 hectares of rice fields and 800 hectares of canals and other waterways within about 12 years.

Could the disaster at Lake Kariba have been avoided? The infestations of *Salvinia* in Ceylon were well known. Scheduled inspections of the shore line where weeds build up and prompt treatment of early infestations seem the only way to prevent massive explosions of vegetation which may by then be too expensive or too dangerous to treat. When man-made lakes are constructed, vegetation management must be integrated with the work of all bureaus and agencies concerned with maintenance of the watershed, the lake, and the dam. Recently, an infestation of *Salvinia* was found on Lake Naivasha in Kenya, and it was dealt with promptly and effectively with herbicides applied from aircraft and boats. Constant vigilance has kept the weed under control.

Water Lettuce

Lake Volta in Ghana deserves special mention. The dam was closed in 1964. When filling is completed, the lake will cover 8125 square kilometers. It will be the largest man-made lake in the world. By 1965 great rafts of water lettuce, *Pistia stratiotes,* some many kilometers long, could be seen floating on the surface. Extensive fringes of the weed covered scores of kilometers of the lake's edge and filled the inlets of small rivers entering the lake.

Water lettuce, very widely distributed in the world, is one of the free-floating aquatic plants that must be viewed with concern. It has pale green leaves which are broad and softly pubescent on both sides.

Figure 6 A permanent mat of *Salvinia* in a bay of Lake Kariba. The weed mat closes in the wake of boats and ships which try to enter.

The leaves occur in rosettes, beneath which are long fibrous roots, and the plants are connected by stolons. There is a tendency to overlook the potential danger from water lettuce on large water bodies such as Lake Volta because, presumably, it cannot tolerate action from large waves.

The most important problem caused by water lettuce is that of disease and the nuisance associated with mosquitoes. Water lettuce serves as a preferred host site for larvae of several species of mosquitoes. One or more of these species of mosquitoes serve as principal vectors of each of several forms of encephalomyelitis and of rural filariasis. The *Mansonia* larvae obtain their oxygen directly from the roots of water lettuce and never surface. The only way to control these mosquitoes is to remove water lettuce (*12*).

In an interesting experiment in which herbicides were used to destroy a 120-hectare infestation of water lettuce, there was also complete control of *Mansonia* mosquitoes for 4 months. Only an occasional mosquito of this species was trapped in the year following treatment (*13*). Demonstrations such as this suggest that suitable methods of destroying water lettuce may provide a means of controlling diseases that affect large numbers of people.

All of the foregoing species of plants are free-floating. There are many more, but perhaps the ones that deserve special mention are those of the duckweed family *Lemnaceae L.* A single frond may be the size of a pinhead, and there are no stems or true leaves. Members of this family propagate vegetatively by producing new individuals at the edge of the frond. They can cover the surface of an entire pond in a few weeks. Duckweeds are a nuisance in rice fields and cause trouble in irrigation systems by entering siphon tubes and pumps and by collecting on trashracks.

The free-floating species have drawn much attention if only because massive infestations are spectacular and also frightening when they begin to move. But these are by no means the most difficult of the aquatic weed problems.

Submersed Weeds

Submersed weeds are perhaps the most serious of all aquatic weed problems, because they cannot readily be sprayed with herbicides and do not easily lend themselves to clearance by machines. Herbicidal treatments must be made to the entire volume of water, depending upon sorption of the chemical by the undesirable species to achieve control. It must be obvious that the submersed weeds drastically reduce the rate of water flow. It is as though the water must pass through an infinite series of inverted combs that create friction and turbulence.

Of all aquatics, submersed weeds cause the greatest problems in the United States. The most troublesome in the west and north are the genera *Potamogeton* and *Elodea*, and in the east include these genera and *Myriophyllum*. In the south, *Najas* and *Ceratophyllum* are the most common. In addition to these, *Egeria* is widespread throughout the United States. The filamentous and branched algae are almost always present. Also present in the United States, but important on a world basis as well, are members of the genera *Ranunculus, Vallisneria,* and *Utricularia*.

Submersed weeds also have a history of rapid invasion of new sites. Many of the waterways of Europe were blocked by *Elodea canadensis* when it was introduced in the 18th century. One of the most alarming plant invaders is eurasian watermilfoil, *Myriophyllum spicatum* L. The plant, apparently in the United States since the 19th century, has become a problem only during the past 10 years. In that period it has invaded more than 80,000 hectares in the Chesapeake Bay, 2,000 hectares in the Tennessee Valley Authority reservoirs, and 26,800 hectares in Currituck Sound.

The weed causes large losses in commercial fishing, smothers shellfish beds, hinders navigation, depresses real estate values, interferes with recreational use, provides mosquito breeding sites, and clogs water intake systems. This is a perennial submersed plant that spreads very rapidly by vegetative reproduction and probably by seed. The leaves are finely dissected, and the flowering spike, without leaves, may extend 10 centimeters above the water. The plant can tolerate salinities up to as much as one-third that of seawater and can thus invade most fresh and estuarine waters. Large acreages in the TVA reservoirs have been treated with 2,4-D (2,4-dichlorophenoxyacetic acid). The chemical has been safely applied at rates of 22 to 44 kilograms per hectare. The results have varied from excellent control in protected embayments to poor control in moving water in these impoundments. The varied results of this work and the inadequacy of the control methods indicate the lack of knowledge to cope with the problem. The need is for concerned public agencies to contain new troublesome plants wherever they are first reported. In one area of Florida, eurasian watermilfoil spread from 80 to 1200 hectares in 2 years; there was no attempt to control the plant or limit it to the infested area (*14*).

The story of the invasion by eurasian watermilfoil is still another example of a species that multiplied rapidly after having been in a certain area for a long time. We do not know whether plant material was moved to a more favorable environment, or whether the habitat was altered to make it more suitable for the spread and growth of plants already present.

Submersed weeds are a menace to irrigation systems. Thousands of freshwater reservoirs, large and small, have been constructed throughout the world in the past two decades, and with them have come large new irrigation schemes. Several of these command more than 400,000 hectares. India, for example, the leading producer of three of the world's major crops, has more kilometers of irrigation canals than any other country. As people occupy and farm the land, the effluent from villages, barns, and animal yards, together with the runoff of fertilizers added to the fields, enriches the waters of the canals and reservoirs. Because the distributaries are often shallow, clear, and slow-moving, the added nutrients ensure the growth of weeds. When the weeds enter the system, water can no longer move at the design rate of flow, with the result that the fields most distant from the reservoir cannot be irrigated on schedule. The reduced rate of flow also encourages seepage from canals, and the losses from evaporation are significantly increased. These are matters for serious concern, because they directly affect food production in an already hungry world. The management of aquatic vegetation for an entire irrigation system may be the most complex of all water weed problems. These include reserves of water held in rivers, ponds, and lakes, a network of canals, and a drainage system. Frequently, the water is used for men, animals, and crops.

It is estimated that over a million additional hectares of arable land will be placed under irrigation before the end of the century. One example may serve to illustrate the futility of planning irrigation systems without, at the same time, making preparations to protect them from aquatic weeds. In one of the countries of Asia, an irrigation scheme with a command area of 560,000 hectares was completed in this decade. One main arm of the canal system is 400 kilometers long and, with its distributaries, totals more than 1600 kilometers. The discharge at the head is equal to that of the Seine River as it flows through Paris. Within 5 years after the system was opened submersed aquatic weeds had cut the flow of water in the main canal by 80 percent.

Man's efforts to increase food production in the United States have resulted in treatment of 105,833 kilometers of irrigation and drainage channels for aquatic weed control in 17 western states (*1*). This represents about 55 percent of the infested channels and about 45 percent of the total length in use. Greater world pressure for food production will increase the demand for even more efficient control of vegetation which interferes with water flow.

Man's quest for beauty and recreation has brought about some of the most catastrophic of our weed problems. Exotic and beautiful aquarium species are transported throughout the world. *Egeria, Cabomba, Elodea* and *Hydrilla* are examples of submersed plants that have been carried from one country to another with dire consequences (Figs. 7 and 8). The problem is twofold. First, in the United States and most other countries there are no limitations on the importation of aquatic plants. Second, most of the aquatic plant dealers grow and harvest the plants in public waters. When a grower desires another species, he simply imports it, places it in several streams or lakes, and then harvests it as it is needed. Many submersed weeds have been introduced in this manner. Failure to restrict their movement has allowed wide distribution of some species.

Emersed Weeds

Emersed aquatic weeds have their roots beneath the water and their stems and leaves above the surface. Many are familiar, including the bulrushes, *Scirpus;* cattails, *Typha;* water lily, *Nymphaea;* spatterdock, *Nuphar;* rushes, *Juncus;* arrowhead, *Sagittaria;* and alligatorweed, *Alter-*

Figure 7 A flood control canal 80 feet wide and 22 feet deep completely filled with *Hydrilla* retarded flow by 90 percent.

nanthera, Where water levels fluctuate these species may survive for short periods as terrestrial plants. Many are the first to invade newly flooded areas, and they will prosper in the backwater areas which are intermittently wet. Emersed weeds are especially troublesome in irrigation and drainage systems. They choke canals, increase silt deposition, and impede water flow. Frequently, the design rate of flow can never be achieved because of encroachment of plants on the shoreline and water lost through transpiration.

Phreatophytes

These are the plants that grow at the water's edge or with their roots reaching into the capillary zone overlying the water table. It is as difficult to determine limits for the species which should be in this group as it is to define the shoreline in a swamp. Woody plants, perennial grasses, and broadleaved plants are prominent in this group. There are an estimated 6 million hectares of stream channels, canals, reservoirs, and river flood plains infested with these weeds in the western part of the United States, and they waste 25 million acre-feet of water annually. It would be practical to save 25 percent of this wasted water by maintenance of stream channels and effective weed control *(15).*

The most common woody plants are salt cedar, *Tamarix pentandra;* willows, *Salix* spp.; cottonwood, *Populus* spp.; baccharis, *Baccharis* spp.; buttonbush, *Cephalanthus occidentalis* L.; velvet mesquite, *Prosopis juliflora;* and grease wood, *Sarcobatus vermiculatis.*

Floating Island Weeds

In accord with many natural growth cycles, large masses of floating dead aquatic vegetation may support progressively larger types of vegetation. As these enter the final phase of this ecological succession, woody species become established. Large islands bearing trees several feet in circumference are impossible to cope with when they lodge in the channels of main streams, for example.

One of the most dangerous and dreaded

Figure 8 This $900,000 resort motel and marina was built on the banks of a crystal-clear river famous for fishing and scuba diving. An invasion of *Hydrilla* prevented the use of the waterway and forced the motel into bankruptcy 18 months after opening.

weeds of the African waterways is papyrus, *Cyperus papyrus* L., of the sedge family Cyperaceae. It is feared because it may encroach on open water by extending from the banks. It is one of the principal plants in the formation of sudd which is a mass of free-floating vegetation. The plant grows upright to a height of 4 to 5 meters. The rhizomes by which it spreads vegetatively are woody and strong and may reach far out into open water or may quickly travel over weed mats or penetrate through them and thus knit them together. These massive and sturdy floating islands of vegetation, when loosed on a river at any season, can be a menance to navigation. Sudd formation in two of Africa's largest swamps, the Okavango in Botswana and the great swamp in the White Nile above Malakal, are dominated by papyrus. A study of the papyrus sudd of the White Nile revealed that 50 percent of the water entering the river was lost through evaporation and transpiration as a result of the activity of this weed and its associated vegetation.

Vossia cuspidata (Roxb.) W. Griff., a robust member of the grass family Gramineae is frequently mixed with the islands of associated plants. The weed is sometimes called hippo grass, because it is grazed by the hippopotamus as well as by cattle. It can have floating or submersed stalks which grow rapidly for considerable distances under favorable conditions. The aerial portion of the stem is strong and erect and may reach a height of 2 meters. The grass is widely distributed in Africa and Asia. The irrigation canals of the Gezira cotton scheme along the Blue Nile are sorely troubled with this grass because it impedes the flow of water.

Chemical Control

Many selective herbicides have been developed to combat aquatic weeds. Some have such a narrow range of specificity that one species of a genus many be controlled without affecting other species of the same genus. Thus certain problem species can be controlled without adversely affecting the desirable flora and fauna in a waterway. Formerly, large quantities of toxic chemicals, such as sodium arsenite, were used indiscriminately; but today the herbecides used in our waters must pass rigid tests on efficacy, toxicity to fauna and flora in and near the waterway, residues in irrigated crops, and many other hazards.

Chemical control of several floating weeds is possible. Water hyacinth is "managed" and kept under control in the United States with 2,4-D (2,4-dichlorophenoxyacetic acid). At one time the infestations of this weed on the White Nile and the Congo rivers in Africa were controlled with the same herbicide, but now political turmoil and economic problems have made it impossible to continue the control operations.

If sensitive crops are grown in agricultural areas adjacent to infested waterways, some herbicides may not be used because of the danger from volatility and spray drift. Recent development of invert emulsion (water in oil) formulations of 2,4-D, water thickening agents, and new innovations in spraying equipment (such as the microfoil boom) give the applicator new tools to control placement of the chemical. Diquat (6,7-dihydrodipyrido [1,2- 2 ,1 -*c*] pyrazidiinium dibromide) has proven effective on water hyacinth at 1.7 kilograms per hectare and can be used with greater safety around ornamentals and crop plants that are very sensitive to 2,4-D. Two applications of diquat usually give about the same control of this weed as would be expected from four treatments with 2,4-D. In Australia, amitrole-T (3-amino-1,2,4-triazole + ammonium thiocyanate) has been used effectively on water hyacinth (*16*).

Diquat is also effectively used to control water lettuce at the same rates as it is used to control water hyacinth. As these two plants often grow together, a single application of diquat can be used to control both species (Fig. 9).

The chemical control of submersed weeds, much more difficult, is achieved through the use of chemicals with several different modes of action. Large quantities of emulsified xylene and other aromatic solvents are used for contact control of vegetation in irrigation canals. The xylene is usually released into the flowing canal, over a 30-minute period, at 6 to 10 gallons per cubic foot per second flow. As the chemical moves down the canal, the plants absorb lethal doses, the treated portion of the water is spilled into a waste area, and the untreated flowing water is then available for immediate use. The treatment kills the vegetation back to the bottom mud, but regrowth may occur in a few weeks.

Acrolein (acrylaldehyde) is applied in flowing water in the same manner as aromatic solvents (*17*). In static water sites such as lakes, acrolein (7 parts per million by volume) is injected directly into the water. *Hydrilla* has been controlled with this method for 8 to 16 weeks in both the United States and Australia. Generally two to four treatments are required per year for satisfactory weed control.

Knowledge of the plant cycle has an important bearing on the control of submersed weeds. Sago pondweed, *Potamogeton pectinatus,* and *Hydrilla* produce hydrosoil propagules. In canals that can be dewatered, fenac (2,3,6-trichlorophenylacetic acid) and dichlobenil (2,6-dichlorobenzonitrile) can be applied to the dry canal bottom. The chemical is then leached into the surface soil and released into the water as growth begins. *Hydrilla* also produces large numbers of hydrosoil turions that sprout as soon as the topgrowth is killed. In areas that cannot be dewatered, successive treatments with fatty acid amine salts of endothall (3,6-endoxohexahydrophtalic acid) and diquat

Figure 9 Diquat was applied at 1.7 kilograms per hectare to control water lettuce (top) and resulted in complete control within 30 days (bottom).

reduce the number of these propagules to the point where regrowth is minimum (*18*).

Foliar treatments with dalapon (2,2-dichloropropionic acid), at 5.6 to 22.4 kilograms per hectare, will afford quite satisfactory control of most aquatic grasses, cattails, and rushes. Often two or more treatments are required in a growing season.

Amitrole-T is another selective herbicide that is effective on grasses. The response of shore weeds to these two herbicides provides us with an example of the interrelation of stage of development and herbicidal activity. Dalapon is most effective on cattails before they flower, while amitrole-T is more effective in the fall, after flowering. A combination or mixture of these two herbicides is often used as a single application that may be effective for several months and which may cover a broader spectrum of weeds. A mixture of dalapon and 2,4-D is sometimes used where control of both grass and broad-leaved species is necessary (*19*).

There are still many troublesome species for which there are no mechanical, chemical, or biological controls. For most of these we lack information on the physiology and the stages of growth and development. The pace has quickened in the search for ideas, chemicals, and methods for the selective control of aquatic vegetation. Modern herbicides may also be used safely to keep our waterways productive and useful. However, application of most herbicides to potable water is restricted. Some are toxic to fish. Information is needed on the disposition and effects of herbicides in water, in crops irrigated with treated water, and in fish. Because the cost is generally only 10 to 20 percent of other control methods, there is frequently no other practical way to manage the vegetation in a stream, a lake, or a power system.

We do not know whether we shall be able to restore our water resources. It is certain that we shall continue to use them, and many of them are destined to be misused for some time to come. Some types of aquatic vegetation will flourish; we shall find it disagreeable, and sometimes we shall be overwhelmed by its luxuriant growth. Until we have become wise enough to appreciate our water, and until we have come to respect it as one of the most precious of all the gifts of nature, aquatic herbicides will be needed —— for they will be one of the few tools that we can afford in the selective management of the vegetation in our waterways. In many places in the world there will be no other choice.

Biocontrol

Natural forces affect aquatic plant growths. These forces are used by man in his biological control programs (biocontrol). Biocontrol, which turns nature against herself, may be the most economical method for the control of portions of this perplexing world problem. Biocontrol has certain advantages, for example, relatively low program costs, ready supply sources, ease of application of techniques which often require no special equipment, minimal training of unskilled personnel, and relative permanence of treatments because of the ability to resist weed reinfestations (*20*). In this age of emphasis on chemical and mechanical control of aquatic weeds, most people are unaware of the progress being made in biocontrol. Emphasis on this area of weed control has increased rapidly during the last decade.

Biocontrol works best with agents of foreign origin. Scientists have traveled to the native homes of certain species of aquatic plants to collect various organisms that attack the plants. Frequently, in its native habitat, the plant never presents the problem that it does in other parts of the world. The reason for this is that certain natural agents, such as insects, diseases, and the chemistry of the water, have controlled the plant. For this reason the biocontrol of aquatic vegetation should not be confused with aquatic plant eradication. The plant and the biocontrol agent are part of the aquatic ecosystem.

In nature, a balance is maintained in the aquatic flora through plant-feeding insects, diseases, nematodes, fungi, bacteria, viruses, fish, snails, and mammals. When a biocontrol agent is selected it must be thoroughly investigated before introduction into a new area. It must not be released if it will attack desirable plants or other organisms.

A large freshwater snail *Marisa cornuarietis* has freed small ponds in the southern United States and Puerto Rico of submersed weeds (Fig. 10). *Marisa* carries no diseases of man and has been used as food in Puerto Rico. The indiscriminate feeding habits of *Marisa* have also made it a biocontrol agent for disease-carrying snails. A major disadvantage is its sensitivity to temperatures below 6°C. It may feed upon certain desirable plants such as rice, watercress, and water chestnuts which are growing in the water. A second freshwater snail, *Pomacea australis,* shows promise as a biocontrol for submersed and floating aquatic weeds. This large ampullarid is widely distributed in Brazil (*21*).

Mammals with aquatic weed-cleaning ability are rare, but the sea cow or manatee, *Trichechus manatus latirostris,* has shown its taste for many types of aquatic weeds. Unofficial reports collected from all areas of the world indicate that this mythological mermaid has no peer in the biological world for the volume of aquatic weeds which can be consumed by one animal. There is only limited use of the manatee because of the scarcity of animals, and an increase in population is uncertain because of the lack of knowledge about its reproduction (*22*).

In contrast to the indiscriminate feeding of snails and manatees is the fastidious taste of the alligatorweed beetle, *Agasicles* sp. In nature this small beetle feeds only on alligatorweed and will starve in its absence. No control technique could be simpler than the release of a handful of beetles into an infestation of alligatorweed. Recent laboratory research has also shown that *Salvinia* may be controlled by a wingless aquatic grasshopper, *Paulinia acriminate. Paulinia* is now being evaluated in field trials at Lake Kariba. Investigations are also underway to search for natural enemies of the water hyacinth in its native South American home. (*23*).

Figure 10 Freshwater snail, *Marisa,* feeding on *Hydrilla.*

Certain freshwater fish consume large quantities of aquatic vegetation. The common carp, *Cyprinus carpio*; Chinese grass carp or white amur, *Ctenopharyngodon idella*; tilapia, *Tilapia* sp.; and silver dollar fish *Metynnia* sp., are used for control of aquatic vegetation in many areas of the world. Many of these species are sensitive to cold weather and, in colder climates, must be overwintered in temperature-control tanks. The most promising of this group is the white amur. Ponds choked with *Hydrilla* in India were cleared of the weed in 2 months after they were stocked with 350 white amur per hectare (*24*). Russia, Poland, Czechoslovakia, China, and other countries have used the white amur as a biocontrol for aquatic weeds.

Ducks, geese, and swans can remove small amounts of vegetation, and they are especially effective in small ponds for the control of duckweed. Controlled grazing of cows, horses, and goats may be used to hold down vegetation along lake shorelines and canal banks (*25*).

Diseases, viruses, fungi, bacteria, and nematodes have received very little attention in aquatic weed biocontrol research. Two diseases are considered to be the cause of the decrease in eurasian watermilfoil in the United States (*26*). The diseases have not been identified and may be only an indirect cause of the decline in the milfoil. A disease was the direct cause of the decline of eel grass along the Atlantic Coast of the United States in the 1930's.

Low-growing species of aquatic plants, several centimeters in height, have been planted in canals to compete with the more undesirable species that may attain lengths of several meters. Removal of nutrients from the water environment by partitioning agents is also being considered as a biocontrol agent. Some success has been obtained in controlled submersed aquatic weeds with selective dyes or black plastic that filter out all, or selective portions of, the sunlight in water.

The employment of biocontrol agents in our aquatic weed programs offers a new approach to the solution of an old and aggravating problem. Combinations of biocontrol agents with chemical and mechanical methods of control may someday be the answer to aquatic vegetation management.

Mechanical Removal

The first effort to control aquatic weeds was with hand tools. Since 1900, many machines have been designed to perform such work, and some are so large that several barges are required for flotation. In the United States today there are still many canals and drainage systems maintained by power shovels and draglines, but these machines work slowly and maintenance costs are high. Emersed weeds are sometimes controlled with the use of underwater mowing machines (*27*). All of these types of mechanical equipment can bring temporary relief from weed infestations and sometimes provide channels through portions of an otherwise inaccessible waterway. When water is used for human or animal consumption, or when valuable crops are in the immediate area, there may be no practical way to control aquatic weeds except by mechanical removal. The ratio of costs for these methods as compared to approved herbicidal methods may be on the order of ten to one. (*8*).

Because large quantities of nutrients accumulate in the tissues of some aquatic weeds, the removal of a heavy stand may be beneficial to the waterway. The yield of such massive quantities of green vegetation raises the question of its value for food, for food amendments, or for improvement of soils. But such questions have already been raised, again and again, as to unwanted terrestrial plants (such as woody species that might be used for lumber and wood products) or as to perennial grasses that might be used for pasture. There are many reports of analytical work on water hyacinth and several other aquatic weeds (*28*). There is no shortage of information on percent dry matter, crude fiber and protein content, or carotene and other special constituents of these plants. Why, then, have aquatic weeds not been more widely used? The answer is that we lack economical ways of harvesting and processing large masses of plants with a very high water content. In certain countries the weed nurseries and other important sources of infestation are almost inacessible. Add to this the probable cost of transport of a finished product over long distances and it becomes easy to see why public and private agencies have not accepted these risks.

Summary

The great focal points of civilization placed their roots along streams and in sheltered harbors because man needed to be near water for navigation and housekeeping and because it provided protection. There was little thought of caring for the bountiful supplies of water which seemed endlessly renewable. Within living memory the supplies of fresh water in the beautiful streams and lakes of North America were legendary. But suddenly in the 20th century we have at last begun to sense that both the water and soil of the earth are limited. We now realize that we cannot run off to a clean new place each time we have fouled our nest. We shall have to learn to manage our affairs and our immediate environment. We are sickened by the spectacle of the trash and refuse of our own activity. We become uncomfortable in our role as stewards of this wonderful resource, for we do not understand how we may both use and protect it.

While we have been engaged elsewhere, the rampant growth of aquatic weeds has come to be one of the symptoms of our failure to manage our resources. We assign values to the depreciation of property, to the pollution of municipal water supplies, to the loss of navigable streams, and to the failure of irrigation and power systems because of aquatic weeds. But we must also judge the worth of clean water for man in other, quite different ways. Some of the loveliest places on earth are at the water's edge. These may be the sites of our dwellings or the places that we choose for rest and renewal. As we spoil these, one by one, we shall know that we have surrendered a great part of our humanness, and we shall be anxious because we cannot trust ourselves.

Now we must abandon our view that streams and lakes are great self-cleansing reservoirs that can receive our wastes forever and return to us always as cool, clear water. Many of the watercourses of Asia and Africa, and of Wisconsin and Florida, are now so fertile and so well innoculated with aquatic weeds that they can no longer correct themselves. In many of these places, it is now too late to talk about an equilibrium or the balance of nature. Human activity and neglect have driven the equation far to the right. Within the combinations of mechanical, biological, and chemical methods of aquatic weed control we can find the tools to help with their restoration. We can keep the waterways open so that they can be useful to man and so that he may enjoy them. We can buy the time we need to learn to manage not only the vegetation but each of the resources in an entire watershed.

References and Notes

1. F. L. Timmons, *U.S. Dep. of Agr. ARS-34-14 (1960).*
2. N. Wollman *et al., The Value of Water in Alternative Uses* (Univ. of New Mexico Press, Albuquerque, 1962).
3. C. A. Evans, *New Sci. 19, 666 (1963);* L. J. Mathews, *Pest Art. News Abstr.* **13, 7** (1967); V. Proctor, *Science* **160, 321** (1968).

4. J. Lebrun, *Bull. Agr. Congo* **50, 251** (1959).

5. E. T. Heinen and Salah el Din Hassan Ahmed, *Publication of the Information Production Center* (Dept. of Agriculture, Khartoum, Sudan, 1964).

6. C. E. Timmer and L. W. Weldon, *Hyacinth Contr. J.* **6, 34** (1967); W. T. Penfound and T. T. Earle, *Ecol. Monogr.* **18,** 447 (1948); R. R. Das, *Proc. Indian Sci. Congr.* **6, 445** (1969).

7. W. E. Wunderlich, *Hyacinth Contr.* **J. 1, 14** (1962); *U.S. House of Representatives, Document No. 91, Water Hyacinth Obstructions* (55th Congress, Third Session, 1899); *U.S. House of Representatives, Document No. 37, Water Hyacinth Obstructions in the Waters of the Gulf and South Atlantic States* (85th Congress, First Session, 1957).

8. D. B. Bogart, *Proc. Soil Sci. Soc. Fla.* **9, 32** (1948); J. C. Stephens, R. D. Blackburn, D. E. Seaman, L. W. Weldon, *Proc. Amer. Soc. Civil Eng. J. Irrig. Drainage Div.* **89, 31** (1963).

9. A. S. Boughey, *Adansonia* **3, 49** (1963).

10. E. R. Hattingh, *Weed Res.* **1, 303** (1961).

11. R. H. Williams, *Trop. Agr.* **33, 145** (1956).

12. J. A. Mulrennan, *Mansonia Mosquitoes,* mimeographed report (Div. of Entomology, Florida State Board of Health, Tallahassee); R. W. Chamberlain, W. D. Sudia, J. D. Gillett, *Amer. J. Hyg.* **70, 221** (1959); M. W. Provost, *Mansonia Studies in Leesburg in 1949* (Div. of Entomology, Florida State Board of Health, Tallahassee, 1949); W. D. Sudia and R. W. Chamberlain, personal communication; E. L. Seabrook, *Rep. Annu. Meet. Fla. Antimosquito Ass.* **21, 1** (1950); C. Chow, *Preliminary Note on Herbicides for Pistia Clearance as a Rural Filariasis Control Measure in Ceylon* (World Health Organization, Geneva, 1953).

13. L. W. Weldon and R. D. Blackburn, *Weeds* **15 (1), 5** (1967).

14. G. F. Beaven, *Summary of the 1962 Interagency Research Meeting on Eurasian Watermilfoil* (Natural Resources Institute, Univ. of Maryland, 1962); T. E. Crowell, J. H. Steenis, J. L. Sincock, *Proc. S. Weed Conf.* **20, 348** (1967); G. E. Smith, *ibid.* **15, 258** (1962); L. W. Weldon, R. D. Blackburn, D. S. Harrison, *U.S. Dep. Agr. Agr. Handb. No. 352* (1969); R. D. Blackburn and L. W. Weldon, *Hyacinth Contr.* **J. 6, 15** (1967).

15. H. C. Fletcher and H. B. Elmendorf, *Yearbook of Agriculture 1955* (U.S. Government Printing Office, Washington, D.C., 1955), pp. 423-429; F. L. Timmons and D. C. Klingman, *in Water and Agriculture,* R. D. Hockensmith, Ed. (AAAS, Washington, D.C., 1960), p. 157.

16. L. W. Weldon, R. D. Blackburn, H. T. DeRigo, R. T. Mellen, *Hyacinth Contr.* **J. 5, 12,** (1966); D. S. Harrison, R. D. Blackburn, L. W. Weldon, J. R. Orsenigo, G. F. Ryan, *Univ. Fla. Agr. Ext. Serv. Circ. No. 219B* (1966); W. T. Parsons, *U.S. Dep. Agr. Vermin Noxious Weeds Destruction Board Bull. No. 3a* (1967).

17. R. D. Comes, R. R. Yeo, B. F. Bruns, J. M. Hodgson, F. L. Timmons, L. W. Weldon, T. R. Bartley, W. D. Boyle, N. E. Otto, D. D. Suggs, *U.S. Dep. Agr. ARS 34-57* (1963); V. F. Bruns, R. R. Yeo, H. F. Arle, *U.S. Dep. Agr. Tech. Bull. No. 1299* (1964); F. L. Timmons, *Proc. Tex. Annu. Indust. Weed Contr. Conf.* **2, 1** (1967).

18. P. A. Frank, *J. Exp. Bot.* **17,** 546 (1966); ——— , R. H. Hodgson, R. D. Comes, *Weeds* **11,** 124 (1963); R. D. Blackburn, L. E. Bitting, L. W. Weldon, *Hyacinth Contr.* **J. 5,** 36 (1966).

19. F. L. Timmons, L. W. Weldon, W. O. Lee, *Weeds* **6,** 406 (1958); J. M. Hodgson, V. F. Bruns, F. L. Timmons, W. O. Lee, L. W. Weldon, R. R. Yeo, *U.S. Dep. Agr. Prod. Res. Rep. No. 60* (1962); F. L. Timmens, V. F. Bruns, W. O. Lee, R. R. Yeo, J. M. Hodgson, L. W. Weldon, R. D. Comes, *U.S. Dep. Agr. Tech. Bull. No. 1286* (1963).
20. J. B. Butler and F. F. Ferguson, *Proc. S. Weed Conf.* **21,** 304 (1968).
21. R. D. Blackburn and T. M. Taylor, *Abstr. Weed Sci. Soc. Amer.* (1968), p. **51;** F. F. Ferguson and J. R. Palmer, *Amer. J. Trop. Med. Hyg.* **7,** 640 (1958); F. F. Ferguson, J. Oliver-Gonzales, J. R. Palmer, *ibid.,* **p. 491;** Anonymous, *Agr. Res.* **16,** 8 (1968).
22. W. H. L. Allsopp, *Nature* **188, 762** (1960); G. C. L. Bertram and C. K. R. Bertram, *ibid.* **196** 1329 (1962); P. L. Sguros, T. Monkus, C. Phillips, *Proc. S. Weed Conf.* **18, 588** (1965).
23. L. A. Andres, *Abstr. Weed Sci. Soc. Amer.* (1968), p. **51;** F. D. Bennett, *Proc. S. Weed Conf.* **19, 497,** (1966); *Pest Art. News Sum. Sec. C* **13,** 304 (1967).
24. J. W. Avault, *Proc. S. Weed Conf.* **18, 145** (1965); M. T. Philipose, *Indian Livestock* **1(2),** 20, 34 (1963).
25. J. Levett, *World Crops* **12,** 58 (1960).
26. S. Bayley, H. Rabin, C. H. Southwick, *Chesapeake Sci.* **9 (3),** 173 (1968).
27. W. E. Wunderlich, *Hyacinth Contr.* **J. 1, 14** (1962); C. F. Zeiger *ibid.,* p. 16; M. E. Grinwald, *ibid.* **7, 31** (1968); W. D. Boyle and D. D. Suggs, *U.S. Dep. Interior Reg. I Bur. of Reclam. Misc. Publ.* (1960); G. C. Klingman, *Weed Control: As a Science* (Wiley, New York, 1961), pp. 323-324.
28. E. Little, *Publication No. PL:CP/20* (Crops Protection Branch, Food and Agricultural Organization of the United Nations, Rome, 1968).

29. Contribution from Department of Horticulture, University of Wisconsin, Madison; and cooperative investigations of the Crops Research Division, Agricultural Research Service, U.S. Department of Agriculture; the Central and Southern Florida Flood Control District; and the University of Florida Agricultural Experiment Station, Fort Lauderdale. Journal Series 3262.

31.

POPULATION RESOURCES AND TECHNOLOGY

HARRISON S. BROWN

Our basic problem, really, is not that of supporting comfortably the distressingly large numbers of people who ... will inevitably inhabit the earth in the decades ahead ... it is that of getting from here to there successfully with a minimum of chaos and human suffering. AUTHOR

In 1954 the United Nations estimated that by 1980 the population of the world would lie between 3.3 and 4.0 billion persons, and it gave 3.6 billion as the most likely figure. By 1965 the population had already reached 3.3 billion and the United Nations demographers were estimating that the 1980 world population would lie between 4.1 and 4.5 billion, with a medium of 4.3 billion. Thus, in the span of but a decade the estimate for the most likely population in 1980 was moved upward by about 20%.

This substantial change in estimate resulted directly from the fact that the world rate of population increase has itself increased to levels considerably higher than were deemed plausible in the early 1950's. During the last half-century the rate of population increase has more than doubled, moving from somewhat less than 1% per year to 2.1% per year today. It is quite conceivable that the rate of increase might grow eventually to about 3% per year, corresponding to a doubling of world population every 23 years.

Reprinted with permission from *Bioscience*, Vol. 18., No. 1, pp. 31-33, January 1968, as it appeared in Vol. II, *Proceedings of the XVIII International Horticultural Congress*, pp. 19-24.

Today, the population of human beings is somewhat over 3.3 billion persons. Projections made on the basis of plausible extrapolations of the rate of population increase indicate that world population might well grow to about 7.5 billion by the turn of the next century, only 34 years from now.

In 1950 many students of the population-resource situation concluded that a major world catastrophe was in the making. Today, when we view the growth of world population in relation to what mankind is doing about the situation, that catastrophe appears a near-certainty. We are faced by the fact that in large regions of the world gains in food production are being more than eaten up by increased population, with the result that we are now experiencing actual reductions in *per capita* food production in many nations in which most people are already hungry. Further, with respect to other aspects of economic development, increased rates of economic growth are being in substantial part wiped out by increased rates of population growth.

Time does not permit our going into the reasons for these ominous changes in any detail today. Instead, I would like to confine my remarks to a brief statement concerning steps the more economically well-off nations might take in the all-too-short time that is still available. I feel justified in doing so, for what we do or don't do in this regard is certainly bound to have profound effect upon the quality of human life.

Clearly, an improvement in the situation would result were we to decrease the rate of population growth. Equally clearly, an improved situation would result were we to increase the rate of economic development, including food production. What many of us have failed to appreciate fully, however, is that these two factors are strongly coupled. We have thought so much in terms of the potential magic that could be wrought by the "perfect contraceptive" that we have failed to appreciate that, in the absence of a greatly improved economic situation in the poorer regions of the world, it is doubtful that even the most perfect contraceptive imaginable would have a truly major effect. Oversimplified, Indian families must have some reasonable assurance of security in old age before they are likely to use a contraceptive, no matter how perfect it might be from the point of view of cost and convenience.

I hope that I am not misunderstood concerning this point. I believe that accelerated research on human fertility is essential. We need better contraceptive techniques and we need to understand far better than we do now the cultural problems involved in the individual decision to limit family size. Almost all of us applaud the recent relaxation within our own government and others which permits greater freedom in discussing contraceptive techniques, in carrying out educational programs concerning birth control, and in helping with the establishement of population policies. This change of attitude has been terribly important. I simply wish to underscore the fact that while all of this is necessary it is by no means sufficient. We cannot look upon contraception as a panacea which, if only the churches and the politicians approve, will enable us quickly and inexpensively to improve the world's lot. To repeat, techniques of birth control can help enormously, but only if coupled with sustained economic development at a rate which is considerably greater than that which the poorer nations of the world are experiencing today.

Economic development can be likened to a vast fabric of interwoven threads of different kinds, all of which are necessary yet none of which by itself is sufficient. To mention but a few of the more obvious components, economic development involves the evaluation and development of land and mineral resources, the construction and operation of power plants and factories, the improvement and expansion of farming, the development of transportation and communications systems, the construction of housing, the development of systems of education, as well as the elimination of certain political, social, and cultural institutions, the modification of others and the introduction of still others.

Here, as in the case of birth control, we must beware of the "quick fix," the gimmick, the magic solution. Too often we have looked to land reform or to increased fertilizer production or to the desalting of water as being "cure-alls" for agricultural illnesses. Similarly, it does little good to educate people unless their talents can be put to use; nor is it useful to build complex factories unless there are adequately trained persons available to operate them. Briefly, the economic development of the poorer nations requires massive, sustained, and balanced efforts over a broad front. Unfortunately, the efforts thus far have not been massive enough, nor have they been sustained, trained persons available to operate nor have they been adequately balanced.

It is absolutely clear that the poorer nations cannot, without help, extricate themselves from the vicious circle in which they now find themselves. That help must come predominantly from the industrial countries of Western Europe, North

America, and the western Pacific which, with less than one-fifth of the world's population, produce and consume more than one-half the world's product. At present, aid from these nations remains at about $6 billion a year.

George D. Woods, President of the World Bank, estimates that about $3 to $4 billion more of development finance could be effectively used by the developing countries at the present time. As time goes by, considerably more even than this could be absorbed. He stresses that the capacity of the industrial nations to provide aid also is rising and points out that the national income in these areas is increasing at the rate of $40 to $50 billion a year.

We are faced, however, by the sad fact that for the past 5 years aid contributions from these countries have remained static. Indeed, as a proportion of their incomes, aid contributions have decreased from 0.8% to 0.6%.

The amount of assistance the industrial countries are providing for the developing countries is far indeed from being commensurate with the magnitude of the crisis which confronts us. In the case of the United States, a part of the inadequacy of our own contribution stems from a failure on the part of our Congress to fully recognize the seriousness of the situation and take appropriate action to place the tactics and strategy of aid on as sound and continuing a basis as our defense organization. Most of us would agree that it would be far easier to use economic and technical aid to prevent new crises than to wait until crises appear and then resort to military force.

Certainly a nation which can afford to spend $50 billion annually for defense and $5 billion annually for space and which has a rapidly rising per capita income ought not shrink from thinking in terms of U.S. aid expenditures which are two, three, five, or even ten times greater than our present expenditure of about $3 billion annually. Nor should we look upon such massive injections of aid overseas as gifts. Quite apart from the effects of such investments on economic growth overseas, were they properly handled they could also have an accelerating effect upon our economy. The production of electrical generators, steel plants, and farm machinery for shipment overseas could keep our own people employed much as the production of military equipment does at the present time.

If aid programs are to be successful, we must appreciate the need for continuity and we must appreciate the time-scale involved. The economics of the poorer countries cannot be brought up to reasonable levels overnight. The changing of cultural and political patterns requires a great deal of time. Indeed, in our planning it is essential that we think in terms of 10, 25, and even 50 years.

Today, in the face of Congressional attitudes, programs are started, then often stopped without being given a reasonable chance. Personnel turnover is rapid. It is extremely difficult to maintain any real continuity either of programs or people. More than in any area of government activity with which I am familiar, the attitudes of Congress inhibit the initiation of imaginative and creative programs. Equally important, it is difficult to plan ahead on a time-scale appreciably longer than the magical 2-year Congressional cycle. There are so many pressures in the direction of making a good showing within a short time, our aid efforts have tended to emphasize short-term projects at the expense of long-term efforts. As a result, our overall showing during the last two decades has been remarkably poor, particularly when compared with the need. I often feel that if we had taken at least a 20-year view of the problem when we first started our assistance efforts abroad, they would now be much further ahead than they actually are.

The problem of adequate balance in our aid efforts is equally critical. A banker tends to think in banker's terms; an agriculturalist tends to think in terms of agriculture; an educator tends to think in terms of education. As a result of the tugging and hauling, our aid programs in some countries are often analogous to the proverbial horse which was designed by a committee and which turned out to be a giraffe.

Here I would like to stress that our Agency for International Development is itself far less to blame for these troubles than is our Congress. I know many dedicated, selfless men in the organization who work day and night trying to keep essential programs going and to make sense of a situation the magnitude of which dulls the senses. But the organization and its predecessors have been beaten, trampled upon, and derided to such an extent over the years, I sometimes wonder that it is able to function at all. Indeed, I find it little short of amazing that it is still able to carry out meaningful programs.

Economic development can be effectively blocked at any one of numerous points by failure to fully anticipate needs. The economy of a given nation might improve with breathtaking speed, only to be halted by a lack of trained manpower able to assume supervisory positions. Agricultural production might increase at a satisfactory rate, only to be halted by lack of adequate agricultural research. We must always be on the lookout for the missing nail which might lead to our losing the shoe, the horse, the king, and eventually the war.

I would now like to take advantage of being with a group of fellow-scientists to draw attention to one of the "nails" which has been almost completely absent from most country-development programs. The missing nail is the development of adequate indigenous scientific-technological competence.

As a nation develops economically, it is certain to be faced by a multiplicity of problems of a technical nature which are peculiar to that country. If those problems are to be solved, the nation must have available its own group of applied scientists, its own facilities for research and development, and administrative organizations which will enable industry and government to mobilize research efforts effectively. This "problem-solving competence" must be backed by competent scientific and engineering staffs in the universities and by a research tradition and a scientific attitude. It must also be backed by an appreciation within the government itself of the need to support and nurture scientific acitvity.

Unfortunately, the need for this "problem-solving competence" is often completely ignored – it being more or less tacitly assumed that technology can be transferred intact and that should problems arise they can be solved in Europe or in the United States. Yet it would appear obvious that the technical problems of developing the Amazon basin must of necessity be solved in Brazil; they certainly can't be solved in Iowa.

Finally, I would like to stress that the economic development of the poorer nations will place a tremendous drain upon the earth's available resources. We have seen that by the year 2000 the population of the world might well grow to 7½ billion persons, about 6 billion of whom will be living in the presently underdeveloped nations of the world. If by some miracle all these persons were to be brought up to the level of living now enjoyed by the people of the United States, we would need to extract from the earth over 50 billion tons of iron, one billion tons of copper, an equal amount of lead, over 600 million tons of zinc, and nearly 100 million tons of tin in addition to huge quantities of other substances. These quantities are several hundred times the present world annual rates of production. Their extraction would virtually deplete the earth of all high-grade mineral resources and would necessitate our living off the leanest of earth substances: the waters of the sea and ordinary rock.

Although depletion of the earth's buried high-grade resources should be a matter of serious concern to us all, it is by no means an immediate threat nor is it an insurmountable problem. Given the necessary technology, which I am convinced can be developed, the earth has ample resources to enable persons the world over to lead abundant lives, even were the population to grow eventually to a considerably higher level than that now anticipated in the year 2000. In principle we can feed, clothe, and support a population of, say 10 billion persons, at a reasonably high level of consumption for an indefinitely long period of time.

Our basic problem, really, is not that of supporting comfortably the distressingly large numbers of people who we now know will inevitably inhabit the earth in the decades ahead. I am convinced that technically this can be done. Our basic problem is that of getting from here to there successfully with a minimum of chaos and human suffering. We are called upon to make the most difficult transition the community of man has yet been called upon to make – that of bringing birth rates in balance with rapidly lowering death rates on a world-wide scale. At the same time we are called upon to rapidly improve the economic well-being of the world's people. Although this transition is feasible from a technological point of view, it might well turn out to be impossible politically.

Here indeed is the great potential tragedy of our age! Our science and our technology have given us the power to create a world in which virtually all people can lead free and abundant lives. We have the power to create a civilization as yet undreamed of in its beauty and its accomplishments. Yet somehow we can't seem to organize ourselves to use that power effectively to solve mankind's basic problems.

Will future generations point to ours as that which made possible the realization of this higher level of human culture? Or will they point to ours as the generation which failed humanity at the most critical period of its history?

I fear that there is no middle ground.

32.

Beyond Family Planning

Bernard Berelson

The fact is that, despite the large forward strides taken in international recognition of the problem in the 1960's, there still does not exist an informed, firm, and constant conviction in high circles that this is a matter with truly great implications for human welfare. AUTHOR

This article rests on four propositions: (i) among the great problems on the world agenda is the population problem; (ii) that problem is most urgent in the developing countries, where rapid population growth retards social and economic development; (iii) there is a time penalty on the problem in the sense that, other things being equal, anything not done sooner may be harder to do later, due to increased numbers; and accordingly (iv) everything that can properly be done to lower population growth rates should be done, now. The question is, what is to be done? There is a certain agreement on the general objective (that is, on the desirability of lowering birth rates, though not on how far and how fast), but there is disagreement as to means.

The first response to too high growth rates deriving from too high birth rates is to introduce voluntary contraception on a mass basis, or try to. Why is family planning the first step taken on the road to population control? Probably because, from a broad political standpoint, it is the most acceptable one; since it is closely tied to maternal and child care it can be perceived as a health measure beyond dispute, and since it is voluntary it can be justified as a contribution to the effective personal freedom of individual couples. On both scores, it ties into accepted values and thus achieves political viability. Moreover, it is a gradual effort and an inexpensive one, both of which features contribute to its political acceptability.

How effective have family-planning programs been as a means toward population control? There is currently some controversy among qualified observers as to its efficacy (*1*), and this is not the place to review that issue. There is agreement, however, that the problem is of such magnitude and consequence that additional efforts are needed to reach a "solution," however that is responsibly defined.

For the purpose of this article, then, let us assume that today's national family-planning programs, mainly based on voluntary contraception, are not "enough" – where "enough" is defined not necessarily as achieving zero growth in some extended present but simply as lowering birth rates quickly and substantially. "Enough" begs the question of the ultimate goal and only asks that a faster decline in population growth rates be brought about than is presently being achieved or in prospect – and, within the range of the possible, the faster the better (*2,3*). Just to indicate roughly the order of magnitude, let us say that the proximate goal is the halving of

Reprinted from *Science*, Vol. 163, pp. 533-543, February 7, 1969.

the birth rate in the developing countries in the next decade or two – from, say, over 40 births per thousand per year to 20 to 25 (*4*). For obvious reasons, both emigration and increased death rates are ruled out of consideration.

What is to be done to bring that reduction about, beyond present programs of voluntary family planning? I address that question in two ways: first, by listing the programs or policies more or less responsibly suggested in recent years for achieving this end; second, by reviewing the issues raised by the suggested approaches.

Proposals Beyond Family Planning

Here is a listing of the several proposals, arranged in descriptive categories. The list includes both proposals for consideration and proposals for action.

A. *Extensions of Voluntary Fertility Control.*

1) Institutionalization of maternal care in rural areas of developing countries: a feasibility study of what would be required in order to bring some degree of modern medical or paramedical attention to every pregnant woman in the rural areas of five developing countries, with professional backup for difficult cases and with family-planning education and services a central component of the program, aimed particularly at women of low parity (*5*).

2) Liberalization of induced abortion (*6;7*,p. 139: *8*).

B. *Establishment of involuntary fertility control.*

1) Mass use of a "fertility control agent" by the government to regulate births at an acceptable level. The "fertility control agent," designed to lower fertility in the society to a level 5 to 75 percent below the present birth rate, as needed, would be a substance now unknown but believed to be available for field testing after 5 to 15 years of research work. It would be included in the water supply in urban areas and administered by "other methods" elsewhere (*9*). A related suggestion is the "addition of temporary sterilants to water supplies or staple food" (*10*).

2) "Marketable licenses to have children", given to women and perhaps men in "whatever number would ensure a reproduction rate of one" (say, 2.2 children per couple). For example, "the unit certificate might be the 'deci-child,' and accumulation of ten of these units, by purchase, inheritance or gift, would permit a woman in maturity to have one legal child" (*11*).

3) Temporary sterilization of all girls by means of time-capsule contraceptives, and of girls and women after each delivery, with reversibility allowed only upon governmental approval. Certificates of approval would be distributed according to national popular vote on desired population growth, and saleable on the open marked (*12*).

4) Compulsory sterilization of men with three or more living children (*13*); a requirement of induced abortion for all illegitimate pregnancies (*6*).

C. *Intensified educational campaigns.*

1) Inclusion of educational materials on population in primary and secondary school systems (*6, 14, 15*).

2) Promotion of national satellite television systems for directly disseminating information on population and family planning and for indirectly promoting acceptance of modern attitudes and practices in general (*7,* p. 162;16; *17,* especially pp. 13-14; *18*).

D. *Incentive programs.* As used here, the term *incentive programs* refers to payments, or their equivalent, made directly to couples who use contraceptives or to couples who do not have children for specified periods. It does *not* refer to payments to field workers, medical personnel, volunteers, and others, for securing acceptance of contraceptive practice.

1) Payment, or the equivalent (for ex-

ample, the gift of a transistor radio), for accepting sterilization (*6, 19-21*) or for the effective practice of contraception (*21-24*).

2) A bonus for child spacing or nonpregnancy (*25-28*); a savings certificate to couples for each 12-month period in which no child is born (*29*); a lottery scheme for preventing illegitimate births among teenagers in a small country (*30*); "responsibility prizes" for each 5 years of childless mariage or for vasectomy before the birth of a third child, and special lotteries, with tickets available to the childless (*7*, p. 138).

E. *Tax and welfare benefits and penalties* that would discourage childbearing rather than encourage it, as present systems tend to do.

1) Withdrawal of maternity benefits, perhaps after the birth of *N* (3?) children (*6,21,26*) or in cases where certain limiting conditions, such as adequate child spacing, knowledge of family planning, or attainment of a given level of income, have not been met (*31*, pp. 130-31).

2) Withdrawal of child or family allowances, perhaps after the birth of *N* children (*6; 26; 31*, pp. 131-36).

3) Levy of tax on births after the *N* th child (*21; 26; 28*, p. 30).

4) Limitation of governmentally provided medical treatment, housing, scholarships, loans, subsidies, and so on, to families with fewer than *N* children (*6, 26*).

5) Reversal of tax benefits, to favor the unmarried and the parents of fewer rather than more children (*6; 7*, pp. 136-37; *21; 26; 31*, p. 137; 32).

6) Provision by the state of *N* years of free schooling, at all levels, to each family, to be allocated among the children as desired (*33*).

7) Pensions for poor parents with fewer than *N* children, as social security for their old age (*21, 34, 35*).

Shifts in social and economic institutions – that is, broad changes in fundamental institutional arrangements that could have the effect of lowering fertility.

1) Raising the minimum age at marriage, through legislation or through imposition of a substantial fee for marriage license (*6, 32*); through direct payment of bonuses for delayed marriage (*25*); through payment of marriage benefits only to parents of brides over 21 years old (*31*, p. 130); through government loans for wedding ceremonies when the bride is over a given age, or with the interest rate inversely related to the bride's age (*36*); through a "governmental 'first marriage grant' . . . awarded each couple in which the age of both [sic] partners was 25 or more" (*7*,, p. 138); or through establishment of a domestic "national service" program for all men for the appropriate 2-year period in order to develop social services, inculcate modern attitudes toward (among other matters) family planning and population control, and delay marriage (*37*).

2) Measures to promote or require the participation of women in the labor force (outside the home), in order to provide roles and interests for women that are alternative or supplementary to marriage (*6, 32, 38*).

3) "Direct manipulation of family structure itself – planned efforts at deflecting the family's socializing function . . . or introducing nonfamilial distractions . . . into people's lives," specifically through employment of women outside the house (*39*); "selective restructuring of the family in relation to the rest of society" (*6*).

4) Promotion of "two types of marriage, one of them childless and readily dissolved, and the other licensed for children and designed to be stable"; marriages of the first type would have to constitute 20 to 40 percent of the total in order to allow free choice of family size for mariages of the second type (*16, 40*).

5) Encouragement of long-range social trends leading toward lower fertility – for example, "improved and universal general

education, or new roads facilitating communication, or improved agricultural methods, or a new industry that would increase productivity, or other types of innovation that may break the 'cake of custom' and produce social foment" (*41*); improvement in the status of women (*42*).

6) Efforts to lower death rates even further, particularly infant and child death rates, in the belief that lower birth rates will follow (*43*).

G. *Political channels and organizations.*

1) U.S. insistence on "population control as the price of food aid," with highly selective assistance based thereon, and exertion of political pressures on governments or religious groups that impede "solution" of the population problem (*7*, pp. 161-66; 44).

2) Reorganization of national and international agencies to deal with the population problem: within the United States, "coordination by a powerful governmental agency, a Federal Department of Population and Environment . . . with the power to take whatever steps are necessary to establish a reasonable population size" (*7*, p. 138; *45*); within India, creation of "a separate Ministry of Population Control" (*46*, p. 96); development of an "international specialized agency larger than WHO to operate programs for extending family limitation techniques to the world. . . charged with the responsibility of effecting the transfer to population equilibrium" (*16*).

3) Promotion of zero growth in population as the ultimate goal, and acceptance of this goal now in order to place intermediate goals of lowered fertility in proper context (*6*).

H. *Augmented research efforts.*

1) More research on social means for achieving necessary fertility goals (*6*).

2) Focused research on practical methods of sex determination (*47*).

3) Increased research directed toward improvement of the contraceptive technology (*48*).

Proposals: Review of the Issues

Here are 29 proposals beyond family planning for dealing with the problem of undue population growth in the developing world. Naturally I cannot claim that these are all the proposals that have been made more or less responsibly toward that end, but my guess is that there are not many more and that these proposals are a reasonably good sample of the total list.

Since several of the proposals tend in the same direction, it seems appropriate to review them against the criteria that any such proposals might be required to meet. What are such criteria? There are at least six:(i) scientific, medical, and technological readiness; (ii) political viability; (iii) administrative feasibility; (iv) economic capability; (v) moral, ethical, and philosophical acceptability; and (vi) presumed effectiveness. In other words, the key questions are: Is the scientific, medical, technological base available or likely? Will governments approve? Can the proposal be administered? Can the society afford the proposal? Is it morally acceptable? And, finally, will it work?

Scientific, medical, technological readiness. Two questions are involved: (i) is the needed technology available? and (ii) are the medical or paramedical personnel needed in order to assure medical administration and safety available or readily trainable?

With regard to temporary contraception, sterilization, and abortion, not only is the needed technology available now, but it is being steadily improved and expanded. The intrauterine device (IUD) and the oral pill have been major contraceptive developments of the past decade, and several promising leads are now being followed up (*49*), though it cannot be said with much confidence that any of the efforts will produce measures suitable for mass use within the next few years (*50*). Improved technologies for sterilization, both male and female, are being worked on, and there has been a recent development in abortion technique, the so-called suction device.

However, neither Ehrlich's "temporary sterilants" nor Ketchel's "fertility control agent" (B-1) is now available or on the technological horizon, though that does not mean that the research task ought not to be pursued against a subsequent need, especially since any such substance could be administered to individuals on a voluntary basis as well as to the population as a whole on an involuntary basis. In the latter case, if administered through the water supply or a similar source, the substance would have to be medically safe and free of side effects for men and women, young and old, well and ill, physiologically normal and physiologically marginal, as well as for animals and perhaps plants. As some people have remarked, the proposal that such a substance be added to a water supply would face far greater difficulties of acceptance, simply on medical grounds, than the far milder proposals with regard to fluoridation to prevent tooth decay.

Though a substantial technology in fertility control does exist, that does not necessarily mean that the techniques can be applied where they are most needed; this is true partly because of limitations in the number of trained personnel. In general, the more the technology requires the services of medical or paramedical personnel (or, what is much the same, is perceived as requiring them), the more difficult it is to administer in the developing countries. In the case of sterilization and abortion, the medical requirement becomes more severe. For example, when the policy of compulsory vasectomy of men with three or more children was first being considered in India (see *13*), it was estimated that the policy would affect about 40 million males: "one thousand surgeons or parasurgeons each averaging 20 operations a day for five days a week would take eight years to cope with the existing candidates, and during this time of course a constant supply of new candidates would be coming along" (*51*) – at present birth rates, probably some 3.5 million a year. A program of large-scale abortion (provided such a program was legal and acceptable) might additionally require hospital beds, which are in particularly short supply in most developing countries. However, the newer abortion technique might not require hospitalization – theoretically,the abortion "camp" may be feasible, as the vasectomy "camp" was, though the problems are substantially greater.

In short, the technology is available for some but not for all current proposals, and the case is similar for properly trained personnel.

Political viability. The "population problem" has been increasingly recognized by national governments and international agencies over the past decade, and policies for dealing with it have been increasingly adopted: national family-planning programs in some 20 to 25 countries; positive resolutions and actions within the United Nations family; large programs of support by such developed countries as the United States and Sweden; the so-called World Leaders' Statement, in which 30 heads of governments endorsed efforts to limit population growth. There is no reason to think that the trend toward population limitation has run its course.

At the same time, the political picture is by no means unblemished. Some favorable policies are not strong enough to support a vigorous program, even one limited to family planning on health grounds; in national politics, "population control" can become a handy issue for a determined opposition; internal ethnic balances are sometimes delicately involved, with political ramifications; national size is often equated with national power, from the standpoint of international relations and regional military balances; the motives behind the support and encouragement of population control by the developed countries are sometimes perceived as neocolonialist or neoimperialist; and on the international front, as represented by the United Nations, there is still considerable reluctance based on both religious and political considerations. In short, ambivalence on the part of the elite and recognition of the issue as a political liability are not absent even in the countries that favor population limitation.

Any social policy adopted by government rests on some minimum consensus concerning goals and means. They need not be the ultimate goals or the final means; the socioeconomic plans of developing countries are typically 5-year plans, not 20- or 40- or 100-year plans. Indeed, the ultimate goal of population policy – that is, zero growth – need not be agreed upon or even considered by officials who *can* agree upon the immediate goal of lowering growth by a specified amount or by "as much as possible" within a period of years. And since there are always goals beyond goals, one does not even need to know what the ultimate goal is – only the direction in which it will be found (which is usually more readily agreed upon). Would insistence *now* on the acknowledgment of an *ultimate* goal of zero growth advance the effort or change its direction?

To start with, the proposal of compulsory controls in India in 1967 (B-4) precipitated "a storm of questions in Parliament" (*52*); the proposal was withdrawn, and the issue resulted in a high-level shift of personnel within the family-planning organization. No other country has seriously entertained the idea. Other considerations aside, in many countries political instability would make implementation virtually impossible.

Social measures designed to affect the birth rate indirectly – for example, tax benefits, social security arrangements, and so on – have been proposed from time to time. In India there have been several such proposals: for example, by the United Nations mission (*53,* chap. 11), by the Small Family Norm Committee (*26*), by the Central Family Planning Council (*54*), and in almost every issue of such publications as *Family Planning News, Centre Calling,* and *Planned Parenthood.*

As Samuel reports, with accompanying documentation (*21*), "the desirability of imposing a tax on births of fourth or higher order has been afloat for some time. However, time and again, the suggestion has been rejected by the Government of India." In some cases action has been taken either by the central government [for example, income tax "deductions for dependent children are given for the first and second child only" (*53,* p.87] or by certain states ["Maharashtra and Uttar Pradesh have decided to grant educational concessions and benefits only to those children whose parents restrict the size of their families" (*55*)].Indicative of political sensitivity is the fact that an order withdrawing maternity leave for nonindustrial women employees with three or more living children – at best a tiny number of educated women – was revoked before it went into effect (*56*). There is a special political problem in many countries, in that economic constraints on fertility often turn out in practice to be selective on class, racial, or ethnic grounds, and thus exacerbate political tensions. Moreover, the promotion of female participation in the labor force runs up against the political problem that such employment would be competitive with men in situations of already high male unemployment and underemployment.

Whether programs for eliminating population growth are or are not politically acceptable appears to depend largely upon whether they are perceived as positive or negative; where"positive" means that they are seen as promoting not only population limitation but other social benefits as well, and where "negative" means that they are seen as limited to population control. For example, family planning programs, as noted above, are often rationalized as contributing to both maternal and child health and to the effective freedom of the individual family; a pension for the elderly would have social welfare benefits as well as indirect impact upon family size, in countries where a large family has been the traditional "social security system"; contraceptive programs in Latin America are promoted by the medical community as a medical and humanitarian answer not to the population problem but to the extensive illegal and dangerous practice of abortion. On the other hand, imposing tax liabilities or withdrawing benefits after the birth of the *N*th child, not to mention involuntary measures, can be attacked as punitive means whose only purpose is that of limiting population.

It would thus require great political courage, joined to very firm demographic convictions, for a national leader to move toward an unpopular and severe prescription designed to cure his country's population ills. Indeed, it is difficult to envisage such a political move in an open society where a political opposition could present a counter view and perhaps prevail.

The governmental decisions about measures to be taken to deal with undue population growth must be made mainly by the countries directly involved; after all, it is their people and their nation whose prospects are most centrally affected. But in an interconnected world, with peace and human welfare at issue, others are properly concerned, for reasons both of self-interest and of humanitarianism — other governments from the developed world, the international community, private groups. What of the political considerations in this connection?

A recommendation (G-1) that the United States exert strong political pressures to effect population control in developing countries seems more likely to generate political opposition abroad than acceptance. It is conceivable that such measures might be adopted here, but it is hardly conceivable that they would be agreed to by the proposed recipients. Such a policy seems likely to boomerang against its own objective, quite aside from ethical or political considerations.

The proposal (G-2) to create an international superagency seems more likely of success, but is not without its difficulties. The World Health Organization, UNICEF, and UNESCO have moved some distance toward family planning, if not population control, but only slowly and in the face of considerable political restraint on the international front (*57*). A new international agency would find the road easier only if its efforts were restricted to the convinced countries. Certainly the international organizations now concerned with this problem would not be expected to abdicate in favor of a new agency. If it could be brought into being and given a strong charter for action, then, almost by definition, the international political climate would be such as to favor action by the present agencies, and then efficiency and not political acceptability would be the issue.

Administrative feasibility. Given technical availability and political acceptability, what can actually be done? This is where several "good ideas" run into difficulties in the developing world, in the translation of a theoretical idea into a practical program.

It is difficult to estimate the administrative feasibility of several of the proposals listed above, if for no other reason than that the proponents do not put forward the necessary organizational plans or details. How are "fertility control agents" or "sterilants" to be administered on an involuntary mass basis in the absence of a central water supply or a food-processing system? How are men with three or more children to be reliably identified in a peasant society and impelled to undergo sterilization against their will; and what is to be done if they decline, or if a fourth child is born? What is to be done with parents who evade the compulsory programs, or with the children born as a result of this evasion? How can an incentive system be honestly run in the absence of an organized network of offices positioned and staffed to carry out the regulatory activity? How can a system of social benefits and penalties, including incentives to postpone or forego marriage, be made to work in the absence of such a network?

These questions are meant only to suggest the kinds of difficulties that must be taken into account if proposals are to be translated into programs. It would seem desirable that every responsibly made proposal address itself to such administrative problems. Some proposals do move in that direction. The feasibility in administration, personnel, and costs of the plan (A-1) to institutionalize maternal care in rural areas, with family planning attached, is currently under study in several developing countries.

The plan (C-1) to include population as a subject in the school curriclum has been carried forward as far as the preparation of educational materials, and in a few cases beyond that (*58*). The plans for incentive programs sometimes come down to only the theoretical proposition that people will do anything for money (in this case refrain from having children), but in some cases the permissible payment is proposed on the basis of an economic analysis, and in a few cases an administrative means is also proposed (*59*). The plan for governmental wedding loans scaled to the bride's age recognizes that a birth-registration system might be needed to control against misreporting of age (*6*).

Thus the *why* of population control is easy, the *what* is not very hard, but the *how* is difficult. We may know that the extention of popular education or an increase in the number of women in the labor force or a later age at marriage would all contribute to population control in a significant way. But there remains the administrative question of how to bring those developments about. In short, several proposals assume workability of a complicated scheme in a country that cannot now collect its own vital statistics in a reliable manner. Moveover, there is a limit to how much administrative burden the typical developing country can carry: it cannot manage many large-scale developmental efforts at a time, either within the field of population or overall. After all, population is not the only effort; agriculture, industry, education, health, communications, the military — all are important claimants. And, within the field of population, a country that finds it difficult to organize and run a family-planning program will find that harder when other programs are added. So, difficult administrative choices must be made.

Economic capability. From the standpoint of economic capability there are two questions: (i) is the program worthwhile when measured against the criterion of economic return, and (ii) if worthwhile, can it be afforded from present budgets?

Most of the proposals probably pass the second screen. If a fertility-control agent suitable for mass administration becomes available and politically and administratively acceptable, such a program would probably not be prohibitively expensive; incorporation of population materials into the school curriculum is not unduly expensive; imposing of taxes or withdrawing of benefits or increasing fees for marriage licenses might even return a net gain after administrative cost.

But a few proposals are costly in absolute if not relative terms. For example, the institutionalization of maternal care (proposal A-1) might cost some \$500 million for construction and \$200 million for annual operation in India, or, respectively, \$25 million and \$10 million in a country with population of 25 million (*5*) (although recent estimates are substantially lower). The plan for a "youth corps" in India would cost upward of \$450 million a year if the participants were paid only \$50 annually. The plan for payment of pensions to elderly fathers without sons could cost from $400 million to $1 billion a year, plus administrative costs (*35*). The satellite television system for India would cost $50 million for capital costs only, on a restricted project (*17*, p. 23), with at least another $200 million needed for receiving sets, broadcast terminals, and programming costs if national coverage were to be secured. All of these proposals are intended to have beneficial consequences beyond population control and hence can be justified on multiple grounds, but they are still expensive in absolute amounts.

The broad social programs of popular education, improved methods of agriculture, and increased industrialization (F-5) already absorb even larger sums, and they could no doubt utilize even more. Here the question is a different one. At present, in such countries as India, Pakistan, South Korea, and Turkey, the funds allocated to family-planning programs constitute less than 1 percent – in most cases, much less – of the total funds devoted to economic development. Would that tiny proportion make a greater contribution to population control, over some specified period, if given over to education or industrialization or road-building than it makes when utilized directly for family planning (*60*)? From what we now know, the answer is certainly "No."

Beyond family planning, the situation is still less clear. On the assumption that some level of incentive or benefit would have a demographic impact, what would the level have to be to cut the birth rate by, say, 20 percent? We simply do not know: the necessary experiments on administration and effectiveness have not been carried out. Let us review what has been proposed with respect to incentives. On the ground that incentives for vasectomy are better than incentives for contraception – since vasectomy is a one-time procedure and is likely to be more effective in preventing births – Pohlman (*20*) proposes for India a range of money benefits depending upon parity and degree of acceptance – from $7 to a father of four or more children if half the villagers in that category enter the program up to $40 to a father of three children if 75 percent enter. If the 50 percent criterion were met in both categories throughout India, the current plan would cost on the order of $260 million in incentives alone, apart from administrative costs. The decline in the birth rate would be slightly over a fourth, perhaps a third – roughly equivalent to $35 to $40 per prevented birth (*61*).

Simon proposes an incentive of half the per capita income "each year to each fertile woman who does not get pregnant" (*23*). Here a special popularises. In a typical developing population of 1000, about 25 to 30 percent of the married women of reproductive age give birth each year: a population of 1000 means from 145 to 165 such women, and a birth rate of, say, 40. Thus, the incentives paid to about three-fourths of the married women of reproductive age would have no effect on the birth rate, since these women would not be having a child that year in any case; thus the cost could be three to four times the amount "needed" for a desired result. Even if the incentive were fully effective and really did prevent a birth, a cut of ten points in the Indian birth rate would cost on the order of $250 million (or 5 million prevented births at $50 each). The cost would be substantially larger if the women (including the nonfecund or the semi-fecund) who would not have had a child that year in any case, could not be screened out effectively.

But these and other possibilities are only speculations: to date we simply do not know whether incentives will lower a birth rate, or, rather, we do not know how large the incentives would have to be in order to do so. These illustrations show only that an incentive program could be expensive. In any case, incentive systems would require a good amount of supervision and record keeping; and, presumably, the higher the incentive (and hence the greater the chance of impact), the greater the risk of false reporting and the greater the need of supervision — which is not only expensive but difficult administratively.

Moral, ethical, and philosophical acceptability. Next, is the proposal not only politically acceptable but considered right and proper — by the target population, government officials, professional or intellectual elites, and the outside agencies committed to aid in its administration.

Coale states (*3,62*), "One reason the policy of seeking to make voluntary fertility universal is appealing — whether adequate or not — is that it is a natural extension of traditional democratic values: of providing each individual with the information he needs to make wise choices, and allowing the greatest freedom for each to work out his own destiny. The underlying rationale is that if every individual knowledgeably pursues his self-interest, the social interest will best be served." But what if "stressing the right of parents to have the number of children they want . . . evades the basic question of population policy, which is how to give societies the number of children they need?" (*6*). The issue rests at the center of political philosophy: how best to reconcile individual and collective interests.

Today, most observers would acknowledge that having a child is theoretically a free choice of the individual couple. However, for many couples, particularly among the poor of the world, the choice is not effectively free in the sense that the individual couple does not have the information, services, and supplies needed to implement a free wish in this regard. Such couples are restrained by ignorance, not only of contraceptive practice but of the consequences of high fertility for themselves, their children, and their country; they are restrained by religious doctrine, even though they may not accept the doctrine; they are restrained legally, as in the case of people who would choose abortion if that course were open to them; they are restrained culturally, as in the case of women subject to a tradition that reserves for them only the childbearing and childrearing roles. Hence effective freedom of choice in the matter of childbearing is by no means realized in the world today, as recent policy statements have remarked (*63*).

To what extent should a society be willing to compromise its ethical standards for the sake of solving a great social problem? Suppose a program for population control resulted in many more abortions in a society where abortion is morally repugnant and where, moreover, abortion by acceptable medical standards is widely unattainable; how much fertility decline would be "worth" the result? What of infanticide under the same conditions? How many innocent or unknowing men may be vasectomized for a fee (for themselves or for others who obtained their consent) before the practice calls for a moral restraint? How large an increase in the regulatory bureaucracy, or in systematic corruption through incentives, or in differential effect by social class to the disadvantage of the poor (*64*) is worth how much decrease in the birth rate? How much association of childbearing with monetary incentive is warranted before "bribing people not to have children" becomes contaminating, with adverse long-run effects on parental responsibility (*65*)? How much "immorality," locally defined as extramarital sex, outweighs the benefits of contraceptive practice (assuming that there is an association)? How much withholding of food aid is ethical, judged against degree of fertility decline? If it were possible to legislate a later age at marriage, would it be right to do so against the will of young women, in a society in which they have nothing else to do? In countries, like our own, where urbanization is a serious population problem, is it right to tell people *where* to live, or to impose heavy economic constraints that in effect "force" the desired migration? Is it right to withdraw educational benefits from children in "too large" families? Such withdrawal would not only be repressive from the standpoint of free education but in the long run would be unfortunate from the standpoint of fertility control. In the balance – and this is a question of great but neglected importance – what weight should be given the opportunities of future generations as against the ignorance, the prejudices, or the preferences of the present one?

Guidance on such ethical questions is needed. For further consideration, these propositions are put forward. (i) "An ideal policy would permit a maximum of individual freedom and diversity. It would not prescribe a precise number of children for each category of married couple, nor lay down a universal norm to which all couples should conform" (*3*). (ii) "An ideal program designed to affect the number of children people want would help promote other goals that are worth supporting on their own merits, or at least not conflict with such goals" (*3*).(iii) An ideal program would not burden the innocent in an attempt to penalize the guilty – for example, would not burden the *N*th child by denying him a free education simply because he *was* the *N*th child of irresponsible parents. (iv) An ideal program would not weigh heavily upon the already disadvantaged – for example, by withdrawing maternal or medical benefits or free education from large families, policies that would tend to further deprive the poor. (v) An ideal program would be comprehensible to those directly affected and hence subject to their response. (vi) An ideal program would respect present values concerning family and children, values which some people may not be willing to bargain away in a cost-benefit anaysis. (vii) An ideal program would not rest upon the designation of population control as the final value justifying all others; "preoccupation with population growth should not serve to justify measures more dangerous or of higher social cost than population growth itself" (*3*).

Presumed of effectiveness. If proposals are scientifically ready to be implemented, politically and morally acceptable, and administratively and financially feasible, to what extent will they actually work in bringing population growth under control? That is the final question.

To begin with, the compulsory measures would probably be quite effective in lowering fertility. Inevitably in such schemes, strongly motivated people are ingenious enough to find ways "to beat the system"; if such people were numerous enough the system could not be enforced except under conditions of severe political repression (*66*). Otherwise, if the scheme was workable, compulsion could have its effect.

What about the proposals for the extension of voluntary contraception? Institutionalizing maternal care in the rural areas, with family planning attached, does promise to be effective within, say, 5 to 10 years, particularly in its potential for reaching younger women and women of lower parity. The International Postpartum Program did have that effect in the urban areas (*67*), and presumably the impact would extend to the rural areas, though probably not to the same degree because of the somewhat greater sophistication and modernization of the cities.

A liberalized abortion system – again, if workable – could also be effective in preventing unwanted births, but it would probably have to be associated with a contraceptive effort; otherwise there might be too many abortions for the system, as well as for the individual woman (who might need three a year to remain without issue).

Free abortion in cases where contraception had failed would probably make for a decline in fertility, but how large a one would depend upon the quality of the contraceptive program. With modern contraception (the IUD and the pill)the failure rates are quite small, but women who only marginally tolerate these two methods could fall back on abortion. Free abortion has certainly lowered fertility in Japan and in certain countries of eastern Europe (*68*) and, where medically feasible, would do so elsewhere as well; as a colleague observes, in this field one should not underestimate the attraction of a certainty as compared to a probability.

The large question of the impact of the various incentive and benefit or liability plans (D and E) simply cannot be answered: we have too little experience to know much about the conditions under which financial factors will affect childbearing to any substantial degree. Perhaps everyone has his price for everything; if so, we do not know what would have to be paid, directly or indirectly, to make people decide not to bear children.

Such as it is, the evidence from the pronatalist side on the effectiveness of incentives is not encouraging. All the countries of Europe have family allowance programs of one kind or another (*69*), most of them legislated in the 1930's and 1940's to raise the birth rate; collectively Europe has the lowest birth rate of any continent. The consensus among demographers appears to be that such programs cannot be shown to have effected an upward trend in the birth rate where tried.

As in the case of abortion for illegitimate pregnancies, several of the benefit or liability proposals would effect only a trivial fraction of people in much of the developing world. However, because the impact of incentive and benefit or liability plans is uncertain and may become important, we need to become better informed on the possibilities and limitations, and this information can come only from experimentation under realistic circumstances and at realistic levels of payment.

A higher age at marriage and a greater participation of women in the labor force are generally credited with effecting fertility declines. In a recent Indian conference on raising the age at marriage, the specialists seemed to differ only on the magnitude of the fertility decline that would result: a decline of 30 percent in the birth rate in a generation of 28 years if the minimum age of the woman at marriage were raised to 20 (*70*), or a decline of not more than 15 percent in 10 years (*71*). I say "seemed to differ" since these figures are not necessarily incompatiable. In either case, the decline is a valuable one. But an increase in the age at marriage is not easy to achieve, and that must come before the fertility effect.

Similarly, an increase in the proportion of working women would have its demographic effect, but could probably come about only in conjunction with other broad social trends like education and industrialization, which themselves would powerfully affect fertility, just as a decline in fertility would assist importantly in bringing these trends about (*72*). Both compulsory education and restrictions on child labor would lower the economic value of children, hence tend to produce a decline in fertility. The question is, how are they to be brought about?

Finally, whether or not research would affect fertility trends depends of course upon its nature and outcome. Most observers believe that, under the typical conditions of the developing society, any improvement in contraceptive technology would lead toward the realization of present fertility goals and might help turn the spiral down. Indeed, several observers believe that this is the single most imporatant desideratum, over the short run. Easy means of determining sex should have some effect upon the "need for sons" and thus cut family size to some extent. Research on the social-economic side would probably have to take effect through programs of the kinds discussed above.

The picture is not particularly encouraging. The measures that would work to sharply cut fertility are politically and morally unacceptable to the societies in question (as with coercion), and in any case unavailable; or they are difficult of attainment in any foreseeable future, as in the case of broad social trends or a shift in age at marriage. The measures that might possibly be tried in some settings, like some version of incentives or benefit or liability plans, give uncertain promise of results at the probable level of operation. Legalization of abortion, where the needed medical facilities are available, would almost certainly have a measurable effect, but acceptability is problematic.

Conclusion

This review leaves us with some conclusions concerning proposals that go beyond family planning.

1) There is no easy way to achieve population control. If this review has indicated nothing else, it has shown how many obstacles stand in the way of a solution to the population problem. Table 1 shows, by way of recapitulation, how the various proposals seem to fit the several criteria (*73*). That is only one observer's judgment of the present situation, but, whatever appraisal is made of specific items, it would appear that the overall picture is mixed.

2) Family-planning programs do not compare unfavorably with other specific proposals, especially when one considers that any *actual* operating program is at a disadvantage when compared with any competitive *ideal* policy. Indeed, on this showing, if family-planning programs did not exist, they would have to be invented; it appears that they would be among the first proposals to be made and the first programs to be tried, given their generally acceptable characteristics.

In fact, when such proposals are made, it turns out that many of them call for *more* family planning, not less, but in a somewhat different form. In the present case, at least a third of the proposals listed above put forward, in effect, simply another approach to family planning, often accepting the existing motivation as to family size. In any case, family-planning programs are established, have some momentum, and, importantly, would be useful as the direct instrument through which other proposals would take effect. So, as a major critic (*74*) acknowledges (*6*), "there is no reason to abandon family-planning programs."

What is needed is the energetic and full implementation of present experience. Much more could be done on the informational side, on encouraging commercial distribution of contraceptives, on the use of paramedical personnel, on logistics and supply, on the training and supervision of field workers, on approaches to special groups of individuals, ranging from women after childbirth to young men drafted into the armed forces. If workers in this field did well what they know how to do, that in itself would in all likelihood make a measurable difference, competitive in magnitude with the probable effects of other specific proposals – not to mention the further impetus of an improved contraceptive technology.

3) Most of the proposed ideas are not new; they have been around for some time. So, if they are not being tried, it is not because they have not been known but because they have not been accepted – presumably, for reasons like those discussed above. In India, for example, several of the social measures being proposed have been, it would seem, under almost constant review by one or another committee for the past 10 to 15 years. So it is not correct to imply that it is only new ideas that are needed; the ideas are there, but their political, economic, or administrative feasibility are problematic.

4) All of the proposers are dissatisfied to some degree with present family-planning efforts, but that does not mean that they

agree with one another's schemes for doing better. Thus, Ohlin believes that "the demographic significance of such measures [maternity benefits and tax deductions for children] would be limited" (*34*). Ketchel eloquently opposes several "possible alternatives to fertility control agents" (*9*). Meier argues against the tax on children on both humanitarian and political grounds (*16*). The U.N. Advisory Mission to India comments (*53*, p. 87), "it is realised that no major demographic effects can be expected from measures of this kind [maternity benefits], particularly as only a small proportion of families are covered . . . but they could contribute, together with the family planning programme, to a general change in the social climate relating to childbearing." Earlier, in supporting a family-planning effort in India, Davis noted that "the reaction to the Sarda Act [the Child Marriage Restraint Act of 1929] prohibiting female marriage [below age 14] shows the difficulty of trying to regulate the age of marriage by direct legislation" (*75*). Myrdal warns against cash payments to parents in this connection and supports social awards in kind to the children (*76*). Kirk believes that "it might prove to be the height of folly to undermine the existing family structure, which continues to be a crucial institution for stability and socialization in an increasingly mobile and revolutionary society" (*77*). Finally, Ehrlich is contemptuous of the professors whose 'ideas of action' is to form a committee or to urge 'more research.' Both courses are actually substitutes for action" (*7*, p. 191).

5) In a rough way, there appears to be a progression in national efforts to deal with the problem of population control. The first step is the theoretical recognition that population growth may have something to do with the prospects for economic development. Then, typically, comes an expert mission from abroad to make a survey and report to the government, as has occurred in India, Pakistan, South Korea, Turkey, Iran, Tunisia, Morocco, and Kenya, among others. The first action program is in family planning, and most of the efforts are still at that level. Beyond that, it apparently takes (i) some degree of discouragement about progress combined with (ii) some heightened awareness of the seriousness of the problem to move the effort forward. To date, those conditions have been most prominently present in India – and that is the country that has gone farthest in the use of incentives and in at least consideration of further steps along the lines mentioned above.

6) Proposals need to be specific—proposals both for action and for further research. It is perhaps too much to ask advocates to spell out all the administrative details of the way their plan is to operate in the face of obstacles and difficulties, or even to spell out how it is to get permission to operate; the situations, settings, opportunities, and personalities are too diverse for that. But it does seem proper to ask for the fullest possible specification of actual plans, under realistic conditions, in order to test out their feasibility and likely effectiveness. Similarly, advocates of further research ought to spell out not only what would be studied, and how, but also how the results might be applied in action programs to affect fertility. Social research is not always readily translated into action, especially into administrative action; and the thrust of research is toward refinement, subtlety, precision, and qualification, whereas the administrator must act in the large. Short of such specification, the field remains confronted with potentially good ideas, such as "raise the age at marriage" or "use incentives" or "substitute pension systems for male children," without being able to move very far toward implementation.

7) Just as there is no easy way, there is no single way. Since population control will at best be difficult, it follows that every acceptable step that promises some

Table 1. Illustrative Appraisal of Proposals, by Criteria.

Proposal	Scientific readiness	Political viability	Administrative feasibility	Economic capability	Ethical acceptability	Presumed effectiveness
A. Extension of voluntary fertility control	High	High on maternal care, moderate-to-low on abortion	Uncertain in near future	Maternal care too costly for local budget, abortion, feasible	High for maternal care, low for abortion	Moderately high
B. Establishment of involuntary fertility control	Low	Low	Low	High	Low	High
C. Intensified educational campaigns	High	Moderate-to-high	High	Probably high	Generally high	Moderate
D. Incentive programs	High	Moderately low	Low	Low-to-moderate	Low-to-high	Uncertain
E. Tax and welfare benefits and penalties	High	Moderately low	Low	Low-to-moderate	Low-to-moderate	Uncertain
F. Shifts in social and economic institutions	High	Generally high, but low on some specifics	Low	Generally low	Generally high, but uneven	High, over long run
G. Political channels and organizations	High	Low	Low	Moderate	Moderately low	Uncertain
H. Augmented research efforts	Moderate	High	Moderate-to-high	High	High	Uncertain
Family-planning programs	Generally high, but could use improved technology	Moderate-to-high	Moderate-to-high	High	Generally high, but uneven, on religious grounds	Moderately high

measure of impact should be taken. The most likely prospect is that population control, to the degree it is realized, will be the result of a combination of efforts – economic, legal, social, medical – each of which has some effect but not an immediately overwhelming one (*78*). Accordingly, it is incumbent upon workers in the professional fields concerned to look hard at various approaches, including family planning itself, in order to screen out what is potentially useful for application. In doing so, it may be the path of wisdom to move with the "natural" progression. Some important proposals seem reasonably likely of adoption – institutionalization of maternal care, population study in the schools, the TV satellite system for disseminating information, a better contraceptive technology, perhaps even liberalization of abortion laws in some settings – and we need to know not only how effective such efforts will be but, beyond them, how large a money incentive would have to be to effect a given amount of fertility control and how effective those indirect social measures are that are morally acceptable and capable of realization. It may be that some of these measures would be both feasible and effective – many observers 15 years ago thought that family-planning programs were neither –and a genuine effort needs to be made. The "heavy" measures – involuntary measures and political pressures – may be put aside for the time being, if not forever.

8) In the last analysis, what will be scientifically available, politically acceptable, administratively feasible, economically justifiable, and morally tolerated depends upon people's perceptions of consequences. If "the population problem" is considered relatively unimportant or only moderately important, that judgment will not support much investment of effort. If it is considered urgent, much more can and will be done. The fact is that, despite the large forward strides taken in international recognition of the problem in the 1960's, there still does not exist an informed, firm, and constant conviction in high circles that this is a matter with truly great implications for human welfare (*79*). Such convictions must be based on sound knowledge. Here it would appear that the demographers and economists have not sufficiently made their case to the world elite – or that, if made, the case has not sufficiently commanded their attention and support. Population pressures are not sharply visible on a day-to-day or even year-to-year basis, nor, short of major famine, do they show themselves in dramatic events. Moreover, the warnings of demographers are often dismissed, albeit unfairly and wrongly, on the basis of past forecasts that were not borne out (*80*). After, all, only a generation ago we were being warned about a decline in population in the West. Asking government leaders to take steps toward population control is asking them to take very substantial steps indeed – substantial for their people as well as for their own political careers – hence the case must be virtually incontrovertible. Accordingly, the scientific base must be carefully prepared (and perhaps with some sense of humility about the ease of predicting or urging great events, for the record is not without blemishes). Greater measures to meet the problem – measures which exclude social repression and needless limitation of human freedom – must rely on heightened awareness of what is at stake, on the part of leaders and masses alike.

What is beyond family planning? Even if most of the specific plans are not particularly new, that in itself does not mean that they are to be disregarded. The questions are: Which plans can be effected, given such criteria? How can they be implemented? What will be the outcome?

This article is an effort to promote the discourse across the professional fields concerned with this important issue. Given the recent stress on family–planning programs as the "means of choice" in dealing with the problem, it is natural and desirable that counter-positions be put forward and reviewed. But that does not in itself settle the critical questions. What can we do now to advance the matter? Beyond family planning, what?

References and Notes

1. See, for example, K. Davis, *Science* **158,** 730 (1967); P. C. Potter, R. Freedman, L. P. Chow, *ibid.* **160** 848 (1968); F. W. Notestein, "Population growth and its control," paper presented before the American Assembly on World Hunger, Fall 1968.
2. See, for example the section on "Goals" in K. Davis, *Science* **158,** 730 (1967).
3. A. J. Coale, "Should the United States start a campaign for fewer births?," presidential address presented before the Population Association of America, 1968.
4. For current targets of some national family-planning programs, see B. Berelson, "National family planning programs: Where we stand," paper presented at the University of Michigan Sesquicentennial Celebration, November 1967; the paper concludes: "By and large, developing countries are now aiming at the birth rates of Western Europe 75 years ago or the United States 50 years ago."
5. H. C. Taylor, Jr. and B. Berelson, *Amer. J. Obstet. Gynecol.* **100,** 885 (1968).
6. K. Davis, Science **158,** 730 (1967).
7. P. R. Ehrlich, *The Population Bomb* (Ballantine, New York, 1968).
8. S. Chandrasekhar, *Population Rev.* **10,** 17 (1966).
9. M. M. Ketchel, *Perspect. Biol. Med.* **11,** 687 (1968); see also Ketchel's article in *Med. World News* (18 Oct. 1968), p. 66.
10. Ehrlich appears to dismiss the scheme as unworkable (7, p. 136), though two pages later he advocates "ample funds" to "promote intensive investigation of new techniques of birth control, possibly leading to the development of mass sterilizing agents such as were discussed above."
11. K. E. Boulding, *The Meaning of the Twentieth Century;The Great Transition* (Harper & Row, New York, 1964), pp. 135-36.
12. W. B. Shockley, in a lecture delivered at McMaster University, Hamilton, Ontario, December 1967.
13. S. Chandrasekhar, as reported in the New York *Times,* **24** July 1967. Just as the present article was being completed. Chandrasekhar proposed *(ibid.,* **21** Oct. 1968) "that every married couple in India deny themselves sexual intercourse for a year. . . Abstinence for a year would do enormous good to the individual and the country." The reader may wish to consider this the 30th proposal and test it against the criteria that follow.
14. S. Wayland, in *Family Planning and Population Programs,* B. Berelson, R. K. Anderson, O. Harkavy, J. Maier, W. P. Mauldin, S. J. Segal, Eds. (Univ. of Chicago Press, Chicago, **1966,** pp. 353-62; ——— , in "Family Planning Programs: Administration, Education, Evaluation," J. Ross and J. Friesen, Eds., in preparation; "Teaching Population Dynamics: An Instructional Unit for Secondary School Students" (Columbia University,

New York, 1965); "Critical Stages in Reproduction: Instruction Materials in General Science and Biology" (Columbia University, New York, 1965). The two last-named publications are pamphlets prepared under Wayland's direction at Teachers College.

15. P. Visaria, *Economic Weekly* (8 Aug. 1964), p. 1343.
16. R. L. Meier and G. Meier, "New Directions: A Population Policy for the Future," unpublished manuscript.
17. *Preparatory Study of a Pilot Project in the Use of Satellite Communication for National Development Purposes in India* (UNESCO Expert Mission, 1968).
18. W. Schramm and L. Nelson, *Communication Satellite for Education and Development – The Case of India* (Stanford Research Institute, Stanford, Calif., 1968). pp. 63-66.
19. S. Chandrasekhar, as reported in the New York *Times,* 19 July 1967.
20. E. Pohlman (Central Family Planning Institute, India), "Incentives for 'Non-Maternity' Cannot 'Compete' with Incentives for Vasectomy," unpublished manuscript.
21. T. J. Samuel, J. *Family Welfare* **13, 11** (1966).
22. J. Simon, "Money Incentives to Reduce Birth Rates in Low-Income Countries: A Proposal to Determine the Effect Experimentally," unpublished manuscript; "The Role of Bonuses and Persuasive Propaganda in the Reduction of Birth Rates," unpublished manuscript.
23. ——— , "Family Planning Prospects in Less-Developed Countries, and a Cost-Benefit Analysis of Various Alternatives," unpublished manuscript.
24. S. Enke, *Population Rev.* **4, 47** (1960).
25. M. Young, "The Behavioral Sciences and Family Planning Programs: Report on a Conference," *Studies in Family Planning,* **No. 23** (1967), **p.10.**
26. D. Bhatia, "Government of India Small Family Norm Committee Questionnaire," *Indian J. Med. Educ.* **6,** 189 (1967). As the title indicates, this is not a proposal but a questionnaire soliciting opinions on various ideas put forward to promote "the small family norm."
27. S. Enke, "The Gains to India from Population Control," *Rev. Econ. Statist.* **42,** 179, 180 (1960).
27a. J. W. Leasure, *Milbank Mem. Fund Quart.* **45,** 417 (1967).
28. J. J. Spengler, "Agricultural development is not enough," paper presented before the Conference on World Population Problems, Indiana University, May 1967.
29, M. C. Balfour, "A Scheme for Rewarding Successful Family Planners," *Population Council Mem.* (1962).
30. W. P. Mauldin, "Prevention of Illegitimate Births: A Bouus Scheme," *Population Council Mem.* (1967).
31. R. M. Titmuss and B. Abel-Smith, *Social Policies and Population Growth in Mauritius* (Methuen, London, 1960)
32. A. S. David, *National Development, Population and Family Planning in Nepal* (1968), pp. 53-54.
33. J. Fawcett, personal communication.

34. G. Ohlin, *Population Control and Economic Development* (Development Centre of the Organisation for Economic Co-operation and Development, New York, 1967), p. **104.**

35. W. P. Davison, personal communication. Davison suggests a good pension (perhaps $400 a year) for men aged 60, married for at least 20 years, with no sons.

36. K. Davis, personal communication.

37. B. Berelson and A. Etzioni, brief formulations, 1962 and 1967, respectively.

38. P. M. Hauser, in "The Behavioral Sciences and Family Planning Programs: Report on a Conference," *Studies in Family Planning,* No. 23 (1967), p. **9.**

39. J. Blake, in *Public Health and Population Change: Current Research Issues,* M. C. Sheps and J. C. Ridley, Eds. (Univ. of Pittsburgh Press, Pittsburgh, 1965), p. **62.**

40. For the initial formulation of the proposal, see R. L. Meier, *Modern Science and the Human Fertility Problem* (Wiley, New York, 1959), chap. **7.**

41. P. M. Hauser, *Demography* **4, 412,** (1967).

42. "Family Planning and the Status of Women: Interim Report of the Secretary-General" (United Nations Economic and Social Council, Commission on the Status of Women, New York, 1968), especially p. **17** ff.

43. R. Revelle, quoted by M. Viorst, *Horizon* (summer 1968), p. 35; D. M. Heer and D. O. Smith, "Mortality level and desired family size," paper presented before the Population Association of America, April 1967.

44. Ehrlich makes the same point in *New Scientist* (14 Dec. 1967), p. 655: "Refuse all foreign aid to any country with an increasing population which we believe is not making a maximum effort to limit its population. . . . The United States should use its power and prestige to bring extreme diplomatic and/or economic pressure on any country or organization (the Roman Catholic Church?) impeding a solution to the world's most pressing problem."

45. In an earlier article Ehrlich calls for a "Federal Population Commission with a large budget for propaganda," presumably limited to the United States.

46. S. Chandrasekhar, in *Asia's Population Probblems,* S. Chandrasekhar, Ed. (Allen & Unwin, New York, 1967), p. 96; Chandrasekhar cites a suggestion made in 1961 by Julian Huxley.

47. S. Polgar, in "The Behavioral Sciences and Family Planning Programs: Report on a Conference," *Studies in Family Planning,* No. 23 (1967), p. **10.**

48. *The Growth of World Population* (National Academy of Sciences, Committee on Science and Public Policy, Washington, D.C., **1963),** pp. 5, 28-36. This recommendation has of course been made on several occasions by several people. For an imaginative account of the impact of biological developments, see P. C. Berry-, appendix to *The Next Thirtyfour Years: A Context for Speculation* (Hudson Institute, Croton-on-Hudson, New York, 1966).

49. See, for example, S. J. Segal, "Biological aspects of fertility regulation," paper presented at the University of Michigan Sesquicentennial Celebration, November **1967.**
50. It is worth noting that such expectations are not particularly reliable. For example, in **1952-53** a Working Group on Fertility Control was organized by the Conservation Foundation to review the most promising "leads to physiologic control of fertility," based on a survey conducted by Paul S. Henshaw and Kingsley Davis. This group did identify a "lead" that became the oral contraceptive (then already under investigation) but did not mention the intrauterine device. It was searching specifically for better ways to control fertility because of the population problem in the developing world, and considered the contraceptive approach essential to that end: "It thus appears imperative that an attempt be made to bring down fertility in overpopulated regions without waiting for a remote, hoped-for transformation of the entire society. . . . It seems plausible that acceptable birth control techniques might be found, and that the application of science to developing such techniques for peasant regions might yield revolutionary results" [*The Physiological Approach to Fertility Control, Report of the Working Group on Fertility Control* (Conservation Foundation, New York, 1953)]
51. A. S. Parkes, *New Scientist* **35,** 186 (1967).
52. New York *Times* (17 Nv. 1967). The then Minister had earlier suggested a substantial bonus (100 rupees) for vasectomy, the funds to be taken from U.S. counterpart funds, "but both Governments are extremely sensitive in this area. Yet in a problem this crucial perhaps we need more action and less sensitivity"[S. Chandrasekhar (46)]
53. *Report on the Family Planning Programme in India* (United Nations Advisory Mission, New York, 1966).
54. *Implications of Raising the Female Age at Marriage in India* (Demographic Training and Research Centre, Chembur, India, 1968), p. 109; *Centre Calling* (May 1968), p. **4.**
55. *Planned Parenthood* (Mar. 1968), p. **3.**
56. *Ibid* (Apr. 1968), p. **2.**
57. For a review of this development see R. Symonds and M. Carder, *International Organisations and Population Control* (1947-67) (Institute of Development Studies, Univ. of Sussex, Brighton, England, 1968).
58. At present, population materials are being included in school programs in Pakistan, Iran, Taiwan, and elsewhere.
59 See, for example, Balfour (*29*), Mauldin (*30*). and Pohlman (*20*) and, for the economic analysis, Enke (*27*) and Simon (*22*).
60. For the negative answer, see Enke (*27*) and Simon (*22*). Data are from family-planning budgets and national development budgets contained in 5-year development plans.
61. E. Pohlman, "Incentives in birth planning," in preparation.
62. Coale, however, does point out that "it is clearly fallacious to accept as optimal a growth that continues until overcrowding makes additional births intolerably expensive."
63. See, for example, the World Leaders' Statement [*Studies in Family Planning,* **No. 26** (1968)] and the Resolution of the International Conference on Human Rights on "Human Rights Aspects of Family Planning," adopted **12** May 1968, reported in *Population Newsletter,* **No.**

2 (issued by the Population Division, United Nations) (1968), p. **21** ff. Incidentally, the issue of population policy was apparently a live one in classical times, and resolved by the great philosophers in ways not fully consonant with modern views. Plato, in the *Republic* (Modern Library edition, pp. **412, 414,** says, "the number of weddings is a matter which must be left to the discretion of the rulers, whose aim will be to preserve the average of population and to prevent the State from becoming either too large or too small" – to which end certain marriages have "strict orders to prevent any embryo which may come into being from seeing the light; and if any force a way to the birth, the parents must understand that the offspring of such a union cannot be maintained, and arrange accordingly." Aristotle, in *Politics* (Modern Library edition, p. 316) says, "on the ground of an excess in the number of children, if the established customs of the state forbid this (for in our state population has a limit), no child is to be exposed, but when couples have children in excess, let abortion be procured before sense and life have begun. . . ."

64. After noting that economic constraints have not been adopted in South Asia, though often proposed, Gunnar Myrdal continues: "The reason is not difficult to understand. Since having many children is a main cause of poverty, such measures would penalize the relatively poor and subsidize the relatively well off. Such a result would not only violate rules of equity but would be detrimental to the health of the poor families, and so of the growing generation" [*Asian Drama: An Inquiry into the Poverty of Nations* (Pantheon, New York, 1968), vol. 2, pp. 1502-03].
65. F. W. Notestein, in *Family Planning and Population Programs,* Berelson *et al.,* Eds. (Univ. of Chicago Press, Chicago, 1966), pp. 828-29: "There is a real danger that sanctions, for example through taxation, would affect adversely the welfare of the children. There is also danger that incentives through bonuses will put the whole matter of family planning in a grossly commercial light. It is quite possible that to poor and harassed people financial inducements will amount to coercion and not to an enlargement of their freedom of choice. Family planning must be, and must seem to be, an extension of personal and familial freedom of choice and thereby an enrichment of life, not coercion toward its restriction."
66. In this connection see the novel by A. Burgess, *The Wanting Seed* (Ballantine, New York, 1963). At the same time, Myrdal, a long-time observer of social affairs, remarks that "the South Asian countries. . . can, to begin with, have no other principle than that of voluntary parenthood. . . . State direction by compulsion in these personal matters is not effective. . ." [G. Myrdal, *Asian Drama: An Inquiry into the Poverty of Nations* (Pantheon, New York, 1968), p. 1501]
67. G. I. Zatuchni, "International Postpartum Family Planning Program: Report on the First Year," *Studies in Family Planning, No.* **22** (1967), p. **14** ff.
68. For example, the repeal of the free abortion law in Rumania resulted in an increase in the birth rate from 14 in the third quarter of 1966 to 38 in the third quarter of 1967. For an early report, see R. Pressat, *Population* **22,** 1116 (1967).

69. See *Social Security Programs Throughout the World, 1964* (U.S. Department of Health, Education, and Welfare, Washington, D.C., 1964).
70. S. N. Agarwala in *Implications of Raising the Female Age at Marriage in India* (Demographic Training and Research Centre, Chembur, India, 1968), p. **21.**
71. V. C. Chidambaram, *ibid* **p. 47.**
72. Actually, recent research is calling into question some of the received wisdom on the prior need of such broad institutional factors for fertility decline. If further study supports the new findings, that could have important implications for present strategy in the developing countries. See A. J. Coale, in *Proc. U.N. World Population Conf.* (1965), vol. 2, pp. 205-09, and — — — , "The decline of fertility in Europe from the French Revolution to World War II," paper presented at the University of Michigan Sesquicentennial Celebration, 1967.
73. As the roughest sort of summary of Table 1, if one assigns values from 5 for "high" to 1 for "low," the various proposals rank as follows: family-planning programs, 25; intensified educational campaigns, 25; augmented research efforts, 24; extension of voluntary fertility control, 20; shifts in social and economic institutions, 20; incentive programs, 14; tax and welfare benefits and penalties, **14;** political channels and organizations, **14;** establishment of involuntary fertility control, **14.**
74. Davis was a strong advocate of family planning in India, and quite optimistic about its prospects even in the pre-IUD or pre-pill era. See K. Davis, in The Interrelations of Demographic, Economic, and Social Problems in Selected Underdeveloped Areas (Milbank Memorial Fund, New York, 1954). Davis concludes (pp. 87-88): "Although India is already well-launched in the rapid-growth phase of the demographic transition, there is no inherent reason why she should long continue in this phase. She need not necessarily wait patiently while the forces of urbanization, class mobility, and industrial development gradually build up to the point where parents are forced to limit their offspring on their own initiative and without help, perhaps even in the face of official opposition. . . . Realistically appraising her situation, India has a chance to be the first country to achieve a major revolution in human life —— the planned diffusion of fertility control in a peasant population prior to, and for the benefit of, the urban-industrial transition."
75. K. Davis, in *The Interrelations of Demographic, Economic, and Social Problems in Selected Underdeveloped Areas* (Milbank Memorial Fund, New York, 1954), p. **86.**
76. G. Myrdal, *Asian Drama: An Inquiry into the Poverty of Nations* (Pantheon, New York, 1968), p. 1503.
77. D. Kirk, "Population research in relation to population policy and national family planning programs," paper presented before the American Sociological Association, August 1968.
78. It begins to appear that the prospects for fertility control may be improving over the decades. Kirk, after reviewing several factors that "favor a much more rapid [demographic] transition than occurred in the West" – changed climate of opinion, religious doctrine, decline of infant mortality, modernization, fertility differentials, grass-roots concern, and improved contraceptive technology – shows, in a remarkable tabulation, that the later a country began the reduction of its birth rate from 35 to

20 births per thousand, the shorter the time it took to achieve this reduction: from 73 years (average) for the period 1831-60, for example, to 21 years after 1951; the trend has been consistently downward for over a century [D. Kirk, "Natality in the developing countries: recent trends and prospects," paper presented at the University of Michigan Sesquicentennial Celebration, 1967]

79. Nor, often, does such a conviction exist among the general public. For example, in midsummer of 1968 a national sample of adults was asked in a Gallup poll, "What do you think is the most important problem facing this country today?" Less than 1 percent mentioned population growth (Gallup release, 3 Aug. 1968, and personal communication).

80. For an old but enlightening review, see H. Dorn, *J. Amer. Statist. Ass.* **45, 311** (1950).

33.

Infectious Diseases and Human Ecology

A. Macdonald

It is against this background of explosive population growth, declining mortality and food production barely keeping pace with minimal needs that we have to consider the importance of infectious disease in the modern world. AUTHOR

Infectious disease is one of the manifestations of the interaction of human or animal hosts with their microbial parasites. Human ecology, however it may be defined, is concerned with the mutual relations of man and his environment. We are thus dealing with biological phenomena and increasingly it has become necessary for microbiologists and others to study infectious disease from the point of view of evolution in which both the parasite and the host are concerned. We can only do this if we appreciate that clinically detectable disease is but one of the manifestations of infection with potentially dangerous micro-organisms. In many, perhaps in most, infectious diseases symptomless or unrecognized infection is as common as overt illness. Poliomyelitis is the best-known example of a condition in which very large numbers of children – it may even be as many as 1,000 – are infected and only one develops the typical paralytic disease.

New techniques, particularly in virology, have greatly increased our knowledge and widened our understanding both of the host/parasite relationship and of its effects on the community. Many years ago Theobald Smith, the American microbiologist, emphasized that in the natural evolution of an infectious disease man and microbe tend to move towards a state of amicable co-existence and this is certainly true of most diseases which are confined to the human host. Syphilis, which in recent years, until about 1958, had been becoming a rarer as well as a comparatively mild disease, was an acute killing disease when first introduced into Europe. So severe and widespread was the great pox, in comparison with smallpox, that more than forty treatises on the subject had already been produced by the end of the sixteenth century in Europe. But it is dangerous to generalize about infectious disease, and classical smallpox and typhoid fever are examples of human infections which show little diminution in virulence despite a long and intimate dependence on man for their survival. The biological fact of the uniqueness of the individual person is true in great measure of the infectious diseases to which he is subject.

While the conditions of human existence obviously affect the incidence of infectious disease it is often difficult to identify and relate particular environmental changes to alterations in community health. We know that in recent times the application of new methods in agriculture, medical science and public health have been followed by dramatic reductions in

Reprinted by permission from *Impact of Science on Society,* Vol. XVII, No. 3, 1967.

mortality. We also know that the so-called demographic transition in the more developed countries – the change from high mortality and high birth-rate to low mortality and low birth-rate – has not occurred in the underprivileged parts of the world. There the revolution in morbidity and mortality has taken place very rapidly following the use of cheap and efficient methods for the prevention and cure of certain diseases. But the lower mortality has not yet brought in its train any decrease in the fertility rate. In several countries of Africa there are birth-rates of 55 and 60 per 1,000 inhabitants and over almost all regions of Africa, Asia and Latin America the average rate is about 35. This contrasts with a birth-rate of twenty-five or under in the wealthier countries of North America, the Soviet Union and Western Europe. It is against this background of explosive population growth, declining mortality and food production barely keeping pace with minimal needs that we have to consider the importance of infectious disease in the modern world.

Rapidly increasing numbers of mouths to feed and insufficient food are the standard ingredients of social misery which, in the past, has always been fertile soil for infectious disease. It is unlikely, however, that the great epidemic diseases such as plague and smallpox and major endemic diseases such as malaria and trypanosomiasis, which decimated whole countries and held populations in check, will be permitted to go on playing a major demographic role. This does not mean that they can be discounted: indeed diseases like tuberculosis remain among the greatest of all causes of morbidity in the underprivileged countries. Even in the United States it has recently been estimated that some 35 million living Americans have been infected at one time with the tubercle bacillus and there were still some 120,000 cases under active treatment at the end of 1960.

During the great era of bacteriology beginning in the 1880s it appeared to many people that single specific causes of most infectious diseases would soon be identified. It was a simple but erroneous view to equate a single germ with a single disease and it still persists in some quarters today. The concept of disease as an interplay between host and parasite, of epidemics as crowd diseases rather than the summation of infections, is relatively new. Even much modern public health epidemiology is concerned almost exclusively with the epidemic episode, with modes of spread of the infection from case to case, with the probability of chance of contracting the disease, and other statistical data. We have learned a great deal from such studies, but as Professor René Dubos and others have emphasized we need an even broader approach to the study of the whole natural history, or ecology, of disease.[1] We do not understand tuberculosis unless we take into consideration how susceptible people live, their hours and type of work, their housing and nutrition, together with genetic, social and psychological factors which all influence the outcome of infection with the tubercle bacillus. In a rapidly changing world such factors are also continually in flux and we would be unwise to assume that change is always for the better.

Environmental and social factors may affect not only man's chances of becoming infected but also the micro-organism's potentialities of causing disease. In many instances the two effects are inextricably intertwined and preventive medicine must be concerned both with the countless interactions among and between people as well as with the private and communal life of germs. Environment includes physical environment, for example, geography and climate as well as the biological and socio-

[1] R. J. Dubos and J. G. Hirsch, *Bacterial and Mycotic Infections of Man*,1965, p 20.

economic milieu. For the pathogenic organism man himself consitutes the environment during the period of actual human infection. This may be an indispensable phase in the parasite's life-cycle or it may be an entirely irrelevant episode during its survival. The natural histories of the numerous infectious disease agents are so varied that it is difficult to categorize and dangerous to generalize.

One further general point may be raised before considering some of the infectious diseases in more detail. There is a widespread and entirely erroneous belief among people in many Western countries that the conquest of infectious diseases has been finally achieved and that little is to be done except mop up a few pockets of resistance. This belief is often fostered by medical and other authorities who are anxious to direct attention and funds to problems which they consider of more pressing importance. But as Dubos has said, "surprisingly enough, this euphoria has not yet been dampened by the increasing awareness that the morbidity rates of infection have not decreased significantly and in some cases have actually increased. Despite so much oratory on the conquest of microbial diseases the paradox is that the percentage of hospital beds occupied by patients suffering from infection is now as high as it was fifty years ago. Today, as in the past, disorders of the respiratory and the digestive tracts with a microbial aetiology are the most frequent causes of absenteeism from school, factory or from training in the armed forces." Dubos is referring primarily to the so-called endogenous microbial diseases which are ubiquitous and are caused by micro-organisms which are usually present in the body without causing obvious harm except when the infected person is under physiological stress. These microbial diseases are different in their origin and manifestations from the classical infectious diseases and require a different approach for their understanding and treatment. We still know very little about the phenomena concerning the interrelationships between man and his biological environment and until we do endogenous infection will remain one of the major hazards.

In the developed countries the endogenous infections are now more important than most of the classical infectious diseases, but we are still a long way from abolishing epidemic diseases. Indeed, the very technological advances which have reduced mortality from the infectious fevers bring in their train serious potential hazards. Safe water supplies, for example, are almost taken for granted in most developed countries yet when a breakdown occurs the number of individuals affected may be large, as at Zermatt in Switzerland in 1963 when an explosive outbreak of typhoid fever involved 437 people. In Aberdeen, Scotland, a single 6 lb can of corned beef infected in South America gave rise in 1964 to 469 cases of typhoid fever. In this instance, and it is by no means unique, a technical failure in the preparation of canned meat at source initiated an outbreak which might have been small but for the opportunities for contamination and spread which occurred in a modern supermarket serving very large numbers of people. These are perhaps isolated outbreaks but it is important to realize that the environment of modern urban society does not automatically preclude large outbreaks of exogenous infectious disease. Indeed the frequent high prevalence of bacillary dysentery in most cities of the world shows that we may be far from fully understanding let alone controlling the spread of such infections.

Bacterial Infections

Cholera

It may be useful to discuss in some detail one or two of the main bacterial infections, and choice must necessarily rest on those which illustrate essential environmental problems and, at the same time, are of serious concern to the world health authorities. Cholera certainly comes within this category although in Western countries, where it has long been eradicated, it seems to arouse little general interest. The 'conquest' of cholera was one of the earliest triumphs of scientific medicine and if predictions had been fulfilled its history should have ended soon after the close of the nineteenth century. Today we are, or should be, less sanguine about the eradication of almost any infectious diseases because we know a great deal most about them. Cholera is an extremely acute diarrhoeal disease which in the space of six to eighteen hours can kill by vascular collapse. A severely affected cholera patient may have a cumulative loss of fluid equivalent to the total weight of the body within a period of five days. With modern methods of treatment, however, including fluid and electrolyte replacement and antibiotic therapy, mortality should be very small.

We have learned a good deal about the epidemiology of the disease since John Snow's classical study in London in 1849 but his proof of the importance of contaminated water is still regarded as the most important single fact in the transmission of the disease. With all the information available there tended to be some complacency about cholera, which up till about ten years ago seemed to be confined to its endemic home in the Ganges and Brahmaputra valleys. In 1961 however, a strain, slightly different from the classical organism and known as the El Tor vibrio, began to spread from its endemic areas and invaded practically all countries in the western Pacific and South-East Asia regions, most of which had been free from the disease for many years. By 1965 cholera El Tor had spread northwards and westwards to Afghanistan, Nepal, Iran and Uzbekistan in the U.S.S.R. After decades of freedom cholera was once again at the gates of Europe and Africa.

One of the great unsolved problems about the cholera vibrio itself is just how it contrives to multiply in the human intestinal tract and produces such violent physiological effects. It is always assumed that human beings acquire the organism in drinking water or food and that somehow the organism passes the acid barrier of the normal human stomach and then multiplies enormously in the intestine. Yet we can show that the vibrio is extraordinarily sensitive to the adverse environment of acid and that under the reducing conditions prevalent in the gut it seems to multiply slowly. Moreover there is evidence that the organisms may not thrive very well in the complex and changing ecological environment of the competitive intestinal bacterial flora. Despite all this there is no doubt that in the disease itself the organisms are passed in the faeces in enormous numbers but they never invade tissues outside the gut. One other unexplained feature is the total lack of gross pathological changes in the wall of the bowel, as occurs, for example, in such conditions as enteric fever or bacillary dysentery. The vast outpouring of fluid into the intestinal tract occurs in the presence of an anatomically intact intestinal epithelial lining and it may be that some specific choleraic moiety affects the microcirculation of the gut mucosa which appears to be the prime target organ. Whatever the nature of this agent, it is so potent that as many as 60 per cent of untreated patients may die.

One further fact of ecological importance for the organism may be cited. Non-cholera vibrios are frequently isolated both from cholera patients and from patients with milder cholera-like infections, sometimes as the only bacteria present. There are at least seventy-five immunological types of these non-cholera vibrios and their relationship to true cholera vibrios may well be very important. Non-cholera vibrios have been isolated in the laboratory from well-established clones of cholera vibrios and the reverse is also true. It seems entirely likely that such phase variation occurs in nature and if so it could have an important bearing on the whole cholera problem.

We can now return to the reasons for the recent resurgence of cholera. Unfortunately we do not know exactly how changes in the organism or in man's habits or in the environment altered a situation which appeared to be in a steady state. Epidemiological findings have confirmed that polluted water from wells remains a very frequent source of cholera, but there is more to it than this. The disease occurs where there are dense populations, poverty and poor sanitation especially in the rural areas, in low-lying river valleys. Water for a variety of uses is taken from the rivers and streams, from canals and artificial ponds or tanks. In other areas shallow wells are often placed close to dwellings and these may be preferred because of local prejudices, because the water is believed to have a better taste or because they are more secluded than safer wells – a point of special importance for women in a cloistered society. In many rapidly growing Asiatic cities water is supplied intermittently, perhaps twice a day, and this practice results in pressure changes which may permit the contamination of drinking water by sewage. Cholera organisms can frequently be isolated by laboratory means from such sources of from murky tanks and ponds during the occurrence of an outbreak or for a few days thereafter. It is, however, not always clear what the temporal relationship of the organism in the supply is to the occurrence of the disease in people who drink the water.

As one might expect, food is frequently contaminated with the cholera organisms and it has been shown in Bengal that the most commonly used and least expensive variety of rice cooked with salt and stored overnight in the traditional manner provides an ideal medium for the multiplication of the organisms which cause the disease.

Cholera, however, does not always occur as the explosive outbreak so characteristic of food-or water-borne infection. In investigations carried out in the outbreak in the Philippines during 1961-62, where there had been no case for over a quarter of a century, many of the incidents seemed to involve person-to-person spread since single individuals in a household were often the only persons with symptoms of the disease. There is not much evidence to suggest that convalescent carriers of the organism are important in initiating new outbreaks of the disease but, as in so many other infections, the symptomless or healthy carrier may be very important. It is not now believed that malnourished or debilitated people are any more prone to the disease than the healthy. In Bengal there are often two waves of the disease each year, the autumn outbreaks being smaller and isolated, compared with those in the spring which are much more serious and widespread. Such epidemics are frequently terminated with the arrival of the monsoon.

Epidemiological facts such as those just mentioned do not offer an adequate understanding of the ecological relationship of the cholera organism and its victims although we do know that a population adequately served with environmental sanitation does not suffer from the disease. The control of cholera, however, is another matter: increasing populations, growing opportunities for travel from the local villages and towns, limited educational facilities and skills, poor housing and services are all part of the total environment in areas when the disease is endemic and from which it may spread. The control of cholera is not so much a medical problem or, at least, not so dependent on further medical advances as on social, educational and economic improvement in large areas of the world. The value of immunization against the disease, long accepted on faith, has only recently been shown to be reasonably effective in a controlled field trial, but with most preparations protection is only for a limited period of some months and may not eliminate the danger of the symptomless carrier. Even if more effective vaccines were available they might only provide an answer in special and limited circumstances. The control of cholera will never be securely based until the educational and technological environment of more affluent societies is shared by the underprivileged countries of the world. As Dr. Robert Gordon has recently said, "until that Utopia arrives, however, cholera remains as a distant, but potentially real, concern for all ".[2]

The Venereal Diseases

The venereal diseases are often considered separately from the other infectious diseases but there are very good reasons why this distinction should not be followed here. Superficially they might be considered a special and clear-cut problem since the bacteria concerned are entirely human-dependent and their mode of spread, by sexual intercourse, is almost invariable. Nevertheless the ecological factors which determine the incidence of these diseases are complex, involving not only the physical environment but the social and moral climate as well. Moreover they affect every country in the world and at the present time there are many signs that their prevalence is increasing when methods of treatment are highly effective.

After the Second World War the incidence of venereal diseases declined rapidly and with the widespread use of penicillin and the end of the social disturbance there were high hopes that these diseases might become as rare as diphtheria. If the causative organism of syphilis, the *Treponema pallidum,* is not destroyed by early treatment an infected individual usually ceases to become infective for other people within about a year but the organism persists in the body and in about a quarter of cases late manifestations of the disease appear. These include severe damage to the heart and blood vessels and, although nowadays much less often, damage to the brain and central nervous system. In addition a young mother infected with syphilis may transmit the infection to her unborn child. The organism of gonorrhoea, the other main venereal disease, survives only for a brief time outside the human body. Gonorrhoea is primarily a disease of the mucous membranes of the genito-urinary tract and, although much less severe than syphilis, still causes serious complications including sterility and occasionally damage to the eyes of the new-born baby. Both of these venereal diseases still carry an emotional content of guilt and shame which other infectious diseases lack and this enormously adds to the difficulty of prevention.

In all countries, gonorrhoea is far commoner than syphilis but both diseases declined rapidly after the end of the First World War. In 1946 in England and Wales, for example, there were 47,300 reported cases but by 1954 the number was reduced to 17,500. Since 1954 the incidence has continued to increase and by 1962 there were 35,400 cases. In many countries neither gonorrhoea nor syphilis is notifiable and even when morbidity statistics are published, the true incidence of infection may be several times higher than represented. More disturbing still is the great increase of these diseases in young people. In 1964 the WHO Expert Committee on the Health Problems of Adolescence reported that in 1961 in Denmark 43 per cent of all female patients were under 25. About 1 in every 100

[2] "Cholera–Combined Clinical Staff Conference at the National Institutes of Health," *Annals of Internal Medicine,* June 1966, p. 1328.

young people between 15 and 19 were estimated to be suffering from the disease. In Canada the rate for boys and girls combined in the 15-19 age group rose from 145 per 100,000 population in 1959 to 171 in 1963. Similar figures could be obtained from many European countries although it is claimed, without detailed evidence, that in the U.S.S.R. in 1960 the over-all rate was only 57 per 100,000 population. In parts of South-East Asia, the eastern Mediterranean and western Pacific regions the rates of infection are said to exceed 1,000 for every 100,000 adults and even higher rates have been given for some centres in Africa and the Americas.

A WHO Expert Committee in 1962 made an important but little known summary of the situation, which included the following statement: "There is no evidence that the advent and wide use of 'ideals' antibiotics over the last 20 years, particularly penicillin and streptomycin, had made an indent on the reservoir of gonococcal infections. The infectiveness of the gonorrhoea, its short incubation period, the mode of transmission and other factors has not made it possible to evolve epidemiological methods effective against the very rapid spread of the infection. It had nowhere been possible to bring a sufficiently large number of cases and contacts to treatment quickly enough to overtake the rapid spread of the infection in the community."

This statement is as true today as it was in 1962.

The recent history of infectious syphilis is rather similar to that of gonorrhoea. The infection rates are considerably lower but the potential severity of the disease is much greater. In the United States of America the numer of early syphilis cases reported in 1957 was 3.8 per 100,000 of the population; by 1964 the rate was 12.1, giving a total of 22,730 reported cases. This is more than a threefold increase. There is reason to believe that something similar has occurred in other countries in the Western world although the U.S.S.R. claims only 1.4 cases per 100,000. According to the World Health Organization the highest recent rates have been reported from Costa Rica, Italy, Greece and the United States while the United Kingdom, Norway, Sweden and Canada have had considerably lower rates of increase. In 1955 in most countries the reported incidence of syphilis was the lowest ever recorded but since then it has climbed steadily.

How can we begin to explain this dramatic and almost world-wide increase in venereal disease? Ecological changes may throw some light, particularly if among them we include patterns of behaviour and the whole socio-economic environment, changes in the distribution of population density, population mobility, transport facilities, the stage of political, educational and economic organization and much else. The World Health Organization cited the following factors as important in the increase of syphilis but they apply equally to gonorrhoea:

1. Less fear of the disease with the advent of easy penicilin treatment.
2. Increasing urbanization.
3. Migration of labour groups from country to country.
4. Ease of international transmission.
5. Revoluation of 'mores' and 'new standards for a new age'. The factors here involve the increases in syphilis in the very young and in homosexuals, the decline in the importance of prostitution and the increase in consumption of alcohol.

A committee of the British Medical Association appointed to review the problem of venereal disease in young people in the United Kingdom gave a similar list of possible causes but they included other factors.[3] Among these they claim are the modern emphasis on sexuality in books, films, television and advertisements, declining religious faith and ignorance of the nature and meaning of sex. They noticed that the onset of menstruation in young girls in Great Britain has become progressively earlier in the past 100 years and now usually occurs at the age of 13. They did not think, however, that this earlier maturity was a cause of promiscuity. Their conclusion includes the following: "It is obvious that the problem has its medical aspects but we believe it to be primarily a social and moral problem. The increase in promiscuity among young people is a symptom of change in society as a whole, for which all sections of the community are responsible."

From the microbiological point of view the problems associated with the venereal diseases are relatively simple but the potentially relevant environmental factors in determining spread are exceedingly complex. Patterns of social behaviour, moral and psychological changes operate simultaneously and we have yet to learn how to measure let alone control them.

Viral Diseases,

Infectious diseases caused by viruses are, in general, technically more difficult to study than bacterial infections. Only one or two of the larger viruses can be seen with optical equipment, none grows on inanimate media. The dependence of the human pathogenic viruses for multiplication within tissue cells involves an intimacy of association which makes the study of ecological relationships particulary important. Some viruses, for example, smallpox, measles and chickenpox, are exclusive human parasites whereas other such as yellow fever and encephalitis have lower vertebrate and invertebrate vectors. Rabies and psittacosis viruses are essentially parasites of animals or birds and man is only accidentally infected. Although viruses can only multiply within living cells some of them survive for considerable periods in the external environment but their relative vulnerability to environmental forces is an important factor limiting transmission.

In the past, outbreaks of smallpox in the United Kingdom have been traced to imported and infected cotton bales from the Middle East. Within recent times air travel has provided a new and much more rapid means for the spread of smallpox from endemic areas such as Pakistan. The incubation period of many virus diseases is such that apparently healthy individuals can travel across the world between the time of contraction of a disease and the development of its first symptoms. Many viruses survive for long periods in sewage-polluted waters and are highly resistant to disinfectants; the 1955 outbreak of infectious hepatitis, which caused more than 35,000 cases in Delhi, was attributed to sewage-contaminated drinking water which, it is claimed, had been so intensively chlorinated that all other bacterial and viral pathogens were destroyed. Thus despite their highly restricted multiplication sites many viruses have efficient and often man-made means for survival and long-range transport.

Influenza

Of all the human infectious diseases influenza appears to have the greatest capacity for explosive, world-wide spread. This is perhaps surprising in a virus which so far as we yet know is essentially confined to the human host. Measles, another entirely human respiratory-borne and highly infectious disease, behaves in quite a different manner. Except in very isolated areas such as Greenland or some of the Pacific Islands measles is endemic in all countries with local epidemics, every two or three years, which seem to depend on

[3] British Medical Association, *Veneral Disease and Young People*, London, 1964.

climate and season and on the number of susceptible individuals. In the United States, for example, the peak incidence occurs in children between 3 and 5 years of age whereas in Nigeria the disease is much more serious and by the age of 2 half the children are already infected. Population density and chance of exposure are said to account for these differences but it may be that other environmental factors are also concerned in the fairly constant pattern of measles infection in any community.

For influenza there are claims that animal reservoirs, pigs or horses or other domestic animals might play a part in the initiation of the major outbreaks but these are speculations rather than epidemiological facts. We now have substantial evidence about the world-wide prevalence of influenza A virus infections since 1940 but the more recent epidemics caused by the Asian or A_2 virus have been most closely followed. Beginning in the Kweichow province of China in 1957 it reached Hong Kong by April of that year and quickly spread to other Asiatic countries particularly India and Japan where there were major epidemics by May. In August, Africa, Australia and South America had outbreaks which often began in coastal or port areas and then spread rapidly. In the northern parts of the world the major outbreaks did not occur until late September or October.

From the ecological point of view two aspects of this pandemic spread may be considered. As the virus passes readily from one person to another and repeated outbreaks are common in all countries of the world, most individuals must meet the virus and become infected repeatedly. We would therefore expect a level of immunity to the virus which theoretically should militate against pandemic spread. We now know that in a period of perhaps ten years a given strain of influenza virus creates such widespread immunity in the population that further general spread is unlikely.

However, the influenza virus is genetically highly variable and under the pressure of population immunity selection of a new variant may occur possibly from one of a number of different antigenic forms. The result is that this new variant virus may become dominant if it is sufficiently distinct to elude the antibody resources of most individuals in a community. In the initiation of pandemics it is also possible that by rapid passage from one individual to another the communicability of the virus may be so enhanced that infection could occur with smaller infective doses. Fortunately there is no fixed association of virulence, or the capacity to cause severe disease, with the capacity to spread; the 1957 outbreak, for example, although it travelled round the world, caused only a small fraction of the deaths of the great 1918 pandemic which is alleged to have killed 20 million people.

Nevertheless we can think of each new pandemic of influenza A virus infection as essentially a new disease to the current world population. It is clear that biologically influenza virus has the inherent potentialities to adapt to a wide range of environmental conditions as well as to variations in human, ciliated, respiratory epithelium where it multiplies and exerts its selective injury.

The second aspect of ecological importance concerns the human environment itself. Increasing urbanization and crowding must obviously be considered and many of the initial outbreaks in the 1957 pandemic began in schools or other institutions. Indeed school life seems to play an important part in almost all respiratory-borne outbreaks; in Japan in 1957 the first outbreak of A_2 influenza occurred in May and then decreased in July when the schools were closed for holidays only to resume when they were reopened in September. In Great Britain in the autumn of 1957, in ten large towns, the incidence of the infection was estimated to have been never less than 55 per cent. In the week beginning 13 October 1957 the United States National Health Survey estimated that 12 million people were affected and in the two peak months of the outbreak about 70 million people in all contracted the disease.

Ports are often the initial foci of an outbreak and from there spread to the surrounding areas may be rapid. Seasonal influences undoubtedly affect outbreaks of influenza infection particularly in the cooler regions of the world but whether this is by lowering physiological resistance of individuals to infection, by increasing aggregation or facilitating spread of the virus in some other way, is still not understood. Indeed, despite the vast volume of work carried out at the basic biological level on the virus itself and despite the considerable volume of information on the epidemiology of the disease, we know far too little about the ecological relationship of this very common parasite and its human host.

The Arboviruses

A second group of viruses, the arboviruses, provides quite different ecological problems from the respiratory virus infections of man. The arboviruses, so-called because they are arthropod-borne, circulate in the blood stream of one or more animal species, multiply or are carried in the arthropod which feeds on animal blood, and in this way the infection is later transmitted to man. New viruses belonging to this group are constantly being recognized and at present there are over 200 which have been allocated to twenty-one different groups but only two, Groups A and B, have been intensively studied. So far about fifty of the 200 are known to cause disease in man. Four families of blood-sucking arthropods are concerned in transmission: the Culicidae, Ceratopogonidae, Psychodidae and Ixodidae. The natural history of some of the arboviruses is particularly obscure since in their natural cycle they may be almost entirely silent. Moreover the ecology of these infections includes a study of virus, vertebrate host and arthropod vector, the behaviour of each depending on a variety of environmental factors, which applies also to the potential human victim of the disease. It is necessary to find out the range of naturally infected species, how they are infected and the dose of virus required, whether multiplication occurs in the arthropod, and then to show whether the vector is capable of transmitting the virus by biting. The task would be daunting even if conditions remained constant but with so many critical variables it will take many years before such details can be adequately studied.

Human infection with an arbovirus is usually a blind-alley transmission in a cycle which basically involves vector and sylvan host. Man is not essential for the continued existence of the virus, except in a few infections in urban centres where he is the only preferred source of blood. A great variety of illnesses is caused by the arboviruses, including many undifferentiated febrile conditions, dengue fever, liver disease with haemorrhages and encephalitis. Almost certainly most of the infections are symptomless and are only recognized by laboratory investigations. A few of the main arthropod-borne viral infections can be controlled by vaccination or by protecting man from attack by known vectors and these provide the main methods of prevention. Attempts to break the basic cycle of vector and sylvan host obviously demand far more information than we usually possess. Fortunately, for the most feared of all the arboviruses, yellow fever, there is a highly effective and long-lasting protective vaccine prepared from the 17D strain. In consequence yellow fever is less often the dreaded man-*Aedes aegypti* cycle but an incidental sylvatic infection of male agricultural workers, since permanent epizooties exist among the monkeys of jungle regions such as the Orinoco or Amazon river basins.

Large outbreaks of yellow fever, however, still occur and one of exceptional ecological interest took place recently in Uganda involving two separate mosquito vectors. The first, *A. africanus*, a mosquito living high in the forest canopy, maintains the true enzootic cycle among monkeys. The virus was conveyed to man in a second cycle in which monkeys while raiding crop plantations infected the local mosquitoes, *A. simpsoni*, which breed around the huts of farm-workers. *A. simpsoni* then not only infected the farm-workers but could maintain the man-mosquito-man cycle.

The significance of these findings is obvious and is underlined by the story of a very large outbreak of yellow fever in Ethiopia in 1960 when *A. simpsoni* was shown to be the man-to-man vector. The biological problems of *A. simpsoni* were shown to be even more complex when it was demonstrated that two distinct forms exist, only one of which sucks human blood while both may occur in the same area. There is thus a continuing need for the constant monitoring of mosquito vectors and in particular of *A. aegypti* which thrives in close proximity to man.

Even when yellow fever has been absent for many years a very large outbreak can readily occur, as in the case of the Diourbel region of Senegal in 1965. Not a single case had been reported there in the previous twelve years but *A. aegypti* was abundant and a constant menace.

Understanding the total ecological relationships of any arbovirus infection is virtually impossible but on limited information preventive action can be taken and even a few predictions can be made with some degree of confidence. It is most unlikely, for example, that the United States could again become a yellow fever area since it lacks the necessary combination of sylvan mosquitoes and susceptible vertebrates. This, despite the fact that it is close to reservoirs of the infection in some of the Caribbean Islands, India with its abundance of vectors and suitable vertebrates has lived, and will continue to live, precariously free from the disease presumably in large measure because of its remoteness from sources of infection. However man is constantly and drastically changing his environment – he fells forests, introduces new animals, exploits newly discovered natural resources and in many areas increasingly penetrates to the reservoirs of infection. In these circumstances prediction becomes increasingly precarious.

We are on more certain grounds if we forecast that very many more of the 200-odd arboviruses will be found to infect man. The first case of O'nyong-nyong virus disease was only recognized in 1959 but by 1962 it had caused an estimated 2 million East Africans to suffer from what the Acholi tribe graphically described as the 'joint-breaker'. It is probable that St. Louis encephalitis first appeared in the United States in 1932, to be followed shortly afterwards by Eastern Equine encephalitis. West Nile virus fever in North Africa, Chikungunya virus infection in Africa and Asia, Kyasanur forest disease in India are only three further examples of numerous new diseases which may well have been brought to prominence by some environmental changes as much as by modern methods of virological investigation.

Trachoma

A disease formerly classified among the true virus infections is trachoma which is of great antiquity, having been clearly described in the Ebers papyrus, 3,500 years ago. Trachoma is caused by a member of a group of agents named the TRIC agents, which differ from the true viruses in that they have metabolic activities which link them to bacteria and they also contain not one but both of the nucleic acids, DNA and RNA. The disease affects only man and is of world-wide prevalence and it has been estimated by WHO that over 400 million people are currently affected and that 20 million have been rendered blind or incapable of normal work because of the disease. It is believed that the disease as it occurs in Mediterranean countries and in the Orient has changed very little over the centuries. Although effective methods of treatment have been available since sulphonamides were introduced in 1938, and although much has been achieved by antibiotic prophylaxis and therapy, the main course of treatment is long and difficult under field conditions. Moreover, according to WHO there is some evidence of diminished rates of cure in some communities that have been exposed to antibiotic treatment over a number of years. For these reasons the search for new drugs and for effective vaccines is still one of the major preoccupations of WHO.

Trachoma is highly infectious in the acute stage but much less so in the chronic form when the symptoms may be so mild that the disease is ignored until major complications develop. Transmission appears to occur mainly in the home during early life and in the Middle East the peak incidence of active disease occurs in children in the first year or two at school. Virtually the whole population of some North African countries is affected by the time they become adults. Poverty, overcrowding and poor hygiene are almost the rule for trachoma countries just as they are with leprosy, tuberculosis and dysentery. In countries where bacterial infections of the eye are as common as in the Middle East or North Africa there is increased severity and increased transmission of trachoma. In other countries such as Yugoslavia and Spain trachoma is largely free from secondary bacterial infection and the incidence of the disease is much less. It is highly significant that among doctors, visitors or teachers in areas of high trachoma incidence, the disease is almost unknown. Probably the customs and habits of the affected people as well as poor education and poverty are of very great importance in spreading the disease. Lack of water, the shared towel, shared cosmetic pencil, the use of the mother's veil in Mohammedan countries to wipe first the mother's own eyes and then those of her children have all been incriminated.

While the trachoma agent is very sensitive to drying and high temperatures, geographical factors appear to play a major role in the incidence of the disease with the dusty, dry, desert areas of the world most severely affected. This was not exclusively so in the past when trachoma was widespread throughout the whole of Europe apart from Scandinavia. Until recently trachoma was common in Russia, Finland and Ireland and in 1860 36 per cent of surgical eye patients in Helsinki were said to be suffering from the disease. In South Africa on the other hand the incidence of trachoma among the Bantu seems to correspond closely to the humidity of the areas they inhabit; the same is true of the desert tribes in New Mexico and Arizona. Trachoma is said to be remarkably uncommon among American Negroes despite their undoubted poverty and poor conditions, whereas in the West Indies it is much more frequent.

It would appear, therefore, that while we know some of the factors which characterize areas where trachoma is prevalent the cultural patterns of the people and other factors must operate in addition to those most often considered by the epidemiologists. In the rapid movements of large numbers of peoples made possible by air transport, trachoma could perhaps readily be introduced to countries from which it has long been eradicated. The history of trachoma in the past, for example the widespread distribution of the disease which occurred in France and England after the return of Napoleon's armies from Egypt, suggests that the risk of extension of trachoma is not negligible.

Natural Selection and Virus Disease – Myxomatosis

J. B. S. Haldane in 1949 was the first scientist to argue cogently that infectious diseases must have played an important part in the evolution of animals and plants. This view has not attracted the attention it deserves but Fenner and his Australian colleagues, working on myxomatosis, have shown that such indeed is the case.[4] While myxomatosis is a virus infection almost entirely confined to the rabbit and not necessarily a close model for human disease, the Australian work has very important biological implications. It will be remembered that myxoma virus has its natural reservoir in the South American rabbit, *Sylvilagus brasiliensis*, in which it causes a skin tumour localized to the site of injection. The tumour serves as a source of virus for mosquitoes which transfer it mechanically. Thus a presumably long-established disease causes either immunity or localized disease but not florid disease in the countries when it is endemic. After several ineffective attempts myxomatosis was introduced successfully into Australia where it had not been known before, in order to control a vast and enormously destructive population of the European rabbit, *Oryctolagus cuniculus*. During a few summer months of 1950-51 the disease spread from its source with astonishing rapidity and soon an area of about a million square miles was affected. Transportation of the virus over these vast areas was by highly mobile mosquitoes and by mid-1953 the distribution of the disease was co-extensive with the rabbit in Australia. The over-all reduction in the rabbit population and the immediate benefits to agriculture were enormous and though difficult to estimate in detail it is now believed that, today, the rabbit population in Australia is about one-fifth of the pre-myxomatosis level.

An important point of general ecological interest is the change that has occurred in the virus causing the disease. During the 1950-51 period it was uniformly virulent, killing an estimated 99.8 per cent of all infected animals. But very soon in 1952 instead of only 2 survivors in every 1,000 infected rabbits the figure had changed to 100 survivors per 1,000 infected animals. In other words from a highly lethal virus there emerged mutant strains with much less virulence. There is good evidence that the selection of these less virulent strains is favoured by mosquito transmission. In Britain since 1953, a somewhat similar process has occurred and although the insect vector, *Spilopsyllus caniculus*, the rabbit flea, is quite different, highly virulent strains of the virus do not predominate as they did earlier but co-exist with strains of much lower virulence. In Australia, Fenner and his colleagues have shown not only that the virus has changed but that the rabbit itself has also undergone important evolutionary changes. During a period of seven years increased genetic resistance to the disease has meant that the case mortality among rabbits has decreased from 90 per cent to 25 per cent and this is one of the few unequivocal qualtitative demonstrations of innate resistance emerging in a single species of wild animal to an infectious disease. The fact that as yet there is little evidence of the emergence of similar resistance among English rabbits is probably also concerned with the selective effect of the environment and of different modes of spread of the disease. The important point, however, is that over a brief span of years there is conclusive quantitative evidence of evolutionary changes both in an animal host

[4] F. Fenner and F. N. Ratcliffe *Myxomatosis,* 1965, Cambridge University Press

and in its parasite and these changes have been determined by the selective forces encountered in particular environments.

Even wider ecological changes have also taken place as a result of the introduction of myxomatosis – changes involving birds and many other animals and above all in the plant life of the areas when it has been prevalent. Changes in the life habits of rabbit predators would be expected and this has been documented for buzzards and foxes, while in Australia certain species of cypress pines which previously could only be grown with difficulty because of the grazing by rabbits have now become something of a noxious weed. In England the reduction in the number of rabbits has allowed, for example, woody plants, including oak and ash to begin to cover the age-old treeless turf of the chalk downlands. The introduction of an infectious disease of animals has thus changed man's environment, but the traditional fears and attitudes of man to such diseases change more slowly. Despite the widely publicized and unequivocal evidence that myxomatosis has never been known to infect human beings or indeed any other animal other than an occasional hare, there was a widespread belief in 1951 that it had been responsible for a severe outbreak of encephalitis caused, in fact, by an arbovirus, which occurred among people living in the Murray Valley of Australia. Concern was not allayed until three of Australia's most distinguished scientists inoculated themselves with living myxoma virus and thus demonstrated to the public that it could not affect man.

Environmental Change and the Control of Viral Infections

The belief that man can alter his environment in such a way as to abolish infectious and other diseases is a basic concept of preventive medicine. There is, indeed, some good evidence to support this belief but nearly all the best examples occur among the mechanically transmitted bacterial infections and a few of the parasitic diseases.

The story of malaria eradication, for example, is a heartening one with new countries achieving freedom from this infection almost every year. Bulgaria, for example, which had 400,000 cases between 1946 and 1955, Taiwan where 5,000 people died every year. Trinidad, Tobago and Jamaica have all recently been declared free from the disease. On the other hand such freedom may not be a permanent state and in Ceylon, for example, spraying to destroy mosquitoes had recently to be resumed to deal with focal outbreaks involving a population of 10,000 people. Malaria remains one of the great scourges of mankind despite the efforts of WHO which at the end of 1965 had twenty-nine pre-eradication programmes and surveys in operation, most of them in African countries.

With the virus infections the achievements are not so much in environmental methods of control as by altering the susceptibility of man by immunization or by creating barriers between him and the sources of infection. The prevention of smallpox, for example, a disease spread by direct or indirect contact with such things as clothing and utensils, is by vaccination or by the strict isolation of infected individuals. Rabies has been kept out of countries like Great Britain by effective methods of dog quarantine. The elimination of urban yellow fever by anti-*A. aegypti* measures in the Americas and elsewhere is one of the comparatively few clear-cut examples of environmental control of a virus infection.

The complexity of the ecological relationship of man and his pathogenic viruses is such that it is difficult to predict what will be the end result of attempts to modify disease incidence by interference with the environment. The story of poliomyelitis is a reminder that progress in environmental public health may actually increase the susceptibility of a population to severe outbreaks of disease. In peoples living under poor and crowded housing conditions with poor sanitation almost every child contracted symptomless poliomyelitis in the first year or two of life and outbreaks of paralytic disease affected the very young, usually the 1-4 age group. Where the environment had been altered to ensure high standards of living, exposure to the disease is delayed and many individuals reach adult life without the immunity which in former times they would have acquired through symptomless infections in infancy. The result is that very severe epidemics now attack older children and adults, who are more prone to paralysis, if virulent strains of the virus are introduced into such communities.

The striking inverse relationship between infantile mortality and the incidence of poliomyelitis reflects the close relationship between epidemics and social conditions. When sanitary and social conditions improve and infant mortality rates fall to about 70 per 1,000, the stage is set for the first epidemic of poliomyelitis to occur. This pattern of the evolution of poliomyelitis has been experienced in many different parts of the world since the first large outbreaks were described in Sweden towards the end of the last century. Once this pattern of epidemicity begins it is apparently irreversible and as older children and increasingly adults are affected by the disease it becomes more severe with a higher mortality. This has already happened in most Western countries, in Australia, New Zealand and South Africa and increased travel facilities have added to the likelihood of the dissemination of virulent strains of virus into non-immune populations. These first epidemics of poliomyelitis are still occurring in many tropical and semi-tropical countries as children are now beginning to grow up without the security of the immunity formerly developed in insanitary surroundings. Fortunately the widespread use of a highly effective vaccine, in particular an oral vaccine, has altered the whole situation. Had this vaccine not become available there is every likelihood that paralytic poliomyelitis would have increased as never before in history.

One further possible example of the risks inherent in improved environmental health may be given. As is well known, German measles or rubella is a minor virus disease of children which almost always runs a benign course. However, about one-third of women who contact rubella during the first two or three months of pregnancy infect their unborn foetus with the result that stillbirth or spontaneous abortion may occur or the infant may suffer from what is termed the "rubella syndrome". This includes anomalies of the eye, ear, heart or central nervous system of which cardiac and ear lesions are the most common and severe. Damage to the foetus may not be apparent at birth and indeed can take some years to develop so that a careful watch on all such children must be maintained.

It is believed that one attack of rubella whether symptomless or overt is followed by very long-lasting immunity – most of the so-called second attacks being due to confusion with other viral exanthemata. This being so, girls who contract rubella early or at least before marriage will be immune from the risk of having children who suffer from the rubella syndrome. In many Western countries particularly in the middle and upper economic groups great efforts are made to protect children from infections of all types but in the case of rubella this is quite misguided. Whether this factor may have played some part in the recent massive outbreak in the United States is open to doubt, but the outbreak was the most severe for thirty years.

Recent observations both in the United States and in Iceland suggest that rubella may, for quite obscure reasons, now be affecting younger age groups than in former years which, if true, would be against the hypothesis that increased protective care of children in an affluent society may account for an increasing number of susceptibles in later life. In the eastern United States it has been estimated that about one in seven young adults has no immunity to the disease and certainly in the 1963-65 outbreak in the United States there were some 20,000 foetal deaths and more than 30,000 congenitally affected infants born to young women throughout the country.

The most striking of the recently discovered ecological facts about the rubella syndrome is that it is a chronic infection in complete contrast to the acute self-limiting infection of post-natal life. The persistence of the virus and its excretion for perhaps months after an affected child is born, when there are demonstrable high levels of neutralizing antibody in the child's own blood, is an anomaly that has still to be explained. It could also be one of the possible means whereby rubella virus persists during inter-epidemic periods.

The shedding of rubella virus by congenitally affected children months after infection and birth, together with the fact that the virus may be excreted in droplets of saliva about a week before a child develops the typical disease, are obviously matters of epidemiological importance. What the ecologist wishes to know is how rubella and other viruses perpetuate themselves during inter-epidemic periods; whether, in fact, the transfer process in human-to-human infections is efficient enough to account for the perpetuation of the species. Shope, speaking of influenza, poliomyelitis and the viral exanthemata, has said: [5]

"The concept that such viral diseases can be maintained in nature in perpetuity by a mechanism so haphazard and serendipitous seems almost unbelievable." Moreover we have, as yet, no adequate explanation of the seasonal incidence, on a cyclically recurring basis, of many of the infectious diseases, even when we take into account that there is a constant build-up of susceptibles as each new generation of children is born.

Shope's own classic work on swine influenza and hog cholera has unravelled the mystery of the regular recurrence and disappearance of these animal diseases. In swine influenza, the lung worm of swine is an intermediate host for the virus which then exists in a masked form. When the larvae of this nematode is eaten by earthworms, which are in turn again eaten by swine, the virus returns to infect fresh animals but only if they are provoked by some additional physiological stimulus such as cold or inclement weather. Shope has shown that the virus in its masked form is widely seeded before the onset of each outbreak and is provoked by a stimulus common to a wide geographic area.

So far we have no evidence of the existence of ecological models, similar to those described by Shope, for the persistence of human virus infections in man. However the capacity of all viruses to show genetic mutation must always be remembered and their ability to adapt to new hosts has been repeatedly demonstrated in the laboratory. Further, the ability of the nucleic acids of many viruses to become intimately associated with host cellular genetic material suggests the possibility of yet another method of inter-epidemic survival. It may well be that for human, as for bacterial viruses, the main reservoirs of infection in nature are in the form of agents stored in a latent condition within the cells of the human body. During the extensive investigations of mammalian cell cultures in recent years a great variety of virus types have been activated. There is every reason to believe that many more will be discovered in the future.

[5] R. E. Shope, in *Viral and Rickettsial Infections of Man*, 1965, p. 385.

Conclusion

In this survey it has been possible to consider in detail only a few of the main infectious diseases and these have been chosen because they illustrate a diversity of ecological problems and because of their current significance for world health. The total biological environment in which both man and microbe live is far too complex for all but the most superficial attempts at description, let alone understanding. Because man is constantly altering both his physical and socio-economic environment there is very little likelihood of a steady state so far as infectious disease is concerned. Change, which has been more rapid and radical in the present century than in any previous age, is still largely undirected and uncontrolled.

It has already been suggested that the increasing population and urbanization so characteristic of our times carries the risk of easier transmission of those diseases spread by case-to-case infection while the greater ease of travel for man ensures that pathogenic organisms will often be given an unobserved and rapid ride to distant countries. On the other hand increasing urbanization will ultimately increase the chances of providing adequate public health services. In addition improvements in education and a better standard of living should become possible by urbanization and these are vitally important in the control of many infections, particularly those which are mechanically transmitted and are thus most readily subject to environmental control. Just how far away we still are from such control in large areas of the world is amply demonstrated by the re-emergence of cholera, described earlier.

It has been emphasized that education and improved living conditions do not in themselves inevitably lead to improvement in the public health: the paradoxical behaviour of paralytic poliomyelitis is only the most obvious example of an infection where, in the absence of vaccination, the severity of outbreaks has increased along with improvements in the physical and socio-economic environment. It has also been suggested that changes in patterns of life together with more liberal attitudes towards sexual behaviour have been, at least in part, responsible for the recent increase in the venereal diseases. As we have learned how to treat such bacterial infection by modern antibiotics so the old partial barrier of fear has become even less effective. Antibiotics have, of course, had a profound effect on the environment, particularly in the hospital environment where the emergence of antibiotic-resistance strains of bacteria and their spread constitutes one of the 'new' infectious diseases.

As is well known this has been a particular problem with *Staphylococcus aureus* which normally co-exists quite harmlessly with an extremely high proportion of healthy people. Outbreaks of disease in adult hospital wards as well as in newborn nurseries are often caused by one or more bacteriophage types of the *Staphylococcus* which are commonly resistant to one or several of the currently employed antibiotics. The ecology of staphylococcal infection in hospital has been more intensively studied in recent years than almost any other infection yet even the dynamics of acquisition of the organism is as yet far from clear. All we can say is that while far fewer people now die from bacterial infections the incidence of bacterial disease in the aging hospital population has not been significantly reduced.

It is well to remember that some of the claims made for 'wonder' drugs and vaccines are really the result of long-term ecological changes. However effective such modern scientific agents may be they are not responsible for the decline in mortality caused by many of the infectious diseases: as Dubos has pointed out, the leading cause of death in Puerto Rico in 1950 was diarrhoea and enteritis which by 1962 had fallen from 138 to 33 per 100,000 of the population. The reasons for this vast improvement are complex but it was certainly not attributable to the use of antimicrobial vaccines and drugs which were not even available. Similar changes occurred in tuberculosis, measles, whooping cough and scarlet fever before the advent of specific remedies and it may well be that the duration and extent of contact of a particular organism with a particular group of people is of overriding importance.

While the epidemiological features of the virus infections of man have many similarities to his bacterial infections, the ecological problems are affected by the intimate dependence of multiplication of all pathogenic viruses on human cells. This biological restriction of the life-history of viruses is largely compensated for by their versatility, in particular, by their capacity for variation. We have seen how influenza A virus eludes successfully the widespread immunity of the population to a current strain, by the selection of a new antigenic variant. Poliomyelitis virus and indeed all the enteroviruses, including particularly Echo and Coxsackie viruses have multiple antigenic types which, in effect, means that each type is, so far as the human population is concerned, a separate disease. Smallpox and yellow fever are examples of viruses which apparently do not have, or have lost, a similar capacity for variation and this is one of the reasons why vaccination against these diseases is so effective. We have not considered the common cold as an example of an infectious disease but here again the multiplicity of antigenic types of the rhinoviruses must be of great importance for the continuance of these ubiquitous agents and for the difficulties of finding effective methods of prevention.

Viruses which can persist for a very long time in the human body such as the viruses of chickenpox and serum hepatitis present further ecological problems which have also not been mentioned but more information is now available about rubella, formerly considered an almost entirely mild condition. We have also briefly considered the vast reservoirs of arthropod-borne agents which have only recently begun to be studied and it may well be that this is among the most important of all potential virus hazards for man. The ecological relationships of virus, animal host, insect-vector and man in these diseases are exceedingly complex and the difficulties of breaking the natural cycle of transmission are often insuperable. In infectious disease as in all other branches of biology the experimental, quantitative approach provides answers which observation and description alone can seldom give. The introduction of myxoma virus into Australia and its subsequent importation into Western Europe has provided a unique opportunity to study, in the field, the ecology and evolution of an infectious disease by quantitative methods. While there are differences between a disease like myxomatosis in rabbits and most human virus infections there are fundamental biological similarities and many lessons can be learned from the comparatively simple study of animal communities. Man, because of his mobility and because of his enormous power to alter his environment and his whole way of living, poses much more complicated problems.

Whether as the result of these changes there may be increasing opportunities for the emergence of now silent, or entirely new pathogenic organisms, or whether new environmental stresses on the human host will alter radically the insecure and ever-changing balance between micro-organisms and man, only the future can tell. We can only hope that in the arrogance of our accomplishments we shall not forget that, ultimately, the health of the human community depends on the health of the biological community as a whole.

XII. Epilogue

In this time of unrest and uncertainty, we need to hear responsible scientists speaking out, expressing opinions, and offering possible solutions for our ills. The authors of our two closing essays, John P. Platt, associate director of The Mental Health Research Institute, University of Michigan, and Jean Mayer, professor of nutrition, Harvard, both do so with considerable courage and the belief that our very survival depends on how wisely we utilize our science and technology. We have misused both, mainly for lack of better goals.

34.

What We Must Do

John Platt

The task is clear. The task is huge. The time is horribly short. In the past we had had science for intellectual pleasure, . . . for the control of nature . . . and science for war. But today, the whole human experiment may hang on the question of how fast we now press the development of science for survival. **AUTHOR**

There is only one crisis in the world. It is the crisis of transformation. The trouble is that it is now coming upon us as a storm of crisis problems from every direction. But if we look quantitatively at the course of our changes in this century, we can see immediately why the problems are building up so rapidly at this time, and we will see that it has now become urgent for us to mobilize all our intelligence to solve these problems if we are to keep from killing ourselves in the next few years.

The essence of the matter is that the human race is on a steeply rising "S-curve" of change. We are undergoing a great historical transition to new levels of technological power all over the world. We all know about these changes, but we do not often stop to realize how large they are in orders of magnitude, or how rapid and enormous compared to all previous changes in history. In the last century, we have increased our speeds of communication by a factor of 10^7; our speeds of travel by 10^2; our speeds of data handling by 10^6; our energy resources by 10^3; our power of weapons by 10^6; our ability to control diseases by something like 10^2; and our rate of population growth to 10^3 times what it was a few thousand years ago.

Could anyone suppose that human relations around the world would not be affected to their very roots by such changes? Within the last 25 years, the Western world has moved into an age of jet planes, missiles and satellites, nuclear power and nuclear terror. We have acquired computers and automation, a service and leisure economy, superhighways, superagriculture, supermedicine, mass higher education, universal TV, oral contraceptives, environmental pollution, and urban crises. The rest of the world is also moving rapidly and may catch up with all these powers and problems within a very short time. It is hardly surprising that young people under 30, who have grown up familiar with these things from childhood, have developed very different expectations and concerns from the older generation that grew up in another world.

What many people do not realize is that many of these technological changes are now approaching certain natural limits. The "S-curve" is beginning to level off. We may never have faster communications or more TV or larger weapons or a higher level of danger than we have now. This means that if we could learn how to manage these new powers and problems in

Reprinted from *Science*, Vol. 166, pp. 1115-1121, November 28, 1969. Copyright 1969 by the American Association for the Advancement of Science. By permission of the author and publisher.

the next few years without killing ourselves by our obsolete structures and behavior, we might be able to create new and more effective social structures that would last for many generations. We might be able to move into that new world of abundance and diversity and well-being for all mankind which technology has now made possible.

The trouble is that we may not survive these next few years. The human race today is like a rocket on a launching pad. We have been building up to this moment of takeoff for a long time, and if can get safely through the takeoff period, we may fly on a new and exciting course for a long time to come. But at this moment, as the powerful new engines are fired, their thrust and roar shakes and stresses every part of the ship and may cause the whole thing to blow up before we can steer it on its way. Our problem today is to harness and direct these tremendous new forces through this dangerous transition period to the new world instead of to destruction. But unless we can do this, the rapidly increasing strains and crises of the next decade may kill us all. They will make the last 20 years look like a peaceful interlude.

The Next 10 Years

Several types of crisis may reach the point of explosion in the next 10 years: nuclear escalation, famine, participatory crises, racial crises, and what have been called the crises of administrative legitimacy. It is worth singling out two or three of these to see how imminent and dangerous they are, so that we can fully realize how very little time we have for preventing or controlling them.

Take the problem of nuclear war, for example. A few years ago, Leo Szilard estimated the "half-life" of the human race with respect to nuclear escalation as being between 10 and 20 years. His reasoning then is still valid now. As long as we continue to have no adequate stabilizing peace-keeping structures for the world, we continue to live under the daily threat not only of local wars but of nuclear escalation with overkill and megatonnage enough to destroy all life on earth. Every year or two there is a confrontation between nuclear powers – Korea, Laos, Berlin, Suez, Quemoy, Cuba, Vietnam, and the rest. MacArthur wanted to use nuclear weapons in Korea; and in the Cuban missile crisis, John Kennedy is said to have estimated the probability of a nuclear exchange as about 25 percent.

The danger is not so much that of the unexpected, such as a radar error or even a new nuclear dictator, as it is that our present systems will work exactly as planned! – from border testing, strategic gambles, threat and counter-threat, all the way up to that "second-strike capability" that is already aimed, armed, and triggered to wipe out hundreds of millions of people in a 3-hour duel!

What is the probability of this in the average incident? 10 percent? 5 percent? There is no average incident. But it is easy to see that five or ten more such confrontations in this game of "nuclear roulette" might indeed give us only a 50-50 chance of living until 1980 or 1990. This is a shorter life expectancy than people have ever had in the world before. All our medical increases in length of life are meaningless, as long as our nuclear lifetime is so short.

Many agricultural experts also think that within this next decade the great famines will begin, with deaths that may reach 100 million people in densely populated countries like India and China. Some contradict this, claiming that the remarkable new grains and new agricultural methods introduced in the last 3 years in Southeast Asia may now be able to keep the food supply ahead of population growth. But others think that the reeducation of farmers and consumers to use the new grains cannot proceed fast enough to make a difference.

But if famine does come, it is clear that it will be catastrophic. Besides the direct human suffering, it will further increase our international instabilities, with food riots, troops called out, governments falling, and international interventions that will change the whole political map of the world. It could make Vietnam look like a popgun.

In addition, the next decade is likely to see continued crises of legitimacy of all our overloaded administrations, from universities and unions to cities and national governments. Everywhere there is protest and refusal to accept the solutions handed down by some central elite. The student revolutions circle the globe. Suburbs protest as well as ghettoes, Right as well as Left. There are many new sources of collision and protest, but it is clear that the general problem is in large part structural rather than political. Our traditional methods of election and management no longer give administrations the skill and capacity they need to handle their complex new burdens and decisions. They become swollen, unresponsive – and repudiated. Every day now some distinguished administrator is pressured out of office by protesting constituents.

In spite of the violence of some of these confrontations, this may seem like a trivial problem compared to war or famine – until we realize the dangerous effects of these instabilities on the stability of the whole system. In a nuclear crisis or in any of our other crises today, administrators or negotiators may often work out some basis of agreement between conflicting groups or nations, only to find themselves rejected by their people on one or both sides, who are then left with no mechanism except to escalate their battles further.

The Crisis of Crises

What finally makes all of our crises still more dangerous is that they are now coming on top of each other. Most administrations are able to endure or even enjoy an occasional crisis, with everyone working late together and getting a new sense of importance and unity. What they are not prepared to deal with are multiple crises, a crisis of crises all at one time. This is what happened in New York City in 1968 when the Ocean Hill-Brownsville teacher and race strike was combined with a police strike, on top of a garbage strike, on top of a longshoremen's strike, all within a few days of each other.

When something like this happens, the staffs get jumpy with smoke and coffee and alcohol, the mediators become exhausted, and the administrators find themselves running two crises behind. Every problem may escalate because those involved no longer have time to think straight. What would have happened in the Cuban missile crisis if the East Coast power blackout had occurred by accident that same day? Or if the "hot line" between Washington and Moscow had gone dead? There might have been hours of misinterpretation, and some fatally different decisions.

I think this multiplication of domestic and international crises today will shorten that short half-life. In the continued absence of better ways of heading off these multiple crises, our half-life may no longer be 10 or 20 years, but more like 5 to 10 years, or less. We may have even less than a 50-50 chance of living until 1980.

This statement may seem uncertain and excessively dramatic. But is there any scientist who would make a much more optimistic estimate after considering all the different sources of danger and how they are increasing? The shortness of the time is due to the exponential and multiplying character of our problems and not to what particular numbers or guesses we put in.

Anyone who feels more hopeful about getting past the nightmares of the 1970's has only to look beyond them to the monsters of pollution and population rising up in the 1980's and 1990's. Whether we have 10 years or more like 20 or 30, unless we systematically find new large-scale solutions, we are in the gravest danger of destroying our society, our world, and ourselves in any of a number of different ways well before the end of this century. Many futurologists who have predicted what the world will be like in the year 2000 have neglected to tell us that.

Nevertheless the real reason for trying to make rational estimates of these deadlines is not because of their shock value but because they give us at least a rough idea of how much time we may have for finding and mounting some large-scale solutions. The time is short but, as we shall see, it is not too short to give us a chance that something can be done, if we begin immediately.

From this point, there is no place to go but up. Human predictions are always conditional. The future always depends on what we do and can be made worse or better by stupid or intelligent action. To change our earlier analogy, today we are like men coming out of a coal mine who suddenly begin to hear the rock rumbling, but who have also begun to see a little square of light at the end of the tunnel. Against this background, I am an optimist – in that I want to insist that there is a square of light and that it is worth trying to get to. I think what we must do is start runing as fast as possible toward that light, working to increase the probability of our survival through the next decade by some measurable amount.

For the light at the end of the tunnel is very bright indeed. If we can only devise new mechanisms to help us survive this round of terrible crises, we have a chance of moving into a new world of incredible potentialities for all mankind. But if we cannot get through this next decade, we may never reach it.

Task Forces for Social Research and Development

What can we do? I think that nothing less than the application of the full intelligence of our society is likely to be adequate. These problems will require the humane and constructive efforts of everyone involved. But I think they will also require something very similar to the mobilization of scientists for solving crisis problems in wartime. I believe we are going to need large numbers of scientists forming something like research teams or task forces for social research and development. We need full-time interdisciplinary teams combining men of different specialties, natural scientists, social scientists, doctors, engineers, teachers, lawyers, and many other trained and inventive minds, who can put together our stores of knowledge and powerful new ideas into improved technical methods, organizational designs, or "social inventions" that have a change of being adopted soon enough and widely enough to be effective. Even a great mobilization of scientists may not be enough. There is no guarantee that these problems can be solved, or solved in time, no matter what we do. But for problems of this scale and urgency, this kind of focusing of our brains and knowledge may be the only chance we have.

Scientists, of course, are not the only ones who can make contributions. Millions of citizens, business and labor leaders, city and government officials, and workers in existing agencies, are already doing all they can to solve these problems. No scientific innovation will be effective without extensive advice and help from all these groups.

But it is the new science and technology that have made our problems so immense and intractable. Technology did not create human conflicts and inequities, but it has made them unendurable. And where science and technology have expanded the problems in this way, it may be only more scientific understanding and better technology that can carry us past them. The cure for the pollution of the rivers by detergents is the use of nonpolluting detergents. The cure for bad management designs is better management designs.

Also, in many of these areas, there are few people outside the research community who have the basic knowledge necessary for radically new solutions. In our great biological problems, it is the new ideas from cell biology and ecology that may be crucial. In our social-organizational problems, it may be the new theories of organization and management and behavior theory and game theory that offer the only hope. Scientific research and development groups of some kind may be the only effective mechanism by which many of these new ideas can be converted into practical invention and action.

The time scale on which such task forces would have to operate is very different from what is usual in science. In the past, most scientists have tended to work on something like a 30-year time scale, hoping that their careful studies would fit into some great intellectual synthesis that might be years away. Of course when they become politically concerned, they begin to work on something more like a 3-month time scale, collecting signatures or trying to persuade the government to start or stop some program.

But 30 years is too long, and 3 months is too short, to cope with the major crises that might destroy us in the next 10 years. Our urgent problems now are more like wartime problems, where we need to work as rapidly as is consistent with large-scale effectiveness. We need to think rather in terms of a 3-year time scale – or more broadly, a 1- to 5-year time scale. In World War II, the ten thousand scientists who were mobilized for war research knew they did not have 30 years, or even 10 years, to come up with answers. But they did have time for the new research, design, and construction that brought sonar and radar and atomic energy to operational effectiveness within 1 to 4 years. Today we need the same large-scale mobilization for innovation and action and the same sense of constructive urgency.

Priorities: A Crisis Intensity Chart

In any such enterprise, it is most important to be clear about which problems are the real priority problems. To get this straight, it is valuable to try to separate the different problem areas according to some measures of their magnitude and urgency. A possible classification of this kind is shown in Tables 1 and 2. In these tables, I have tried to rank a number of present or potential problems or crises, vertically, according to an estimate of their order of intensity or "seriousness," and horizontally, by a rough estimate of their time to reach climactic importance. Table 1 is such a classification for the United States for the next 1 to 5 years, the next 5 to 20 years, and the next 20 to 50 years. Table 2 is a similar classification for world problems and crises.

The successive rows indicate something like order-of-magnitude differences in the intensity of the crises, as estimated by a rough product of the size of population that might be hurt or affected, multiplied by some estimated average effect in the disruption of their lives. Thus the first row corresponds to total or near-total annihilation; the second row, to great destruction or change affecting everybody; the third row, to a lower tension affecting a smaller part of the population or a smaller part of everyone's life, and so on.

Informed men might easily disagree about one row up or down in intensity, or one column left or right in the time scales,

Table 1. Classification of problems and crises by estimated time and intensity (United States).

Grade	Estimated crisis intensity (number affected X degree of effect)		Estimated Time to Crisis* 1 to 5 Years	5 to 20 Years	20 to 50 Years
1. 2.	10^8	Total annihilation Great destruction or change (physical, biological, or political)	Nuclear or RCBW escalation (Too soon)	Nuclear or RCBW escalation Participatory democracy Ecological balance	(Solved or dead) Political theory and economic structure Population planning Patterns of living Education Communications Integrative philosophy
3.	10^7	Widespread almost unbearable tension	Administrative management Slums Participatory democracy Racial conflict	Pollution Poverty Law and justice	?
4.	10^6	Large-scale distress	Transportation Neighborhood ugliness Crime	Communications gap	?
5.	10^5	Tension producing responsive change	Cancer and heart Smoking and drugs Artificial organs Accidents Sonic boom Water supply Marine resources Privacy on computers	Educational inadequacy	?
6.		Other problems—important, but adequately re-searched	Military R & D New educational methods Mental illness Fusion power	Military R & D	
7.		Exaggerated dangers and hopes	Mind controls Heart transplants Definition of death	Sperm banks Freezing bodies Unemployment from automation	Eugenics
8.		Noncrisis problems being "overstudied"	Man in space Most basic science		

*If no major effort is made at anticipatory solution.

Table 2. Classification of problems and crises by estimated time and intensity (World).

Grade	Estimated crisis intensity (number affected X degree of effect)		Estimated Time to Crisis*		
			1 to 5 Years	5 to 20 Years	20 to 50 Years
1.	10^{10}	Total annihilation	Nuclear or RCBW escalation	Nuclear or RCBW escalation	(Solved or dead)
2.	10^{9}	Great destruction or change (physical, biological, or political)	(Too soon)	Famines Ecological balance Development failures Local wars Rich-poor gap	Economic structure and political theory Population and ecological balance Patterns of living Universal education Communications–integration Management of world Integrative philospphy
3.	10^{8}	Widespread almost unbearable tension	Administrative management Need for participation Group and racial conflict Poverty–rising expectations Environment degradation	Poverty Pollution Racial wars Political rigidity Strong dictatorships	?
4.	10^{7}	Large-scale distress	Transportation Diseases Loss of old cultures	Housing Education Independence of big powers Communications gap	?
5.	10^{6}	Tension producing responsive change	Regional organization Water supplies	?	?
6.	10^{5}	Other problems–important, but adequately researched	Technical development design Intelligent monetary design		
7.		Exaggerated dangers and hopes			Eugenics Melting of ice caps
8.		Noncrisis problems being "overstudied"	Man in space Most basic science		

*If no major effort is made at anticipatory solution.

but these order-of-magnitude differences are already so great that it would be surprising to find much larger disagreements. Clearly, an important initial step in any serious problem study would be to refine such estimates.

In both tables, the one crisis that must be ranked at the top in total danger and imminence is, of course, the danger of large-scale or total annihilation by nuclear escalation or by radiological-chemical-biological-warfare (RCBW). This kind of crisis will continue through both the 1- to 5-year time period and the 5- to 20-year period as Crisis Number 1, unless and until we get a safer peace-keeping arrangement. But in the 20- to 50-year column, following the reasoning already given, I think we must simply put a big "✠" at this level, on the grounds that the peace-keeping stabilization problem will either be solved by that time or we will probably be dead.

At the second level, the 1- to 5-year period may not be a period of great destruction (except nuclear) in either the United States or the world. But the problems at this level are building up, and within the 5- to 20-year period, many scientists fear the destruction of our whole biological and ecological balance in the United States by mismanagement or pollution. Others fear political catastrophe within this period, as a result of participatory confrontations or backlash or even dictatorship, if our divisive social and structural problems are not solved before that time.

On a world scale in this period, famine and ecological catastrophe head the list of destructive problems. We will come back later to the items in the 20- to 50-year column.

The third level of crisis problems in the United States includes those that are already upon us: administrative management of communities and cities, slums, participatory democracy, and racial conflict. In the 5- to 20-year period, the problems of pollution and poverty or major failures of law and justice could escalate to this level of tension if they are not solved. The last column is left blank because secondary events and second-order effects will interfere seriously with any attempt to make longer-range predictions at these lower levels.

The items in the lower part of the tables are not intended to be exhaustive. Some are common headline problems which are included simply to show how they might rank quantitatively in this kind of comparison. Anyone concerned with any of them will find it a useful exercise to estimate for himself their order of seriousness, in terms of the number of people they actually affect and the average distress they cause. Transportation problems and neighborhood ugliness, for example, are listed as grade 4 problems in the United States because they depress the lives of tens of millions for 1 or 2 hours every day. Violent crime may affect a corresponding number every year or two. These evils are not negligible, and they are worth the efforts of enormous numbers of people to cure them and to keep them cured – but on the other hand, they will not destroy our society.

The grade 5 crises are those where the hue and cry has been raised and where responsive changes of some kind are already under way. Cancer goes here, along with problems like auto safety and an adequate water supply. This is not to say that we have solved the problem of cancer, but rather that good people are working on it and are making as much progress as we could expect from anyone. (At this level of social intensity, it should be kept in mind that there are also positive opportunities for research, such as the automation of clinical biochemistry or the invention of new channels of personal communication, which might affect the 20-year future as greatly as the new drugs and solid state devices of 20 years ago have begun to affect the present.)

Where the Scientists Are

Below grade 5, three less quantitative categories are listed, where the scientists begin to outnumber the problems. Grade 6 consists of problems that many people believe to be important but that are adequately researched at the present time. Military R&D belongs in this category. Our huge military establishment creates many social problems, both of national priority and international stability, but even in its own terms, war research, which engrosses hundreds of thousands of scientists and engineers, is being taken care of generously. Likewise, fusion power is being studied at the $100-million level, though even if we had it tomorrow, it would scarcely change our rates of application of nuclear energy in generating more electric power for the world.

Grade 7 contains the exaggerated problems which are being talked about or worked on out of all proportion to their true importance, such as heart transplants, which can never affect more than a few thousands of people out of the billions in the world. It is sad to note that the symposia on "social implications of science" at many national scientific meetings are often on the problems of grade 7.

In the last category, grade 8, are two subjects which I am sorry to say I must call "overstudied," at least with respect to the real crisis problems today. The Man in Space flights to the moon and back are the most beautiful technical achievements of man, but they are not urgent except for national display, and they absorb tens of thousands of our most ingenious technical brains.

And in the "overstudied" list I have begun to think we must now put most of our basic science. This is a hard conclusion, because all of science is so important in the long run and because it is still so small compared, say, to advertising or the tobacco industry. But basic scientific thinking is a scarce resource. In a national emergency, we would suddenly find that a host of our scientific problems could be postponed for several years in favor of more urgent research. Should not our total human emergency make the same claims? Long-range science is useless unless we survive to use it. Tens of thousands of our best trained minds may now be needed for something more important than "science as usual."

The arrows at level 2 in the tables are intended to indicate that problems may escalate to a higher level of crisis in the next time period if they are not solved. The arrows toward level 2 in the last columns of both tables show the escalation of all our problems upward to some general reconstruction in the 20- to 50- year time period, if we survive. Probably no human institution will continue unchanged for another 50 years, because they will all be changed by the crises if they are not changed in advance to prevent them. There will surely be widespread rearrangements in all our ways of life everywhere, from our patterns of society to our whole philosophy of man. Will they be more humane, or less? Will the world come to resemble a diverse and open humanist democracy? Or Orwell's *1984?* Or a postnuclear desert with its scientists hanged? It is our acts of commitment and leadership in the next few months and years that will decide.

Mobilizing Scientists

It is a unique experience for us to have peacetime problems, or technical problems which are not industrial problems, on such a scale. We do not know quite where to start, and there is no mechanism yet for generating ideas systematically or paying teams to turn them into successful solutions.

But the comparison with wartime research and development may not be inappropriate. Perhaps the antisubmarine warfare work or the atomic energy project of the 1940's provide the closest parallels to what we must do in terms of the novelty, scale, and urgency of the problems, the initiative needed, and the kind of large success that has to be achieved. In the antisubmarine campaign, Blackett assembled a few scientists and other ingenious minds in his "back room," and within a few months they had worked out the "operations analysis" that made an order-of-magnitude difference in the success of the campaign. In the atomic energy work, scientists started off with extracurricular research, formed a central committee to channel their secret communications, and then studied the possible solutions for some time before they went to the government for large-scale support for the great development laboratories and production plants.

Fortunately, work on our crisis problems today would not require secrecy. Our great problems today are all beginning to be world problems, and scientists from many countries would have important insights to contribute.

Probably the first stop in crisis studies now should be the organization of intense technical discussion and education groups in every laboratory. Promising lines of interest could then lead to the setting up of part-time or full-time studies and teams and coordinating committees. Administrators and boards of directors might find active crisis research important to their own organizations in many cases. Several foundations and federal agencies already have in-house research and make outside grants in many of these crisis areas, and they would be important initial sources of support.

But the step that will probably be required in a short time is the creation of whole new centers, perhaps comparable to Los Alamos or the RAND Corporation, where interdisciplinary groups can be assembled to work full-time on solutions to these crisis problems. Many different kinds of centers will eventually be necessary, including research centers, development centers, training centers, and even production centers for new sociotechnical inventions. The problems of our time – the \$100-billion food problem or the \$100-billion arms control problem – are no smaller than World War II in scale and importance, and it would be absurd to think that a few academic research teams or a few agency laboratories could do the job.

Social Inventions

The thing that discourages many scientists – even social scientists – from thinking in these research-and-development terms is their failure to realize that there are such things as social inventions and that they can have large-scale effects in a surprisingly short time. A recent study with Karl Deutsch has examined some 40 of the great achievements in social science in this century, to see where they were made and by whom and how long they took to become effective. They include developments such as the following:

Keynesian economics
Opinion polls and statistical sampling
Input-output economics
Operations analysis
Information theory and feedback theory
Theory of games and economic behavior
Operant conditioning and programmed learning
Planned programming and budgeting (PPB).
Non-zero-sum game theory

Many of these have made remarkable differences within just a few years in our ability to handle social problems or management problems. The opinion poll became a national necessity within a single election period. The theory of games, published in 1946, had become an important component of American strategic thinking by RAND and the Defense Department by 1953, in spite of the limitation of the theory at that time to zero-sum games, with their dangerous bluffing and "brinkmanship." Today, within less than a decade, the PPB management technique is sweeping through every large organization.

This list is particularly interesting because it shows how much can be done outside official government agencies when inventive men put their brains together. Most of the achievements were the work of teams of two or more men, almost all of them located in intellectual centers such as Princeton or the two Cambridges.

This list might be extended by adding commercial social inventions with rapid and widespread effects, like credit cards. And sociotechnical inventions, like computers and automation or like oral contraceptives, which were in widespread use within 10 years after they were developed. In addition, there are political innovations like the New Deal, which made great changes in our economic life within 4 years, and the pay-as-you-go income tax, which transformed federal taxing power within 2 years.

On the international scene, the Peace Corps, the "hot line," the Test-Ban Treaty, the Antarctic Treaty, and the Nonproliferation Treaty were all implemented within 2 to 10 years after their initial proposal. These are only small contributions, a tiny patchwork part of the basic international stabilization system that is needed, but they show that the time to adopt new structural designs may be surprisingly short. Our cliches about "social lag" are very misleading. Over half of the major social innovations since 1940 were adopted or had widespread social effects within less than 12 years – a time as short as, or shorter than, the average time for adoption of technological innovations.

Areas for Task Forces

Is it possible to create more of these social inventions systematically to deal with our present crisis problems? I think it is. It may be worth listing a few specific areas where new task forces might start.

1) *Peace-keeping mechanisms and feedback stabilization.* Our various nuclear treaties are a beginning. But how about a technical group that sits down and thinks about the whole range of possible and impossible stabilization and peace-keeping mechanisms? Stabilization feedback-design might be a complex modern counterpart of the "checks and balances" used in designing the consitutional structure of the United States 200 years ago. With our new knowledge today about feedbacks, group behavior, and game theory, it ought to be possible to design more complex and even more successful structures.

Some peace-keeping mechanisms that might be hard to adopt today could still be worked out and tested and publicized, awaiting a more favorable moment. Sometimes the very existence of new possibilities can change the atmosphere. Sometimes, in a crisis, men may finally be willing to try out new ways and may find some previously prepared plan of enormous help.

2) *Biotechnology.* Humanity must feed and care for the children who are already in the world, even while we try to level off the further population explosion that makes this so difficult. Some novel proposals, such as food from coal, or genetic copying of champion animals, or still simpler contraceptive methods, could possibly have large-scale effects on human welfare within 10 to 15 years. New chemical, statistical, and management methods for measuring and maintaining the ecological balance could be of very great importance.

3) *Game theory.* As we have seen, zero-sum game theory has not been too academic to be used for national strategy and policy analysis. Unfortunately, in zero-

sum games, what I win, you lose, and what you win, I lose. This may be the way poker works, but it is not the way the world works. We are collectively in a non-zero-sum game in which we will all lose together in nuclear holocaust or race conflict or economic nationalism, or all win together in survival and prosperity. Some of the many variations of non-zero-sum game theory, applied to group conflict and cooperation, might show us profitable new approaches to replace our sterile and dangerous confrontation strategies.

4) *Psychological and social theories.* Many teams are needed to explore in detail and in practice how the powerful new ideas of behavior theory and the new ideas of responsive living might be used to improve family life or community and management structures. New ideas of information handling and management theory need to be turned into practical recipes for reducing the daily frustrations of small businesses, schools, hospitals, churches, and town meetings. New economic inventions are needed, such as urban development corporations. A deeper systems analysis is urgently needed to see if there is not some practical way to separate full employment from inflation. Inflation pinches the poor, increases labor-management disputes, and multiplies all our domestic conflicts and our sense of despair.

5) *Social indicators.* We need new social indicators, like the cost-of-living index, for measuring a thousand social goods and evils. Good indicators can have great "multiplier effects" in helping to maximize our welfare and minimize our ills. Engineers and physical scientists working with social scientists might come up with ingenious new methods of measuring many of these important but elusive parameters.

6) *Channels of effectiveness.* Detailed case studies of the reasons for success or failure of various social inventions could also have a large multiplier effect. Handbooks showing what channels or methods are now most effective for different small-scale and large-scale social problems would be of immense value.

The list could go on and on. In fact, each study group will have its own pet projects. Why not? Society is at least as complex as, say, an automobile with its several thousand parts. It will probably require as many research-and-development teams as the auto industry in order to explore all the inventions it needs to solve its problems. But it is clear that there are many areas of great potential crying out for brilliant minds and brilliant teams to get to work on them.

Future Satisfactions and Present Solutions

This is an enormous program. But there is nothing impossible about mounting and financing it, if we, as concerned men, go into it with commitment and leadership. Yes, there will be a need for money and power to overcome organizational difficulties and vested interests. But it is worth remembering that the only real source of power in the world is the gap between what is and what might be. Why else do men work and save and plan? If there is some future increase in human satisfaction that we can point to and realistically anticipate, men will be willing to pay something for it and invest in it in the hope of that return. In economics, they pay with money; in politics, with their votes and time and sometimes with their jail sentences and their lives.

Social change, peaceful or turbulent, is powered by "what might be." This means that for peaceful change, to get over some impossible barrier of unresponsiveness or complexity or group conflict, what is needed is an inventive man or group – a "social entrepreneur" – who can connect the pieces and show how to turn the advantage of "what might be" into some present advantage for every participating party. To get toll roads, when highways were hopeless, a legislative-corporation mechanism was invented that turned the future need into present profits for construction workers and bondholders and continuing profitability for the state and all the drivers.

This principle of broad-payoff anticipatory design has guided many successful social plans. Regular task forces using systems analysis to find payoffs over the barriers might give us such successful solutions much more often. The new world that could lie ahead, with its blocks and malfunctions removed, would be fantastically wealthy. It seems almost certain that there must be many systematic ways for intelligence to convert that large payoff into the profitable solution of our present problems.

The only possible conclusion is a call to action. Who will commit himself to this kind of search for more ingenious and fundamental solutions? Who will begin to assemble the research teams and the funds? Who will begin to create those full-time interdisciplinary centers that will be necessary for testing detailed designs and turning them into effective applications?

The task is clear. The task is huge. The time is horribly short. In the past, we have had science for intellectual pleasure, and science for the control of nature. We have had science for war. But today, the whole human experiment may hang on the question of how fast we now press the development of science for survival.

35.

On The Life Sciences: Science to Wisdom

Jean Mayer

We need to decide where we want to go and to whom we are accountable before we decide what means we shall use. AUTHOR

That technology is becoming much more biological – and even behavioral – than it has been in the past is, I believe, a happy development in the history of Man. Biology has occasionally been put to bad use. The history of the development of bacteriological weapons reads like an alphabet of horrors – bacterial diseases: anthrax, brucellosis, cholera, etc., supplemented by viral diseases from dengue to yellow fever and zoopathogens, by rickettsial diseases, fungal diseases, and toxins, but the fact is that the same countries which are ready to spend billions on "ABM's" and "MIRV's" agree on banning "BW." But I believe with Pasteur – with due apology to my illustrious predecessor on this platform – that the study of life, unlike that of physical, mechanical, and chemical phenomena, has ethical implications which educate all but the most insensitive researchers to a sense of the value as well as of the complexity of Life. Biological research, particularly when it deals with the complexity of physiological organization is a very different journey from that taken by the physicist.

Reprinted from *Science, Engineering and the Quality of Life: Proceedings of the Charles Allen Thomas Symposium*, St. Louis, March 5, 1970. By permission of the author and the sponsor, the Monsanto Company.

This may be even truer of pshychological and behavioral research. Again we can site the simple minded perversions exemplified by some propaganda and much of advertising. But poring into the behavior and mental processes of animals or children, into educational methodology, psychiatry or social anthropology produces almost invariably concerned men and women who, while they believe in the development of their science or technology, are aware of its ethical implications, and in more sophisticated terms than most who deal with the technology of inanimate objects. Hence, my view that the massive entry of the life sciences into technology is going to be almost wholly beneficent.

I obviously cannot cover here all important aspects of the life sciences in tomorrow's technology. In fact, I shall not cover some which are dearest to my heart, however promising they are. As a man who has for over twenty years studied the mechanism of regulation of food intake, I am aware of the quantum jump which, I believe in the next few years, the study of regulatory mechanisms is likely to achieve when enough hitherto isolated facts have been accumulated for computer models to be built and tried for fit. I am also aware of the fact that this development will transform our understanding not only of the integrative functions of the central nervous system but of hormones and of enzymatic feedback systems as well. It will transform pharmacology. Again, as one who for many years has correlated behavioral phenomena with physiological and metabolic events, I am unlikely to underestimate the advances which we are likely to see in the relation of neurophysiology and of behavior and as a result, in psychopharmacology. As a professor, and as both an historian and an activist of public health, I entertain high hope that an increase in our knowledge of the theory of learning and of social anthropology will lead to a technology of child and adult education which will give us a rapid increase in knowledge and enjoyment of the arts and the sciences, improve the quality of relations between men and perhaps between nations, and permit much more effective assistance programs in nutrition, health, education and technology than we have achieved so far. In spite of these tempting vistas, I shall not discuss these fields of biological endeavor. I intend to limit myself to only four: genetics and embryology and their technological implications; one area of preventive medicine, that of cardiovascular diseases; the problem of food and population; and one limited set of difficulties concerning agriculture and the new concern for the environment, namely those raised by pesticides. In all four cases, I shall attempt to show how far we are from the simple minded optimism of the second half of the nineteenth century and the pre-World War I twentieth century, the era of Verne, and Spengler, and H.G. Wells when technology was going to solve the problems of society. We are also beyond, I believe, the guilt, despair, and skepticism which followed, for many scientists, the use of chemical warfare agents during World War I and of atomic bombs during World War II. We are entering the era of massive biological technology, conscious of the gigantically complex problems we face, but with some hope and much compassion.

Let us start with genetics, reproduction and embryology, with our understanding and someday indeed our engineering of the development of organisms. This is a field in which we become instantly aware of the fact that we are living in an age of revolution. The rapidity with which this part of the biological sciences is evolving, the sociological, psychological and even ethical consequences of the new knowledge are such that biology is likely to transform our society in the next century as much as the Industrial Revolution has in the past two. Donald Fleming, the Harvard historian justifies the term "revolution" in the context of changes brought about by the biological sciences on the basis of three "revolutionary" characteristics of certain present day biologists, and among them, particularly of geneticists and embryologists. They have, he finds, a distinctive attitude toward the world; a program for utterly transforming it; and an unshakable, not to say fanatical confidence that this program can be enacted – "a world view, a program and a faith." These characteristics are well exemplified by a young Harvard scientist whom I have known since his undergraduate days, James Shapiro. Having made a basic discovery at the age of 26 – the first isolation of a gene, he announced last year his immediate retirement from direct participation in further research to consecrate his full time to work on the political consequences of his and related discoveries; and did so with the approval and active support of many of his colleagues, surely a new attitude in the annals of science.

In the last fifteen years of genetic research, we have learned the structure of DNA, the basic genetic substance. Through Watson and Crick's visualization of the double helix, we have become able to apprehend the general structure of the process by which the chromosomal strands are replicated. We now know the code by which DNA specifies the insertion of amino acids in protein and have discovered in viruses how to achieve the perfect replication of biologically effective DNA molecules. We understand the ability of viruses to invade bacterial and other cells and to insert the genes of the virus into the genome of the host (a phenomenon probably crucial in the etiology of cancer).

As for reproduction and embryology, we have during the same period devised effective hormonal contraception for females and are on the way to do so for males as well; we have learned to bank, for indefinite periods, deep frozen mammalian (bovine) sperm and can draw upon such stock to produce viable offspring; we have discovered in rabbits how to regulate the sex of offspring by removing fertilized ova from the female before they become implanted in the uterine wall, "sexing" the embryos by a technique entailing the removal of a few hundred cells, flushing embryos of the "wrong" sex out and reinserting, if necessary, embryos of the desired sex which develop normally to term.

Herman J. Muller believed, in 1927, when he discovered artificial mutation through X-rays, that he had discovered the tool with which to direct evolution. By the time of his death, forty years later, he had come to realize that the vast majority of mutations by irradiation were actually harmful and been thrown back to a neo-Galtonian position of advocating sperm banks for the sperm of exceptionally gifted and socially useful men, complemented as soon as possible by comparable ova banks.

Actually, biologists are beyond this rejuvenated eugenics and are already laying the foundations of revolutionary bioengineering developments, whereby all or parts of chromosomes and even individual genes will be inserted into the genome or whereby "synthetic" DNA molecules will be inserted into enucleated cultured germ cells. Superior genes will be produced by the hybridization of cells. The fact that we are limited at present to viruses, bacteria and cell cultures has little bearing on these predictions when one remembers the rate at which genetic engineering has progressed.

Joshua Lederberg thinks that these ultimate techniques can be temporarily bypassed for speedier action by what he calls "euphenics." the engineering of embryological development where, by modifying or inserting DNA, new instructions can be coded into the living organism. Lederberg states: 'It would be incredible if we did not soon have the basis of developmental engineering technique to regulate, for example, the size of the brain by prenatal or early postnatal intervention." Robert Edwards' success at Cambridge University with test tube fertilization of human eggs (the "test tube baby" discussed in the past few weeks by biologists and theologians may be the first of a new breed), has opened early development to the view of the microscope and perhaps action by the "interventionists."

At this point, of course, we encounter some of the deepest moral and ethical (not to say theological) problems. Crick states the dilemma at the lowest and most brutal level when he states: "Take the suggestion of making a child whose head is twice as big as normal. There is going to be no agreement between Christians and any humanists who lack their particular prejudice about the sanctity of the individual and who simply want to try it scientifically." The conservative attitude is represented by the Princeton theologian, Paul Ramsey, recently appointed by Georgetown University Medical School the world's first professor of "genetic ethics." Ramsey holds that genetic tailoring would be a "violation of man." In his masterful summation of this new, all-important controversy, Donald Fleming points out well how difficult ethical judgments will be when he states: ". . . immediate practical consequences will probably not present themselves as threatening to the individuals involved – quite to the contrary. Abstract theological speculations about genetic tailoring would be totally lost upon a woman who could be sure in advance that her baby would not be born mentally retarded or physically handicapped. The private anxieties of individuals are likely to diminish rather than increase any effective resistance to the broader consequences of the Biological Revolution." And he goes on to predict "general acquiescence, though occasionally sullen" to the concept of children becoming, in effect, objects of manufacture. I am not so sure that this is the wave of the future. Granted we need to weed out the aggressive instincts of nuclear-armed Man, to improve on the low average intelligence he brings to bear on increasingly interrelated problems, to limit his prolificity, which is no longer balanced by a high enough infant mortality and perhaps may soon be enhanced by enormously increased life spans, I do not see – and I am not sure I wish for – a scientific aristocracy deciding not just the future of Man, but future Man as well. But then mine may be the misgivings of a man too long used to think of the Ancient Greeks, Jesus and the Apostles, the artists of the Renaissance, the thinkers of the eighteenth century and our beautiful wives and daughters as models for the future generations.

And now on to Medicine. In a study on "Longevity and Medical Costs" published in 1967, W. H. Forbes of Harvard pointed out that by 1966 medical care was tied with agriculture as our second largest industry. In that year, American people had spent 44 billion in medical care as compared to 11 billion in 1949; by 1969 the figure was 62 billion . If present trends continue, medical care will soon be *the* largest consumer of manpower.

Surely, with expenditures of this magnitude, it is readily understandable that the public should expect impressive results. In fact, as Forbes pointed out, we are certainly getting "health services" for those tens of billions but, in terms of what we expect from these services – longevity, health and happiness or, as they are often expressed in the negative, postponement of death, freedom from discomfort and disability, and freedom from apprehension and worry, the results are much more doubtful. The last two quantities are difficult to measure or to evaluate – hospital days or absenteeism due to physical or mental disease are influenced by cultural attitudes, prevalence of prepaid medical care, differences in diagnosis – so that longevity is usually found to be the easiest yardstick, particularly if international comparisons are planned. The sad fact then becomes obvious that not only are we, the United States, constantly slipping with respect to the rest of the world (from 11th in 1966 for the average remaining lifetime for men at age 40 to 37th in 1969 to close to 40th now and from 9th to 23rd for women), we are also showing essentially no progress. Our picture for men is particularly dismal – five years behind the leaders. While part of this poor showing can be attributed to poor medical care for some of our population, particularly poor minorities, this differential accounts for only a small part of our problem, and would not explain why our women are doing much better than the men. Looking at specific causes of death, we find that we have a high mortality from violence – several times the death rate from automobile, home and industrial accidents and ten to twenty times the rate from homicides of leading developed countries, that we have also a high – and growing – death rate from lung cancer but that as compared with our main cause of death – cardiovascular disease – these and other factors are minor. Cardiovascular death is our big problem: one million such deaths in each of the past three years, with three quarters of these having been coronaries. That for almost half of our men in the forty-to sixty-year range death is the result of heart disease means that we have returned unwittingly to the days of the great pandemics (like the Great Plagues of the 14th century) when a quarter or a half of the population died of one disease. And this American phenomenon is probably a preview of what will happen in all industrial societies.

We cannot take care of a morbidity of this magnitude through curative medicine. It is true that in the past few years we have discovered in principle – and to some extent in practice – how to repress the immunological defenses of the body and gone on to combine these immunological techniques with surgical procedures by which the heart can be transplanted, with some chance of the recipient surviving for several months. But even if in the future these months become decades, we will still be unable to cope with our coronary mortality in this way: even if warning devices bring patients in the hospital in time for the transplant techniques to succeed, even if we have the thousands of additional thoracic surgeons, the tens of thousands of additional nurses, the billions of dollars worth of surveillance and monitoring instruments, the thousands of additional operating rooms and intensive care units, where shall we find the donors? Obviously, if we take that route, it will make more sense to manufacture at least a million, perhaps two million mechanical hearts (there may have been two nonfatal coronaries or more for each coronary death).

It makes even more sense, however, to try to *prevent* cardiovascular deaths. We are beginning to know a number of the main factors involved in the etiology of cardiovascular disease: hypercholesterolemia, hypertension, obesity, lack of exercise, (and perhaps cigarette smoking, too much caffein and lack of sleep). A number of these factors are amenable to technological attack, others to changes in behavior. Health education could make much better use of new educational techniques, going all the way to changes in mode of life.

Hypercholesterolemia is undoubtedly an important factor. As the cholesterol level of American males rises from 150 to 250 mg per 100 ml, the chances of a coronary triple. It appears that cholesterol can be manipulated to a certain extent (the limits being defined by heredity and mode of life) through changes in the diet, by decreasing the saturated fat content and increasing the polyunsaturated fats. If changes in regulation make it possible, if incentive is created by better labeling laws (as was recommended by the First (1969) White House Conference on Food, Nutrition, and Health), alteration of the food supply could soon make a considerable difference: half of our food or so is now processed industrially before it is consumed, and the nature of the fat used can be changed, especially as "textures" proteins, gel technology and other methods permit a more generous use of polyunsaturated oils without unfavorable change in structure or taste. The use of new feeding techniques, supplemented perhaps by endocrine manipulation, may permit the production of leaner meat which is yet tender and which can be further tenderized by enzymatic techniques. These techniques would be a vast improvement over the massive fat infiltration to "marbelize" the meat used at present to prepare tender meat. The development of fishing techniques, and, better still of fish husbandry, will, I hope, permit the development of a highly palatable source of animal protein rich in polyunsaturated fatty acids. Beyond this, the possibilities of production of "synthetic" foods of a given composition, texture and taste are almost unlimited. Again these should take note of health needs, such as the lowering of cholesterol either in their composition, or through the inclusion of a chemical or physical trap which can combine with cholesterol as it is secreted in the bile to prevent its reabsorption through the intestinal mucosa and its reappearance in the general circulation.

Hypertension is still another basic risk leading to coronaries (and strokes). It may respond to pharmacological research – many leads are being followed from the mechanism of diuresis to the role of cadmium. Certainly there, too, food technology can help by producing palatable foods lower in salt than present-day foods (and perhaps by including sodium traps). Organization of preventive medicine practices which would permit hypertension to be found before any sympton is perceived by the patient so that it can be treated before its effects are irreversible, would obviously help – the same is true of diabetes.

Obesity, a major factor in cardiovascular disease (as well as in diabetes and liver disease) is a complex syndrome where genetic and environmental factors interact in some cases traumatic factors are also involved. Our studies show that it can be due to disturbances in the function of regulatory mechanisms of which the better known are situated in a part of the brain known as the hypothalamus; we are beginning to understand the mechanism of action of these centers well enough to imagine the possibility of a "magic bullet" depressing the action of the "feeding" centers or activating the "satiety" centers without altering the function of the surrounding regulatory centers. Other types of obesity (and perhaps, in particular, that associated with "maturity-onset" diabetes – the diabetes of the fat fiftyish) may be due to enzymatic disorders which again may respond to specific pharmacology. Most obesity, however, appears to be directly the result of our mode of life. I have shown a number of years ago that the mechanism of regulation of food intake works poorly or not at all at very low levels of energy expenditure: as physical work decreases below a certain critical level (determined for each individual by heredity and clearly associated with body type), food intake stops decreasing correspondingly (or decreasing at all) and the individual becomes fat. Obesity is, for many, and probably for most of our "overweight" population, the result of a technology which, by emphasizing door-to-door individual transportation, labor saving devices and televised spectator sports, is affecting unfavorably the health (and appearance) of close to half of our adults

Lack of exercise does more than lead to obesity. Again our research shows that hard and prolonged exercise protects against the hypercholesterolemic effects of a diet high in saturated fat and to a certain extent, against hypertension. In the presence of a significant narrowing of the arteries of the heart, it also leads to the opening of collateral vessels and thus may avert a fatal coronary. It is thus not enough to mention technological approaches to preventing the number one killer (and crippler) of men in industrial societies; it is even more important to understand that the pandemic prevalence of cardiovascular disease is to a large extent the result of our uncontrolled technology. We must associate medicine in the planning of our society, not just use it after we have created new diseases. We have the technology to make cities which are pleasant and safe to walk in (rather than simply to drive and park in). We need planning, we need system analysis which includes effects on health as a variable if we are going to put our technology to good use; in more and more cases we will have to recognize that the answer to our problems is either a more delicate technology – the right kind of urban planning and construction, based on an understanding of physiology and social interactions, rather than simply of structural engineering and transportation, or in some cases, that the answer to technology is not necessarily more technology: until our bodies are changed we may be better off not eliminating physical work altogether.

Public health has, through elimination of microbiological pollution of water and food and through preventive inoculations against contagious disease, become responsible for a gigantic acceleration of the "population explosion." The decrease in death rate, first confined to white populations, but now spreading throughout the world, produces, as long as birth rates do not drop correspondingly, not only an increase in the population, but in the rate of increase as well; demographers say that a decline in "age specific mortality rates" prior to the end of childbearing age has the same demographic effect as an increase in the birth rate. (They mean that, for example, in the United States, 97 out of every 100 white females reach the age of 20; 91 reach the age of 50. In Guatemala, only 70 women out of every 100 reach the age of 20; 49 that of 50. If the death rate in Guatemala falls within the next decade to somewhere near the 1950 United States level, not an unlikely development, this alone will increase the number of women reaching the childbearing age by 85 per cent!) And so we have gone from 250 million people at the time of the birth of Christ to 500 million by 1650 to 1 billion by 1850, 2 billion in 1930 and there shall probably be 4 billion of us by 1975, with the prospect of 8 billion being reached in 30 to 35 years after that unless the birth rates are drastically reduced. If we in the United States maintain the lowest birth rate in our history we will still double our population in a century.

With this picture of ever increasing numbers of people, many people, social scientists and conservationists in particular have decided that the present rate of increase of the world's population is rapidly carrying all of Mankind to starvation. It is my contention that this is not happening and is not going to happen in the foreseeable future. Furthermore, as a nutritionist who has spent a great deal of his life combating hunger and malnutrition in our own country, and malnutrition and famine elsewhere (I was in Biafra a year ago and have been deeply involved since in the organization of the relief to that unhappy region), I can hardly be accused of ignoring the widespread character of severe malnutrition in the world today: caloric malnutrition, kwashiorkor and other protein-calorie malnutrition syndrome, anemias, B-complex vitamin deficiencies, mineral deficiencies and vitamin-A deficiency, an important threat still to life and sight in many areas. Still there is no evidence that the situation is getting worse. The only bases we have for judging the evolution of the race between food and population are the Food Balance Sheetes of FAO, admittedly imperfect instruments. They have imperfections as regards figures for that portion of the food supply that does not move within the channels of trade, such as food grown by the farmer for his family, and for waste. Still they are the best we have. Using population figures of the U.N. which are, again, the best we have, the present production rate for the world at large is increasing faster than the population (though this may not be true for particular subregions). What is more, except in the case of wars or civil disturbances (as in the Nigeria/Biafra conflict) improvement in communications, availability of surpluses in certain countries, the existence of international organizations and the gradual improvement in international morality make large scale famines brought about by natural castastrophies (such as the Irish or the Bengali famine) less likely. The fact is that we are feeding, however badly, a greater and greater number of people.

I believe that this state of affairs will continue for quite some time. Acreage under cultivation is at present 3.4 billion acres, less than 11 per cent of the land mass of the world. Estimates of potentially arable land have ranged from 13 to 19 billion acres. The use of fertilizers, so extraordinarily successful in this country, in Japan, and in Western Europe, has yet to penetrate the underdeveloped nations. It has been estimated that an expenditure of 10 dollars an acre for fertilizers would alone add 50 to 100 per cent to the low yields in the underdeveloped countries. Applying this investment to an area of 1.5 billion acres would be the equivalent of adding at least 750 million acres to the crop areas of these countries, the equivalent of a continent bigger than North America.

There are many other advances in agriculture that have yet to be applied on a large scale. The identification of necessary trace elements and their incorporation into fertilizers and feeds have opened vast areas of cultivation in Australia and elsewhere. This work will no doubt be extended to hitherto poorly studied lands in Asia, Africa and South America. Selective breeding of plants and animals has permitted the development of species with superior hardiness, increased yields and better nutritional value. In the greater part of the world this work has hardly begun. Advances in animal health and nutrition have already permitted the mass production of milk and eggs under indoor conditions in a scale which was unimaginable a few years ago. Computers programmed to calculate the cheapest method of providing a diet of known energy and known content in essential amino acids and other nutrients automatically set the controls that will mix basic staples providing the cheapest adequate poultry or cattle diet as they are informed of the latest commodity prices. Obviously, this is but the beginning of a new technology of enormous promise. Yields will be increased and crops are better generated by herbicides, pesticides and fungicides. Preservation of foods by radiation offers promise. The cultivation of high yield fishes is developing; and beyond these methods, the same genetic and embryologic processes which we have discussed earlier may well be used to create new edible species.

Powerful though these methods of "chemical" agriculture are, I believe that they will, within the life time of at least our younger listeners, be left far behind as methods of food production. Large-scale manufacture of food from petrochemicals, just started during World War II by the Germans to feed forced labor, now developed under more benign auspices by a number of large firms, is approaching the economic range. Carbohydrates, triglycerides, amino acids and fully metabolizable simpler compounds, such as 1,3-butanediol may soon be manufactured at very low cost for human food and animal feed. As abundant and cheap atomic power becomes available, oil and coal will increasingly be seen as raw materials for food and plastics rather than as fuel.

As a potential source of food production, photosynthesis can be used more efficiently in algae than in higher plants. With sufficient power (light), with proper mineral fertilization and with the proper rate of removal of the finished products, one square meter of algae production may support one man, with a high proportion of the calories derived from proteins; vitamins are also produced into the bargain. The problems entailed in passing from the theoretically possible to the economically feasible are formidable, though their solution is likely to be accelerated by plans to include algae production into the life cycle of crews in interplanetary travel of long duration – the space role is, it seems, William James' "moral [and technological] equivalent of war."

I may add that contemplating an increase in "classical" agriculture production which can (albeit at certain costs to the environment) keep up with population growth for the next twenty years, followed by industrial production of food which would remove food as a limiting factor for population growth, does not make me an "optimist" as regards the population problem; quite the contrary. World-wide famine would be a lamentable but effective limit to runaway human growth. I do not think it will occur. If we want to save the beauty of our planet and the social values which we hold dear, if we really value the quality of human life rather than numbers, we shall have to use our wisdom – and our science and our technology – to control population which the technologies of public health and agriculture could multiply almost ad infinitum. In many ways, obsessive concern with the nutriti aspect of the population problem in the future prevents us from looking at the material and social cost of the population explosion now. It hides the fact that food surpluses do not prevent us from having a population problem. And then, there is the risk of crying "wolf" too often.

I would like to end the list of our illustrations by considering the special problems of one limited segment of agriculture – pesticides – one factor among the many I have just mentioned as offering enormous potential for the underdeveloped countries. But this time we shall look at the social cost of that factor and what can be done about it.

Pesticides were at one time considered an unadulterated blessing. We have learned since that, while they are by no means an unmitigated disaster, they, too, exact a price; the nature of the pesticide, the mode of employ and ecological situation determine that price; and for some combinations the price is higher than what we are willing to pay. There are circumstances, notably in tropical countries, where mosquitoes and other insects may be the carriers of dangerous human (and animal) diseases, under which the application of chlorinated hydrocarbons may save many lives; and, indeed it was felt at the time of the invention of the DDT that its creator deserved a Nobel Prize for Medicine. But, more and more, we realize that in less hazardous circumstances, with all biological phenomena so closely interrelated, any action harsh enough to significantly alter one basic aspect of the environment may snowball and drastically change the whole ecology. We understand also that in the process a potentially disastrous impact on man's health may take place. Pesticides are found to act on the food chain in more than one way; they do indeed increase the available food supply by limiting the destruction of crops; but they may cause us to be deprived of the beautiful company of our familiar birds and to have to endure a "silent spring." They also threaten our health. Plants absorb DDT, are eaten by animals which are eaten by other animals and by us. Eventually we carry with us in our fat tissues over 10 parts per million of chlorinated hydrocarbon and our women may secrete over 0.2 parts per million in their milk – well above the amount recognized as the safe limit for meat or cow's milk by the Federal Government of the United Nations.

Similar case studies have been written about nitrates and about phosphates in fertilizers or detergents which are said to be causing the "death" of America's lakes and rivers; about nitrite used as preservatives which may have unfavorable health consequences. Obviously, we have had a serious problem for some time and, at last, we recognize that we have one. Francis Sargent, the Republican Governor of the Commonwealth of Massachusetts, recalling how hard it was just a few years ago to make this now well accepted point, wrote in the Sunday, March 1st, *New York Times:* "The military-industrial complex attempts to discredit its opponents by portraying them as men advocating unilateral disarmament; similarly, the agricultural-chemical complex suggest that the only alternative to massive pesticide use is crops plagued by hordes of insects. And, as the first alliance uses the threat of defeat in nuclear war to justify itself, analogously, the second employs the specter of widespread famine." Let me add that to state this indictment in the present tense and in such general terms makes it partly obsolete. I have found biochemists and other biologists with whom I have been in contact at Monsanto increasingly aware of the environmental problems raised by the use of hard pesticides; and under Dr. Thomas' enlightened leadership, management at various levels has recognized the ecological risks and, I understand, is speeding up the development of substitutes required by our understanding of these ecological risks and the consequent change in national mood.

To agree that the simple, massive, unsophisticated, chemical solution is fraught with risks, and perhaps unacceptable risks, is not to deny the continuing peril to crops (or at least to some crops) of insects or at least of some insects. How do we cope with this problem in the light of our new understanding and the new social morality? I seems to me that four new dimensions become necessary:

First, once again, we must introduce planning in our national (and international) life to a much greater and more detailed extent than we hitherto have. Our life, dependent as it is for its water and support on agriculture, transportation, industry, and public health and for its meaning on values such as the beauty of our country and our planet, cultural undertakings and civilized behavior, is far too complex to be left to chance. Planning, on the scale of, say, the use of pesticides, one firm, one industry, or one department of government is not enough. We must have national planning, leaving scope for individual enterprise. The example of the 1969 White House Conference on Food, Nutrition, and Health shows that such planning does not need to be undemocratic, that indeed, its results are best accepted if all concerned groups in the Nation have had a chance to have an input.

Secondly, we have again encountered, this time in industry and agriculture, the idea that we need system analysis and cost accounting of the most complex order: all large-scale measures we take have a social price which often goes well beyond the financial cost to the manufacturer and to the consumer *qua* consumer of a single article. An automobile or a truck costs more than the dealer's price tag indicates, FOB Detroit or what have you. It uses up land for roads and parking; after polluting the atmosphere and rivers during its manufacture, it pollutes the atmosphere with carbon monoxide, sulphur dioxide, lead organic compounds and incompletely burnt hydrocarbons as it runs; it is eventually abandoned somewhere in our countryside. At this point in time, we make no charge in the price tag for most of these ecological insults. Nor is the costly increase in mobility – ever expanding suburbs consume far more space than villages taxed on mass transportation routes, individual mobility increases familial and social disorganization – accounted for in any way in the price tag of the first, the second, the third family automobile. The same is true for the cost of this or that fertilizer, this or that pesticide, or this or that non-biodegradable container. We must learn to analyze the effects of a new invention or process and determine its cost not only to the manufacturer but to society itself. I may add that I believe that not only is such analysis and cost accounting compatible with free economy, it is only by showing that it can do it that free enterprise will survive.

Third, seen in the perspective of the human community, we find that in industry and agriculture, as in medicine, we are lead to a much more exquisite technology: biology (and even behavioral sciences) will have to replace in many instances the indiscriminate utilization of physical sciences. Let me cite one example, staying on the subject of agriculture: the use of insecticidal traps of sexual attractants, specific for one single zoological species, may be an excellent alternative in some cases to the indiscriminate use of chlorinated hydrocarbons. They may be more expensive in the old fashioned, single problem type of perspective; but they may be very economical of significant ecological change and have no health consequences for man. Similarly the use of opaque 2 corn, a cereal of much bigger biological value than its predecessors, may decrease the need for animal supplements and thus reduce considerably the overall acreage needed.

And finally, once again we must recognize that the answer to problems caused by technology is not always more technology, biological or otherwise. In some cases, the reasonable answer is to renounce certain goods, or to be ready to be satisfied with less than optimal efficiency in obtaining them. Just as I believe that we shall be led to sacrifice some of the desire for individual transportation and accept mass transportation with its acknowledged limitations of mobility as a substitute, we may be led to accept limitations of agricultural efficiency as the price for saving our environment: need we strive always for the highest yields and the highest preservation from pests, only to spend billions every year to limit production? I realize that this type of reasoning is foreign to our acquired reflex, but I believe that we are beginning to look beyond technology to science and beyond science to wisdom. Self restraint is not only the basis for endurable human relations and the functioning of a democratic society, it has become the basis for the perpetuation of life itself.

I would like to close this address by making a plea for utopian thinking. By this term I mean a concentrated effort to define and describe the good life, as Thomas More once set out to do. Technology and social organization are then harnessed to achieve it. This form of writing and thinking has disppeared from contemporary literature and has been replaced by science fiction – that is technology progressing unplanned, with man its servant and its victim instead of its master. We need to decide where we want to go and to whom we are accountable before we decide what means we shall use.